T0258245

EVOLUTIONARY BIOGEOGRAPHY

EVOLUTIONARY BIOGEOGRAPHY
An Integrative Approach with Case Studies

Juan J. Morrone

Professor of Biogeography, Systematics, and Comparative Biology
Facultad de Ciencias, Universidad Nacional Autónoma de México (UNAM),
Mexico City

COLUMBIA UNIVERSITY PRESS NEW YORK

Columbia University Press
Publishers Since 1893
New York   Chichester, West Sussex

Copyright © 2009 Columbia University Press
All rights reserved

Library of Congress Cataloging-in-Publication Data

Morrone, Juan J.
     Evolutionary biogeography : an integrative approach with case studies / Juan J.
Morrone.
       p. cm.
   Includes bibliographical references.
     ISBN 978-0-231-14378-3 (cloth : alk. paper)—ISBN 978-0-231-51283-1 (ebook)
   1. Biogeography.   2. Biogeography--Case studies.   I. Title.

   QH84.M665 2009
   578.09—dc22

                                                                                    2008038927

Columbia University Press books are printed on permanent and durable acid-free paper.
This book is printed on paper with recycled content.
Printed in the United States of America

c 10 9 8 7 6 5 4 3 2

References to Internet Web sites (URLs) were accurate at the time of writing. Neither the author nor Columbia University Press is responsible for URLs that may have expired or changed since the manuscript was prepared.

In memory of my mother, Lidia Lupi (1937–2008), who inspired me to love nature and truth.

There are three things that last forever: faith, hope, and love; and the greatest of the three is love.

—1 Corinthians 13:13

When the miracle occurs, as it sometimes does; when, on one side and the other of the hidden crack, there are suddenly to be found cheek-by-jowl two green plants of hidden species, each of which has chosen the most favourable soil; and when at the same time two ammonites with unevenly intricate involutions can be glimpsed in the rocks, thus testifying in their own way to a gap of several tens of thousands of years suddenly space and time become one: the living diversity of the moment juxtaposes and perpetuates the ages. Thought and emotion move into a new dimension where every drop of sweat, every muscular movement, every gasp of breath becomes symbolic of a past history, the development of which is reproduced in my body, at the same time as my thought embraces its significance. I feel myself to be steeped in a more dense intelligibility, within which centuries and distances answer each other and speak with one and the same voice.

Claude Lévi-Strauss (1955), *Tristes tropiques*

Contents

Preface

The aim of this book is to provide a theoretical and practical guide to evolutionary biogeography for advanced undergraduate and beginning graduate students, academics, and anyone concerned with the study of biogeographic patterns and their evolution. It provides an introduction to evolutionary biogeography, its basic concepts, history, approaches, methods, developments, and case studies. I have assumed that the readers have a basic knowledge of phylogenetic systematics, so cladistic methods are not detailed. Some existing books deal with a specific biogeographic approach or with several methods but fail to provide a coherent framework from which one can choose the most appropriate, leaving the reader uncertain about how to address a particular problem. I discuss the available methods and suggest the appropriate step of a biogeographic analysis in which to use any particular one. The case studies (most from my own research) are intended to help the reader make a rational choice from among the approaches and methods available. I have also included problems, questions for discussion, and a glossary that can be used in the classroom.

In recent decades, biogeography has undergone much debate. Dispersalists, panbiogeographers, cladistic biogeographers, ecological biogeographers, macroecologists, and phylogeographers, among others, have disputed the relative merits of their approaches but have not worked toward their integration. However, a few authors have noted this lack of interaction recently and have stated that the integration of approaches and methods in biogeography is a salient issue. I hope this book can contribute to a future integrative biogeography.

I am indebted especially to my students for their insights and inspiration. Many thanks to my friends and colleagues Roxana Acosta, Alfredo Bueno Hernández, Angélica Corona, Dalton de Sousa Amorim, Malte Ebach, Tania Escalante, David Espinosa Organista, Oscar Flores Villela, John Grehan, Gonzalo Halffter, Michael Heads, Analía Lanteri, Livia León Paniagua, Jorge Llorente Bousquets, Ana Luz Márquez, Susana Magallón, Juan Márquez Luna, Rafael Miranda Esquivel, Adolfo Navarro Sigüenza, Gareth Nelson, Silvio Nihei, Federico Ocampo, Jesús Olivero, Rod Page, Gerardo Pérez-Ponce de León, Paula Posadas, Raimundo Real, Sergio Roig-Juñent, Adriana Ruggiero, Luis A. Sánchez-González, Claudia Szumik, Hernán Vázquez, and Mario Zunino for helpful and stimulating discussions. I am also indebted to Malte Ebach, Tania Escalante, John Grehan, David Hafner, Michael Heads, Analía Lanteri, Rafael Miranda, Silvio Nihei, Federico Ocampo, Jesús Olivero, Lynne Parenti, Paula Posadas, and Adriana Ruggiero for valuable comments on parts of the manuscript. Patrick Fitzgerald (Columbia University Press) and three anonymous reviewers provided very useful suggestions. In recent years my research has been supported by the Universidad Nacional Autónoma de México, the Comisión Nacional para el Conocimiento y Uso de la Biodiversidad, and the Consejo Nacional de Ciencia y Tecnología, México; the National Geographic Society, United States; and the Fundación Carolina, Spain. Parts of this book were written in Paris and Málaga; thanks to my colleagues from the Université Pierre et Marie Curie and the Universidad de Málaga for hosting me during my sabbatical leave. For the last twelve years, Adrián Fortino has provided love, support, and encouragement.

Mexico City, February 20, 2008

EVOLUTIONARY BIOGEOGRAPHY

CHAPTER 1

Introducing Evolutionary Biogeography

Biotas are complex mosaics originated by dispersal and vicariance, having reticulate histories, which should be studied through different methods. Evolutionary biogeography integrates distributional, phylogenetic, molecular, and paleontological data in order to discover biogeographic patterns and assess the historical changes that have shaped them, following a stepwise approach. In this chapter I briefly introduce the steps of this approach.

What Is Evolutionary Biogeography?

One hundred fifty years ago, Charles Darwin published *On the Origin of Species*. The geographic distribution of plant and animal taxa was among the evidence he provided to support evolution. Although the fact that continents have their own distinctive biotas has been known for some time, from Darwin we learned that these biotas evolve, that their composition changes over time. In the nineteenth and twentieth centuries, biogeographers extensively debated the mechanisms leading to biotic evolution. In recent years some authors have concluded that dispersal and vicariance are both relevant processes. When climatic and geographic factors are favorable, organisms actively expand their geographic distribution according to their dispersal capabilities, thus acquiring their ancestral distribution (dispersal). When the organisms have occupied all available space, their distribution may stabilize, allowing the isolation of populations in different sectors of the area and the differentiation of new species through the appearance of geographic barriers (vicariance).

To analyze the resulting complex patterns, we should identify particular questions, choose the most appropriate methods to answer them, and finally integrate them in a coherent framework. Most of the authors who are involved in the theoretical development of biogeography or who apply their methods usually conceive them as representing alternative schools; however, they can be used to answer different questions. Evolutionary biogeography integrates distributional, phylogenetic, molecular, and paleontological data in order to discover biogeographic patterns and assess the historical changes that have shaped them. It follows a stepwise approach (fig. 1.1). Each of its steps is discussed in a different chapter of this book.

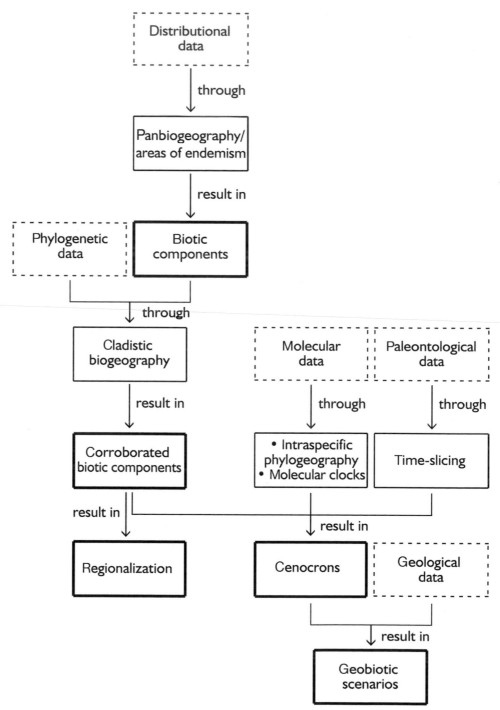

Figure 1.1   Flow chart with the five steps of an evolutionary biogeographic analysis.

Step 1: Identification of Biotic Components

Biotic components are sets of spatiotemporally integrated taxa that coexist in given areas. Their identification is the first stage of an evolutionary biogeographic analysis. There are two basic ways to represent biotic components: generalized tracks and areas of endemism. The former are studied by panbiogeography, whereas the latter are the units of cladistic biogeography.

Panbiogeography emphasizes the spatial or geographic dimension of biodiversity to allow a better understanding of evolutionary patterns and processes (Craw et al. 1999). A panbiogeographic analysis comprises three basic steps: (1) constructing individual tracks for two or more different taxa, (2) obtaining generalized tracks based on the comparison of the individual tracks, and (3) identifying nodes in the areas where two or more generalized tracks intersect. Individual tracks are the basic units of panbiogeography, representing the primary spatial coordinates of species or supraspecific taxa, which operationally correspond to line graphs connecting the different localities or distributional areas of a taxon according to their geographic proximity. Generalized tracks result from the significant superposition of different individual tracks and indicate the preexistence of ancestral biotic components that became fragmented by geologic or tectonic events. Nodes are complex areas where two or more generalized tracks superimpose and are usually interpreted as tectonic and biotic convergence zones.

Areas of endemism are areas of nonrandom distributional congruence between different taxa (Morrone 1994b). Müller (1973) suggested a protocol that has been applied to identify areas of endemism and that consists of plotting the ranges of species on a map and finding the areas of congruence between several species. This approach assumes that the species' ranges are small compared with the region itself, that the limits of the ranges are known with certainty, and that the validity of the species is not in dispute.

Step 2: Testing Relationships Between Biotic Components

Cladistic biogeography assumes a correspondence between the phylogenetic relationships of the taxa and the relationships between the areas they inhabit (Platnick and Nelson 1978). Cladistic biogeography uses information on the cladistic relationships between the taxa and their geographic distribution to postulate hypotheses on relationships between areas. If several taxa show the same pattern, such congruence is evidence of common history. A cladistic biogeographic analysis comprises three basic steps: (1) constructing taxon–area cladograms from the taxonomic cladograms of two or more different taxa by replacing their terminal taxa with the areas they inhabit, (2) obtaining resolved area cladograms from the taxon–area cladograms (when demanded by the method applied), and (3) obtaining a general area cladogram, based on the information contained in the resolved area cladograms.

Taxon–area cladograms are obtained by replacing the name of each terminal taxon in the cladograms of the taxa analyzed with the area where it is distributed. Their construction is simple when each taxon is endemic to a single area and each area has only one taxon, but it is more complex when taxonomic cladograms include widespread taxa, redundant distributions, and missing areas. In these cases, some methods require that taxon–area cladograms be turned into resolved area cladograms (Morrone and Crisci 1995; Nelson and Platnick 1981). General area cladograms based on the information from the different resolved area cladograms represent hypotheses on the biogeographic history of the taxa analyzed and the areas where they are distributed.

Step 3: Regionalization

Because the geographic distributions of taxa have limits, and these limits are repeated for different taxa, they allow the recognition of biotic components. Once they have been identified, they may be ordered hierarchically and used to provide a biogeographic classification. This stage of the analysis takes place before cenocrons are elucidated and a geobiotic scenario is proposed.

Biogeographic regionalization implies the recognition of successively nested areas for which classically the following five categories have been used: realm, region, dominion, province, and district. Sometimes it is more difficult to determine the exact boundaries of two realms or regions, and authors have described transition zones. These zones represent events of biotic hybridization, promoted by historical and ecological changes that allowed the mixture of different biotic components.

Step 4: Identification of Cenocrons

Cenocrons are sets of taxa that share the same biogeographic history, constituting identifiable subsets within a biotic component by their common biotic origin and evolutionary history from a diachronic perspective. After biotic components are established, time slicing, intraspecific phylogeography, and molecular clocks can help establish when the cenocrons assembled in the identified components, incorporating a time perspective in the study of biotic evolution.

Events of biogeographic convergence produce reticulated area histories that decrease the chances of establishing area relationships through congruence. The solution to problems posed by instances of biogeographic convergence is time slicing (Upchurch and Hunn 2002). Whereas assessments of faunal similarity usually are undertaken with faunas of successive geological ages, cladistic biogeography has used only data on organism relationships and spatial distributions on a single time plane (usually the present). Time slicing may reconcile the use of time and a synchronic approach. Ideally, it is possible to use a synchronic approach for each time slice identified.

Intraspecific phylogeography studies the principles and processes governing the geographic distribution of genealogical lineages, especially those within and between closely related species, based on molecular data (Avise et al. 1987). Once the population genetic structure has been assessed based on mitochondrial DNA (mtDNA), it is possible to obtain a network or cladogram of haplotypes, which allows us to analyze historical patterns and the processes that shaped them (e.g., dispersal, vicariance, range expansion, and colonization), sometimes under a statistical framework. This knowledge can suggest when recent cenocrons incorporated into a biotic component.

Cladograms based on molecular data may be used as raw data in cladistic biogeography and intraspecific phylogeography. In addition, the assumption that the rate of molecular evolution is approximately constant over time for proteins in all lineages allows the inference of a clock-like accumulation of molecular changes (Zuckerland and Pauling 1962), where the "ticks" of the clock, which correspond to mutations, do not occur at regular intervals but rather at random points in time (Gillespie 1991). This time is measured in arbitrary units and then calibrated in millions of years by reference to the fossil record or geological data (Magallón 2004; Sanderson 1998), giving minimum estimates of the age of a clade, which in turn may help elucidate the relative minimum ages of the cenocron to which it belongs.

Step 5: Construction of a Geobiotic Scenario

Once we have identified the biotic components and cenocrons, we may be able to construct a geobiotic scenario. By accounting biological and nonbiological data, we can integrate a plausible scenario to help explain the episodes of vicariance or biotic divergence and dispersal or biotic convergence that have shaped the evolution of the biotic components analyzed.

Biogeographers have classified geographic features in terms of their impact on dispersal and vicariance. The most important are barriers (geographic features that hinder dispersal) and corridors (geographic features that facilitate dispersal). In dealing with long-term changes in the biotic distributional patterns, continental drift may be a relevant factor (Briggs 1987; Cox and Moore 1998). Not only do the splitting and collision of landmasses directly affect distributional patterns, but also new mountains, oceans, and land barriers change the climatic patterns on the landmasses.

How to Read This Book

There are different ways to read this book. You may select an individual chapter to address your particular interests (e.g., chapter 2 if you are interested in conceptual issues, chapter 3 if you are interested in the history of biogeography, chapter 4 if you are interested in panbiogeography, chapter 5 if you are interested in cladistic biogeography, or chapter 7 if you are a paleontologist or a phylogeographer). You may also choose to skip a particular chapter (e.g., chapter 2 if you are not philosophically inclined, chapter 3 if you think the history of a discipline is uninteresting, chapters 4 and 5 if you do not feel motivated by the study of biogeographic patterns, or chapters 7 and 8 if you are pattern oriented). But if you are really interested in evolutionary biogeography or are curious about the possibilities of integrating diverse approaches, I suggest that you read all chapters.

CHAPTER 2

Basic Concepts

Evolutionary biogeography integrates distributional, phylogenetic, molecular, and paleontological data in order to discover biogeographic patterns and assess the historical changes that shaped them. To elucidate the ontology of evolutionary biogeography, several complex issues should be understood. In this chapter I discuss the relationship between ecology and history, the relevance of the genealogical and ecological hierarchies, biogeographic patterns and processes, biotic components, and cenocrons. I also provide a general introduction to the available biogeographic methods of evolutionary biogeography.

Biogeography

Biogeography is the study of the geographic distribution of taxa and their attributes in space and time (Hausdorf and Hennig 2007). In addition to recognizing distributional patterns of plants, animals, and other organisms, biogeographers identify natural biotic units to provide a biogeographic regionalization of Earth, postulate hypotheses about the processes that may have shaped such patterns, and, on the basis of discovered patterns, help predict the consequences of global planetary changes and select areas for biodiversity conservation (Morrone 2004a). There are several interesting questions posed by biogeographers, but in essence, they can be reduced to two basic questions: Where are organisms distributed, and why are they distributed there?

Space, time, and form (*form* refers not only to the structure of organisms but also to all their characters, be they structural, functional, or behavioral) are the three dimensions of biodiversity (Croizat 1964). Although space is basic to biogeography and, in general, to evolutionary biology, it has been stated that "an adequate concept of space, that is, spatial differentiation in its temporal context, has been the most elusive factor in the history of systematics, one rivaling if not exceeding Darwin's 'mystery of mysteries'—the origin of species" (Nelson 1977:450). In particular, we lack a relativistic spatiotemporal perspective that might represent a break with the traditional organism–environment dichotomy (Craw and Page 1988). This new perspective might allow reintroduction of biogeography into evolutionary biology.

Biogeography is a peculiar discipline because, despite the existence of books, journals, symposia, and courses on the subject, few people are employed primarily as biogeographers (Nelson 1978a, 1985). This may be because biogeography occupies an intermediate area between geography, geology, and biology, being practiced by systematists, ecologists, paleontologists, anthropologists, naturalists, and geographers, among others. We may consider biogeography a synthetic discipline (Brown and Lomolino 1998) or even interdisciplinary (fig. 2.1). For this reason, biogeography is heterogeneous in its principles and methods, lacking the conceptual unity of other sciences (Morrone 2004a).

In the past two decades biogeography has undergone an extraordinary theoretical and methodological renovation. Morrone and Crisci (1995) held that biogeography is passing through a revolution concerning its foundations, basic concepts, methods, and relationships with other disciplines. Andersson (1996) detected a problem with the ontology of biogeography because there is no consensus about which phenomena should be considered biogeographic. Crisci (2001) referred to external and internal forces that characterize this revolution; the former include the paradigm of plate tectonics in Earth sciences, cladistics as the basic language of biology, and biologists' perceptions of biogeography, whereas the latter include the proliferation of methods and a seemingly endless philosophical debate. I think that other important developments in molecular biology, informatics, geographic information systems, ecology, and geology should be added. Humphries (2004) sadly noted that rather than creating an arena for discussion, biogeographers are becoming politically balkanized, even misrepresenting the ideas of others. Until quite recently, there has been little communication between biogeographers practicing different approaches and too little integration of their findings, even though the need for integration has been recognized for some time (Lomolino et al. 2006). Despite this complex and confusing situation, interest in biogeography is growing. A quantitative measure of this interest can be easily seen by searching the subject "biogeograph*" in recent years in the Science Citation Index (ISI Web of Science, http://www.isinet.com/isi/).

Figure 2.1   Interdisciplinary situation of biogeography, at the intersection of 6 different disciplines.

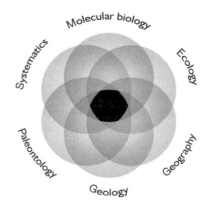

Ecological and Historical Biogeography

In the past two centuries several general approaches and theories have been developed in biogeography. Some of them are regional biogeography (De Candolle 1820; Sclater 1858), dispersalism (Axelrod 1963; Darwin 1859; Matthew 1915; Mayr 1946; Simpson 1965; Wallace 1876), chorology (Haeckel 1868), phylogenetic biogeography (Brundin 1966; Hennig 1950), paleobiogeography (Hallam 1973; Simpson 1953), panbiogeography (Craw et al. 1999; Croizat 1958b, 1964), island biogeography (Carlquist 1974; MacArthur and Wilson 1967; Whittaker 1998), vicariance biogeography (Croizat et al. 1974), Pleistocene refugia (Haffer 1969), dynamic biogeography (Hengeveld 1990; Udvardy 1969), geographic ecology (MacArthur 1972), areography (Rapoport 1975), quantitative biogeography (Crovello 1981), cladistic biogeography (Humphries and Parenti 1999; Nelson and Platnick 1981), systematic biogeography (Morain 1984), evolutionary biogeography (Blondel 1986; Ridley 1996), analytical biogeography (Myers and Giller 1988a), balanced biogeography (Haydon et al. 1994), intraspecific phylogeography (Avise 2000; Avise et al. 1987), macroecology (Brown 1995), comparative phylogeography (Bermingham and Moritz 1998), applied biogeography (Spellerberg and Sawyer 1999), and the unified neutral theory of biodiversity and biogeography (Hubbel 2001). It has been suggested that these approaches broadly constitute two subdisciplines: ecological and historical biogeography. Ecological biogeography analyzes patterns at the species or population level, at small spatial and temporal scales, accounting for distributions in terms of biotic and abiotic interactions that happen in short periods of time. Historical biogeography analyzes patterns of species and supraspecific taxa, at large spatial and temporal scales, being more interested in processes that happen over long periods of time. This distinction between ecological and historical biogeography is rather artificial because it implies a continuum whose extremes are easily identifiable as ecological or historical, but in the middle it is more difficult to justify such division, as occurs with Pleistocene refugia (Cox and Moore 1998; Myers and Giller 1988b). The development of phylogeography and macroecology has also challenged the ecological–historical dichotomy. Some authors (Rousseau 1992; Wiley 1981) have considered a main division between descriptive and causal biogeography, the latter encompassing both historical and ecological biogeography. Other authors (Crisci 2001; Crisci et al. 2000; Morrone and Crisci 1995; Vargas 1993, 2002) accepted the ecological–historical division as merely conventional.

The history–ecology dichotomy is analogous to other dichotomies opposing internal (originating from inside the organism) and external (originating from the environment) factors that have been proposed in biology and other disciplines (table 2.1). Despite being inadequate, these dichotomies are preserved because of the sociology of academic life (Oyama 2000). In biogeography, several authors have criticized the lack of interaction between historical and ecological biogeography and have discussed the possibility of integrating them into a unified research program (Gray 1989; Henderson 1991; Holloway 2003; Morrone 1993a; Riddle 2005; Smith 1988; Wiens and Donoghue 2004). This integration of the historical and

Table 2.1

Disciplines	Internal Factors	External Factors
Epistemology	Innate knowledge	Acquired knowledge
History	Internalism	Externalism
Anthropology	Biology	Culture
Psychology	Nature	Nurture
Genetics	Genotype	Phenotype
Ethology	Innate behavior	Acquired behavior
Evolution	Orthogenesis and phylogenetic constraints	Natural selection
Biogeography	History	Ecology

ecological approaches to biogeography was expressed some decades ago by Os-valdo Reig and Gonzalo Halffter. Reig (1962) coined the term *cenogenesis* to refer to the evolution of biotic associations through time, considering that biogeographic inquiries are historical and, at the same time, should try to explain the development of communities, not of isolated taxa. Halffter (1987) held that biogeographic patterns should be inferred from taxa with similar evolutionary and macroecological trends.

From a different perspective, Ebach and Goujet (2006) and Williams (2007a) also challenged the division between historical and ecological biogeography. They proposed a distinction between a chorological approach that uses evolutionary models to trace distributional pathways and a systematic approach that describes, compares, and classifies biotas. Chorology may date to Buffon's law, although its name was originally coined by Haeckel (Williams 2007a). Systematic biogeography has its beginnings in the classification of the French flora (Lamarck and de Candolle 1805). Chorology and systematic biogeography may be equally historical and ecological.

Hierarchies and Scales in Biogeography

In recent decades, some authors have claimed that a hierarchical approach may allow improvement of evolutionary theory (Eldredge 1985; Eldredge and Salthe 1984; Lieberman 2003b; Morrone 2004c). This hierarchical vision implies that nature is structured in entities that are ordered hierarchically, with smaller entities nested within larger ones. Each level of a hierarchy has its emergent properties and some autonomy. It is not a mere assemblage of smaller entities (Eldredge 1985; Mahner and Bunge 1997). Hierarchical thinking is a way to structure the study of evolutionary patterns and processes, an approach more than a theory (Cracraft 1985). Two main hierarchies have been identified in nature (fig. 2.2). The genealogical hierarchy consists of entities named replicators, which contain information, reproduce in similar entities, and evolve as genes, chromosomes, organisms, and clades. The ecological hierarchy includes entities named interactors, which are involved in the matter–energy interchange as molecules, cells, organisms, populations, and biotas. Organisms are the only members that are common to both

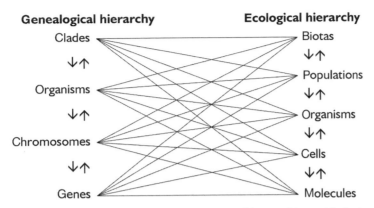

Figure 2.2   Entities of the genealogical and ecological hierarchies, with their connections.

hierarchies. There are interactive processes between the members of a hierarchy. Additionally, each member of a hierarchy interacts with the members of the other hierarchy. For this reason, the collapse of a biota in the ecological hierarchy may cause the extinction of a clade or a specific gene in the genealogical hierarchy.

Biogeography is a complex discipline because it addresses geographic patterns exhibited by entities belonging to both hierarchies (Lieberman 2003b). Patterns of the genealogical hierarchy, studied by evolutionary biogeography, involve historical entities ruled by evolutionary processes. Patterns from the ecological hierarchy, studied by ecological biogeography, involve entities that transfer matter and energy. At the upper level of both hierarchies, the entities contrast markedly, so their study usually entails different data and methodological tools, whereas the distinction between entities at lower levels may not always be clear. Some biogeographic analyses may imply both hierarchies simultaneously. When we correlate the richness of a taxon with the latitude along a continent, we are studying genealogical entities (clades), but it is a basic problem of ecological biogeography. When we compare the distributional patterns of different clades inhabiting the same areas, even when biotas are present, we are analyzing clades, which belong to the genealogical hierarchy (Morrone 2004c). This shows the fuzzy limits between evolutionary and ecological biogeography and allows one to speculate on the possibilities of integration.

In order to build a unified biogeography, it would be possible to limit the scopes of ecological and historical biogeography based on scales: ecological explanations for local, short-term spatiotemporal scales and historical or evolutionary explanations for larger, long-term spatiotemporal scales. However, it has been noted that there is no obvious point at which to divide the domain of ecology from that of history and that some ecological factors may have wide geographic effects, whereas some historical factors may be responsible for local distributional patterns (Gray 1989). In fact, ecology and history are not independent variables acting on different spatiotemporal scales. They act together at all times. Historical changes are mediated ecologically, and ecological changes are historically contingent

Figure 2.3 Balanced biogeography, as envisioned by Haydon et al. (1994), based on the complementary roles of ecology, history, and chance.

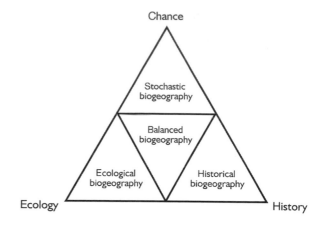

because organisms experience historical factors such as a tectonic change not directly but indirectly through changes in altitude, climate, and so on. After analyzing the possibility of isolating ecological and historical determinants of Andean bird distributions, Vuilleumier and Simberloff (1980) concluded that ecology and history have played significant roles together and at all times. Gray (1989) suggested that a new biogeographic vision should emerge in which different ecological and historical factors are integrated in a network of interacting processes.

An alternative conceptualization (Haydon et al. 1994) highlights the complementary roles of ecology, history, and chance (fig. 2.3). In this model, a balanced biogeography results from the interaction of ecological, historical, and stochastic processes.

Biogeographic Patterns

Biogeographic patterns are nonrandom, repetitive arrangements or distributions of organisms and clades in geographic space. The study of certain specific patterns constitutes the scope of particular biogeographic approaches; for example, specific richness patterns and distribution of life forms are studied in ecological biogeography, chorological patterns are studied in areography, structural and functional patterns of ecological systems are studied in macroecology, and biogeographic homology patterns are studied in evolutionary biogeography (Espinosa Organista et al. 2002).

Biogeographic homology is the basic concept of evolutionary biogeography (Morrone 2004a). In its more general form, *homology* means equivalence of parts and constitutes a sorting procedure used to establish meaningful comparisons within a hierarchical system (de Pinna 1991; Nelson 1994; Rieppel 2004; Williams 2004). Biogeographic homology allows one to identify biotic components, namely, the sets of spatiotemporally integrated taxa that coexist in given areas. If the analogy between systematics and biogeography is accepted, we may consider that the distributions of individual taxa are the statements about biogeographic homology that are being compared. Homology is the relationship between the homologues (in biogeography, the biotic components) rather than the homologues themselves (Nelson 1994; Rieppel 1991; Williams 2004).

Several authors have recognized two stages in the proposition of homologies, which have been named primary and secondary homology by de Pinna (1991). Primary homology, which corresponds to the stage of generation of the hypotheses, represents a conjecture on the correspondence between parts of different organisms. Secondary homology, which corresponds to the stage of legitimation of the hypotheses, represents a test of such conjecture by congruence with similar statements in the cladogram. In biogeography, both stages have been implicitly recognized by several authors (Donoghue et al. 2001; Hausdorf and Hennig 2003; Morrone and Crisci 1990, 1995; Riddle and Hafner 2006).

Primary biogeographic homology (figs. 2.4a and 2.4b) is a conjecture on a common biogeographic history, which means that different taxa, even when they have completely different means of dispersal, are spatiotemporally integrated in a biotic component (Morrone 2001c, 2004a). A panbiogeographic analysis allows comparison of individual tracks in order to detect generalized tracks. In addition to sorting distributions of the analyzed taxa into generalized tracks, it is possible to detect smaller units or areas of endemism within them. Both areas of endemism and generalized tracks represent biotic components. Secondary biogeographic homology (figs. 2.4c and 2.4d) is the cladistic test of the formerly recognized biotic components. A cladistic biogeographic analysis allows one to compare area

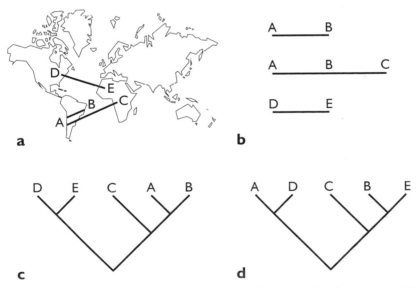

Figure 2.4   Biogeographic homology. (a, b) Primary biogeographic homology, with biotic components drawn as generalized tracks; (c, d) secondary biogeographic homology (general area cladograms): (c) the general area cladogram corroborates the hypothesis of primary biogeographic homology; (d) the general area cladogram falsifies the hypothesis of primary biogeographic homology.

cladograms—obtained by replacing terminal taxa in taxon–area cladograms by the areas of endemism they inhabit—in order to obtain a general area cladogram.

A biogeographic analysis by Donoghue et al. (2001) followed this approach when analyzing disjunctions of several taxa in the Northern Hemisphere. These authors initially assigned the taxa analyzed to two generalized tracks (Atlantic and Pacific) and then undertook the cladistic biogeographic analysis of the taxa belonging to each of the tracks separately. Riddle and Hafner (2006) presented similar arguments to develop an approach for the analysis of historical relationships that combines parsimony analysis of endemicity, phylogeography, and cladistic biogeography.

Biogeographic Processes

Biogeographic processes are those that shape the geographic distribution of taxa. There are three basic biogeographic processes: dispersal, vicariance, and extinction (fig. 2.5). Once patterns have been discovered, explanations on the processes that have shaped them are sought, and the hypotheses can be tested until robust theories become accepted (Brown and Lomolino 1998).

Dispersal is the expansion of the distributional area of a taxon, covering all types of geographic translocation (Myers and Giller 1988b). For classic dispersal-

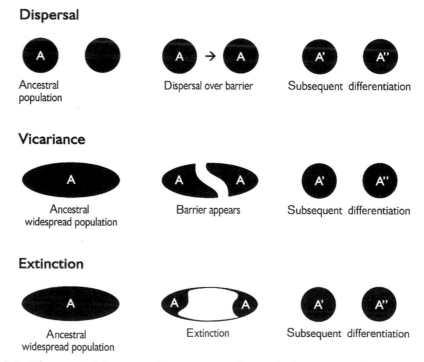

Dispersal

Ancestral population — Dispersal over barrier — Subsequent differentiation

Vicariance

Ancestral widespread population — Barrier appears — Subsequent differentiation

Extinction

Ancestral widespread population — Extinction — Subsequent differentiation

Figure 2.5   Three main biogeographic processes: dispersal, vicariance, and extinction.

ists (e.g., Darwin 1859; Matthew 1915; Wallace 1876), it meant the movement by active migration or passive transfer of a species from its center of origin, usually crossing a preexisting barrier, allowing it to colonize a new area and, eventually, differentiate into new taxa. More recent authors usually imply not a precise center of origin but the ancestral area where the taxon evolved (Bremer 1992, 1995). The term *dispersal* has been used with different and ambiguous meanings because it appears to be an explanation of a pattern in terms of ideas about a process (Eldredge 1981). *Dispersal* has been commonly used to describe processes acting on different temporal scales, such as the routine transport of propagules (short-term or biological timescale), the chance crossing of barriers (short- to long-term scale), and the change of the distributional area of a taxon (short- to long-term or evolutionary timescale). It is useful to distinguish between the movement of an organism within its area of distribution, named dispersion (Platnick 1976), organismic dispersal (Wiley 1981), or intrarange dispersal (MacDonald 2003), and extrarange or biogeographic dispersal (MacDonald 2003).

There are four explanatory models of biogeographic dispersal (Cecca 2002; Heads 2005b; Lieberman 2000; Lieberman and Eldredge 1996; MacDonald 2003; Miranda-Esquivel et al. 2003; Nelson and Platnick 1981; Pielou 1992; Ronquist 1997a, 1997b; Upchurch and Hunn 2002):

- Jump dispersal (also known as long-distance dispersal, waif dispersal, founder effect dispersal, or random dispersal): random movement of organisms through barriers that allows the successful establishment of the species in very distant areas. It was the most popular model in earlier dispersalism, and it has recently received some support to explain disjunct distributions on widely separated areas that apparently were never in contact (Cowie and Holland 2006; De Queiroz 2005; Waters and Roy 2004).
- Diffusion (also known as range expansion): gradual movement of populations crossing adjacent suitable habitats over several generations. There are several examples of species that expanded their distribution in North America with the warming of the climate and the retreat of ice at the end of the Pleistocene (MacDonald 2003).
- Secular migration: movement over a short distance that occurs so slowly that the species evolves in the meantime.
- Geodispersal (also known as mass coherent dispersal, biotic dispersal, concerted dispersal, or predicted dispersal): simultaneous movement of several taxa due to the effacement of a barrier, followed by the emergence of a new barrier that produces subsequent vicariance. An example of geodispersal occurred during the Pleistocene, when the Bering Strait connected North America and Asia, allowing the dispersal of several taxa, including *Homo sapiens,* to the Americas (Lieberman 2004). As a result of geodispersal, biogeographic convergence occurs (Hallam 1974). Its main consequence is the reticulated, nonhierarchical evolution of biotic components.

Vicariance is the appearance of a barrier that allows fragmentation of the distribution of an ancestral species, after which the descendant species may evolve in isolation. The appearance of the barrier causes the disjunction, so both species have the same age. Area fragmentation is not the only way to cause vicariance. The process known as dynamic vicariance (Zunino and Zullini 1995) implies that climatic changes may act by displacing a biotic component gradually in a certain direction, which finally finds a barrier that causes vicariance. One example is the Mediterranean area, where the climatic oscillations of the Pleistocene induced vicariance of formerly widespread taxa in different European peninsulas (Zunino 2003). Another is the South American Chacoan subregion, which developed gradually during the Tertiary, splitting the former continuous Amazonian–Parana forest and leaving a central diagonal of open vegetational formations (Morrone 2006).

Extinction is the local extirpation or total disappearance of a taxon (Morain 1984). It has the potential to obscure biogeographic patterns because biotas may appear different simply because one region has experienced differential extinction (Lieberman 2003a, 2005). Although extinction is a fact, mechanisms explaining it usually do not concern biogeographers because it does not form patterns. Different taxa may actually be biogeographically congruent but not appear so in a cladistic biogeographic analysis because of pruning caused by extinction (Lieberman 2004). Simulation studies conducted by Lieberman (2002) have shown that cladistic biogeographic studies of extant taxa that ignore extinct taxa may be predisposed to artificial biogeographic incongruence. Wiley and Mayden (1985) have also shown the confounding effects of extinctions in the study of extant biotas. Two other possibilities may be invoked to explain the absence of a taxon in an area: primitive absence (it never lived in the area) or pseudoabsence (it lives or lived there but has not been discovered yet).

It has been debated extensively whether dispersal or vicariance represents the most relevant process to explain biogeographic patterns (Humphries and Parenti 1999; Nelson 1978a; Platnick and Nelson 1978). In the nineteenth century and the first decades of the twentieth century, dispersalism emphasized dispersal through a stable geography from centers of origin to explain the distribution of organisms (Darwin 1859; Matthew 1915; Wallace 1876). In the second half of the twentieth century, usually associated with the acceptance of plate tectonics, vicariance arose as a more appropriate explanation than dispersal (Croizat 1964; Croizat et al. 1974). However, intraspecific phylogeography (Avise 2000) shows that dispersal continues to be relevant in explaining the distribution of organisms.

In the mid-nineteenth century, Hooker (1844–1860) discovered that, quite paradoxically, both dispersal and vicariance could explain the same disjunctions. For example, a species may inhabit two disjunct areas, which could be due to a widespread ancestral distribution when both areas were united (vicariance explanation) or to evolution in one area and then dispersal to the other (dispersal explanation). Can one choose between them? The solution to the vicariance–dispersal opposition consists not of choosing one process or the other but of adopting a different reasoning where vicariance includes dispersal, although the latter occurs before the geographic barrier appears (Andersen 1982; Brooks and McLennan 2001; Colacino 1997; Croizat 1958b, 1964; Grehan 1991; Morrone 2004a; Savage 1982). According to this dispersal–vicariance model, geographic distributions evolve in two steps (fig. 2.6):

1. Dispersal (figs. 2.6a–2.6c): When climatic and geographic factors are favorable, organisms actively expand their geographic distribution according to their dispersal capabilities or vagility, thus acquiring their ancestral distribution or primitive cosmopolitism.

2. Vicariance (figs. 2.6d and 2.6e): When organisms have occupied all available geographic or ecological space, their distribution may stabilize. This allows the isolation of populations in different sectors of the area (subspecies, races, or varieties) and the differentiation of new species through the appearance of geographic barriers.

Considering dispersal and vicariance as alternative processes has engendered the impression that they are in opposition (e.g., Avise 2004; Cowie and Holland 2006; MacDonald 2003; Santos 2007; Voelker 1999; Whittaker 1998). According to Parenti and Ebach (personal communication, 2007), the terms *vicariance* and *dispersal* are both explanatory and descriptive, and this is the cause of much confusion in determining which is the most likely explanation for a particular disjunct distribution. Descriptive vicariance represents a way to identify disjunct distributions, showing where biotic components vicariated, not when the isolation occurred. It may function as a general statement of geographic distribution (Humphries and Ebach 2004). Descriptive dispersal identifies the extent to which organisms are able to move within their distributional area. Explanatory dispersal and vicariance are ad hoc mechanisms that are invoked to explain some disjunction, using different models that are particular to each taxon. In evolutionary biogeography, identification and testing of biotic components rely basically on descriptive vicariance.

Biotic Components and Cenocrons

Areas of endemism traditionally have been considered the basic biogeographic units (Humphries and Parenti 1999; Nelson and Platnick 1981; Platnick 1991);

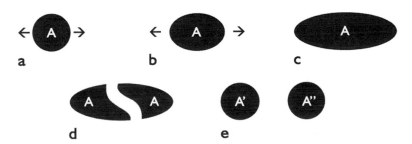

Figure 2.6   Stages of the dispersal–vicariance model: (a–c) dispersal; (d, e) vicariance.

however, Henderson (1991) and Andersson (1996) found that focusing on areas instead of biotas was reductionist, and panbiogeographers (Craw et al. 1999; Croizat 1958b, 1964) considered areas of endemism to be artificial units, preferring instead to recognize generalized tracks. Hausdorf (2002) proposed that biotic elements are more appropriate biogeographic units. Units proposed by other authors include lineages (Jeannel 1942; Ringuelet 1957, 1961), horofaunas and cenocrons (Reig 1962, 1981), chorotypes (Baroni-Urbani et al. 1978; Zunino 2005), and dispersal or distributional patterns (Halffter 1978, 1987). All these concepts basically refer to two different entities: biotic components and cenocrons. Biotic components are sets of spatiotemporally integrated taxa that coexist in given areas, representing biogeographic units, from a synchronic or proximal perspective. Cenocrons are sets of taxa that share the same biogeographic history, constituting identifiable subsets within a biotic component by their common biotic origin and evolutionary history, from a diachronic perspective.

Biotic components are basically like other perduring biological entities (Boniolo and Carrara 2004). They may remain the same despite the possible transformations they could incur (e.g., extinction of particular clades, dispersal of species from other biotic components), they may split into two or more independent biotic components as a result of vicariance, they may mix into a new biotic component as a result of biogeographic convergence, and eventually they may become extinct. From a diachronic perspective, cenocrons allow one to represent how the convergence of biotic components occurs during biotic evolution (fig. 2.7). If a biotic component evolves in isolation (fig. 2.7a), taxa integrated within it will represent a biotic unit, within which it could be possible to track the cenocrons that have contributed to it. When there is biotic convergence (fig. 2.7b) as a result of geodispersal, two or more cenocrons are combined into a single component. Vicariance (fig. 2.7c) splits a biotic component into two or more descendant components. Information about fossils, intraspecific phylogeography, and molecular clocks helps identify cenocrons.

The concepts similar to biotic components (Morrone 2004a; Real et al. 1992) are the lineages (Jeannel 1942), generalized tracks (Croizat 1958b, 1964), horofaunas (Reig 1962, 1981), and areas of endemism (Nelson and Platnick 1981), whereas dispersal and distributional patterns (Halffter 1978, 1987) are similar to cenocrons. Biotic elements, defined as groups of taxa whose ranges are significantly more similar to each other than those of taxa of other such groups (Hausdorf 2002; Hausdorf and Hennig 2007), and chorotypes, defined as sets of species with a coincidence in their spatial distribution that is greater than expected at random (Báez et al. 2004; Baroni-Urbani et al. 1978; Gómez-González et al. 2004; Zunino 2005), represent interesting concepts. Although they seem to correspond to biotic components, I find that they may help identify cenocrons because they may allow identification of biotic units even when substantial dispersal affected the distributions of the taxa analyzed.

Prediction and Retrodiction

Predictions are not usually formulated in evolutionary biogeography because, like other historical disciplines such as systematics, paleontology, or geology, evolu-

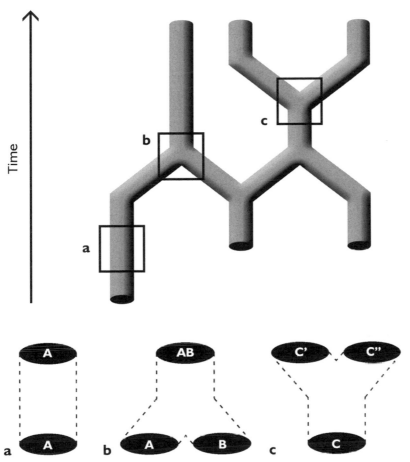

Figure 2.7   Model showing the stages of biotic evolution: (a) stable biotic component;
(b) biotic convergence as a result of geodispersal; (c) vicariance.

tionary biogeography studies unique events that occurred in the past (Mahner and Bunge 1997). Alternatively, we may use biogeographic patterns to make retrodictions, that is, to "predict" past events (Morrone 1997, 2004a). For example, if a panbiogeographic analysis led one to identify an area as a node or a cladistic biogeographic analysis showed an area with conflicting relationships with other areas, one can infer that such area is composite or "hybrid" from a tectonic viewpoint. The geological analysis of such an area may allow one to falsify our retrodiction.

Discovering biogeographic patterns, which provide an organizing framework within which we interpret biological data, makes evolutionary biogeography an empirical science. Patterns are the nearest thing biogeographers have to repeatable experiments; nevertheless, they still represent retrodictive reconstructions (Humphries and Ebach 2004). The validity of a biogeographic hypothesis may be measured by its retrodictive power (Andersen 1982). Well-corroborated biogeographic patterns inform phylogenetic studies by postulating where a sister taxon may occur, reinforce conservation studies by identifying areas of endemism and hotspots, and simplify our understanding of patterns of biodiversity by proposing common causes of our observations (e.g., life and Earth evolve together) rather than a series of unrelated events (Parenti and Humphries 2004).

Biogeographic Approaches and Methods

Evolutionary biogeographic approaches are classified as dispersalism and vicariance biogeography (Morrone 2004a, 2005a). Dispersalism locates centers of origin or ancestral areas and then uses dispersal from them to explain the biogeographic history of particular taxa. Vicariance biogeography looks for the correlation of distributional patterns of different unrelated taxa. Vicariance biogeography includes panbiogeography and cladistic biogeography. The aim of panbiogeography is to identify generalized tracks (primary biogeographic homology), whereas cladistic biogeography deals with general area cladograms (secondary biogeographic homology). Variants of dispersalism include classic dispersalism, phylogenetic biogeography, ancestral areas, and intraspecific phylogeography. Methods implementing panbiogeography include the minimum-spanning tree method, connectivity and incidence matrices, track compatibility, parsimony analysis of endemicity, endemicity analysis, and nested areas of endemism analysis. Methods of cladistic biogeography include the reduced consensus cladogram, ancestral species maps, component analysis, quantitative phylogenetic biogeography, Brooks parsimony analysis, component compatibility, quantification of component analysis, three area statement analysis, tree reconciliation analysis, paralogy-free subtree analysis, dispersal–vicariance analysis, vicariance events analysis, area cladistics, and phylogenetic analysis for comparing trees.

Crisci et al. (2000) recognized nine basic approaches (dispersalism, phylogenetic biogeography, ancestral areas, panbiogeography, cladistic biogeography, parsimony analysis of endemicity, event-based methods, phylogeography, and experimental biogeography) and about thirty techniques for implementing them. Unfortunately, this classification uses different demarcation criteria, with the consequence that the nine approaches are not mutually exclusive (van Veller 2004). Humphries (2004) found that some of the approaches and techniques are intended to resolve problems of conflicting data, some are explanations of patterns, and some are variations on cladistic biogeography implementing different optimizing principles, quantitative methods, and spatial models. The main differences with my classification (Morrone 2004a, 2005a) are as follows:

- Phylogenetic biogeography, ancestral areas, and intraspecific phylogeography are part of dispersalism, sharing the primary objective of estimating the ancestral areas of the taxa analyzed. It may be argued that phylogenetic biogeography also takes into account vicariance, but its use of the progression rule makes it dispersalist. Comparative phylogeography is very different from intraspecific phylogeography; it is aimed toward finding general patterns based on the comparison of the cladograms of different taxa, so it corresponds to cladistic biogeography.

- Parsimony analysis of endemicity is included in panbiogeography, having the same objective: to identify primary biogeographic homology.
- Event-based methods, which include tree reconciliation and dispersal–vicariance analysis, are part of cladistic biogeography.
- One technique of cladistic biogeography ("integrative method") and an event-based method ("combined method") are not techniques in themselves but approaches combining methods.

Evolutionary Biogeography

As a consequence of the frequent episodes of geodispersal, biotic evolution is rarely divergent, resulting in a reticulate rather than branching structure (Brooks 2005; Hovenkamp 1997; Riddle and Hafner 2006; Upchurch and Hunn 2002). To analyze this complexity, we should try to discover the instances of vicariance and those where biotic convergence occurred. A stepwise approach may allow one to identify particular questions, choose the most appropriate methods to answer them, and finally integrate them in a coherent framework. Most authors involved in the theoretical development of biogeography and those who apply their methods usually see them as representing alternative schools; however, they can be used to answer different questions, which can be different steps of an evolutionary biogeographic analysis (fig. 1.1).

This stepwise approach comprises five steps, each corresponding to particular questions, methods, and techniques. Panbiogeography and methods for identifying areas of endemism are used to identify biotic components, which are the basic units of evolutionary biogeography. Cladistic biogeography uses phylogenetic data to test the historical relationships between these biotic components. On the basis of the results of the panbiogeographic and cladistic biogeographic analyses, a regionalization or biogeographic classification may be achieved. Intraspecific phylogeography, molecular clocks, and fossils are incorporated to help identify the cenocrons that become integrated in a biotic component. Finally, the geological and biological knowledge available can be integrated to construct a geobiotic scenario that may help explain the way the biotic components evolved.

This stepwise approach does not imply that every biogeographer must follow all the steps, but anybody may articulate a specific biogeographic question and choose the most appropriate method to answer it. Given some time, as the different analyses accumulate, they can be integrated to formulate coherent theories. I defend this approach within the philosophical framework of integrative pluralism (Mitchell 2002). It is not an eclectic or "anything goes" approach, but the different methods are compatible because they give partial solutions when answering particular questions. Integrative pluralism with respect to methods coexists with the objective of integration in order to explain biotic evolution.

For Further Reading

Andersson, L. 1996. An ontological dilemma: Epistemology and methodology of historical biogeography. *Journal of Biogeography* 23:269–277.

Crisci, J. V. 2001. The voice of historical biogeography. *Journal of Biogeography* 28: 157–168.

Morrone, J. J. and J. V. Crisci. 1995. Historical biogeography: Introduction to methods. *Annual Review of Ecology and Systematics* 26:373–401.

Vuilleumier, F. 1999. Biogeography on the eve of the twenty-first century: Towards an epistemology of biogeography. *Ostrich* 70:89–103.

For Discussion

1. Carefully read the following articles:

   Andersson, L. 1996. An ontological dilemma: Epistemology and methodology of historical biogeography. *Journal of Biogeography* 23:269–277.

   Vuilleumier, F. 1999. Biogeography on the eve of the twenty-first century: Towards an epistemology of biogeography. *Ostrich* 70:89–103.

   a. List the basic problems of biogeography that the authors identify.

   b. Discuss the similarities and differences between both authors' approaches.

2. Carefully read the following articles:

   Craw, R. C. 1983. Panbiogeography and vicariance cladistics: Are they truly different? *Systematic Zoology* 32:431–438.

   Grehan, J. R. 1988. Biogeographic homology: Ratites and the southern beeches. *Rivista di Biologia—Biology Forum* 81:577–587.

   Morrone, J. J. 2001. Homology, biogeography and areas of endemism. *Diversity and Distributions* 7:297–300.

   Nelson, G. 1994. Homology and systematics. In *Homology: The hierarchical basis of comparative biology,* ed. B. K. Hall, 101–149. San Diego: Academic Press.

   a. Identify the authors' ideas about biogeographic homology.

   b. Discuss their similarities and differences.

CHAPTER 3

A Brief History of Evolutionary Biogeography

Biogeography has had a long history woven into natural history, evolutionary biology, systematics, geology, and ecology. In this chapter I highlight some authors whose works focus on evolutionary aspects of biogeography. Many works that might have been included have been omitted because this chapter is not intended to be an exhaustive historical treatment. For historical accounts of biogeography, see Hofsten (1916), George (1964), Browne (1983), Larson (1986), Papavero and Balsa (1985), Bowler (1989, 1996), Papavero (1990, 1991), Papavero et al. (1997), and Lomolino et al. (2004). My approach is largely archaeological (in the sense of Foucault 1966, 1969), trying to identify strategies and practices, in order to reconstruct the space where evolutionary biogeography has developed. Additionally, I have paid special attention to the way more recent authors have interpreted these ideas (metahistory).

The Beginnings of Biogeography

The questions posed by modern biogeographers are already present in prescientific writings. The concepts of centers of origin and dispersal may be recognized in the biblical accounts of the Garden of Eden, Noah's Ark, and the Tower of Babel (Papavero and Balsa 1985; Papavero et al. 1997). In the Middle Ages some progress was made in the study of life and Earth. In his *Confesiones* and *De Civitate Dei*, Saint Augustine (354–430) analyzed the Deluge and the dispersal of animals preserved in Noah's Ark. Muslim scholars Abu al-Rayhan Mohamed ben Ahmad al-Biruni (973–1050) and Avicenna (980–1037) analyzed the expansion of geographic distributions and the interpretation of fossil distributions. In *Summa Contra Gentiles* Saint Thomas Aquinas (1225–1274) defended the physical existence of the Garden of Eden and discussed the secondary dispersal from Mount Ararat, where Noah's Ark landed after the Deluge.

In the sixteenth century, new attitudes toward knowledge developed during the Enlightenment, when the invention of printing by movable metal type and global exploration provided better conditions for the development of biogeography. Spanish Jesuit Joseph d'Acosta (1540–1600) tried to explain the presence of the human species in the New World. In *Historia Natural y Moral de las Indias* (1590) he examined alternative explanations and concluded that the Americas should be connected, in some place, to the Old World, postulating the existence of the Bering Strait, discovered in the eighteenth century (Papavero 1991). German Jesuit priest Athanasius Kircher (1602–1680) calculated the dimensions of Noah's Ark necessary to accommodate a pair of the 310 animal species that were known at that time. By the beginning of the nineteenth century, the idea of the Ark was almost completely abandoned because of the increasing numbers of species that were discovered and described, but the concept of the Deluge was still entrenched (Briggs and Humphries 2004). Matthew Hale (1609–1676) erected theoretical land bridges to explain how animals had reached the New World from the Old. These bridges subsequently disappeared, and the New World species were transformed after living for some time in a new area (Briggs 1995).

Classical Biogeography

Carl Linnaeus (1707–1778) inaugurated classical biogeography. He was one of the first authors to provide an explanation of the geographic distribution of living beings in accordance with the book of Genesis. In his *Oratio de Telluris Habitabilis Incremento,* Linnaeus (1744) situated the Garden of Eden on a tropical island, under the equator, which was the only land emerging from the primordial sea. All the animals and plants inhabited this paradisiacal island, which bore a variety of ecological conditions arranged in elevational and climatic zones; those needing a cold climate lived near the peak of a high mountain, and those needing a warmer climate inhabited the lowlands. After the Deluge, as the waters receded and the lands expanded, species dispersed to the areas where they have remained. Linnaeus's theory clearly comprises two fundamental biogeographic ideas: a small center where species appear and their movement to other areas. Nelson and Platnick (1981) analyzed these assumptions, finding that this generalized center of origin and dispersal explanation was impossible to falsify, either with reference to empirical observations of species, their distributions, and their relationships or with reference to causal processes of species origin and dispersal. To be proper, a theory should suggest additional observations that would confirm or refute it. Whatever Linnaeus might have intended as the implications of his theory, his contemporaries and followers deduced that if he were correct, different areas of the world should be inhabited by the same species (Nelson and Platnick 1981).

George-Louis Leclerc, Comte de Buffon (1707–1788), was the first author to falsify Linnaeus's explanation. Distributional problems occupy so central a position in Buffon's writings that he may be viewed as the founder of evolutionary biogeography (Larson 1986; Mayr 1982; Nelson 1978a). In his monumental *Histoire Naturelle, Générale et Particulière* (1749–1788), Buffon observed that different tropical areas of the world, even when having some similar climatic and environmental conditions, were inhabited by completely different mammal species. This discovery was subsequently named Buffon's law (Briggs 1987; Nelson 1978a). Nelson (1978a) held that the history of biogeography may be subdivided in two parts: one dealing with the development of Buffon's law and the other with the development of causal explanations of the law. In addition to being an early evolutionist, Buffon suggested a vicariance model when he accounted for the origin of disjunct distributions by reference to the times when continents were not yet separated (Briggs and Humphries 2004), although his statements on vicariance and evolution may be considered topics for speculation, not finished doctrines. However, Buffon's advocacy of a historical approach to natural history and his emphasis on factual evidence marked an important change in the history of biogeography (Larson 1986). Pierre Latreille (1762–1833) studied insects, and Georges Cuvier (1769–1832) studied reptiles of Africa and South America, both corroborating Buffon's law.

Johann Reinhold Forster (1729–1798) was a German naturalist who emigrated to England. Together with his son Georg, Forster accompanied Captain Cook on his second expedition to the southern seas. On their return, Forster (1778) published his *Observations Made During a Voyage Round the World,* where he described various natural regions, showing how floras replaced one another as the physical characteristics of the environment changed, and noted the relationships of the vegetation and the animals found in each region (Briggs 1995). Forster understood that biotas are living communities, characteristic of certain geographic areas, giving rise to the concept of natural biotic regions (Briggs 1987). He found the tropics to be the richest areas, where nature reached its highest expression (Browne 1983).

German zoologist Eberhardt August Wilhelm von Zimmermann (1743–1815), in *Geographische Geschichte* (1778–1783), proposed that species were created in the areas where they are distributed today. The climate that prevails today in these areas is the same that prevailed in the time of the species creation. Zimmermann held that naturalists should avoid fruitless speculations and concentrate on solving more modest problems (Larson 1986).

Karl Willdenow (1765–1812) was a German botanist and head of the Berlin Botanical Garden. In *Grundriss der Kräuterkunde,* Willdenow (1792) outlined the elements of phytogeographic regionalization. To account for the differences between botanical provinces, he envisioned an early stage of many mountains surrounded by a global sea. Different plants were created on their peaks and then spread downward, as the water receded, to form present phytogeographic provinces (Briggs 1987, 1995). Willdenow insisted that many features of present distribution were products of historical development (Larson 1986).

German naturalist Alexander von Humboldt (1769–1859), one of Willdenow's students, studied Latreille's and Cuvier's works and generalized Buffon's law to include plants and most terrestrial animals (Brown and Lomolino 1998). His *Essai sur la Géographie des Plantes* (Humboldt 1805) was the result of extensive field work on Mount Chimborazo in the Andes. Humboldt invented "botanical arithmetic," a phytogeographic technique that enabled naturalists to reduce the absolute number of species from different regions into statements of proportions, which could be arranged with others in a table. After calculation of the ratios of species of one plant family to another, the predominant taxa of a region could be discovered objectively. Humboldt, Robert Brown, and de Candolle made "botanical arithmetic" an extremely useful tool, and although it was short lived, it was instrumental in making the studies of geographic distribution a scientific exercise (Browne 1983). Humboldt's (1815) *Personal Narrative of Travels to the Equinoctial Regions of America* was a key element in captivating a generation of naturalists, including Charles Darwin, Alfred R. Wallace, and Joseph D. Hooker.

Swiss botanist Augustin Pyrame de Candolle (1779–1841) emphasized the distinction between "stations" (habitats) and "habitations" (botanical provinces):

> By the term *station* I mean the special nature of the locality in which each species customarily grows; and by the term *habitation*, a general indication of the country wherein the plant is native. The term *station* relates essentially to climate, to the terrain of a given plant; the term *habitation* relates to geographical, and even geological, circumstances. . . . The study of stations is, so to speak, botanical topography; the study of habitations, botanical geography. (de Candolle 1820:383)

According to this author, explanations for the former depend on physical causes that are acting in the present, whereas those for the latter depend on causes that existed in the past. De Candolle believed that the distinction between stations and habitations was important and that the confusion between them limited the study of the geographic distribution of plants (Nelson and Platnick 1981). Nelson (1978a) identified them with ecological and historical biogeography, respectively.

Darwinian Biogeography

British naturalist Charles Lyell (1797–1875) was the first to articulate the classic dispersalist model developed later by Darwin and Wallace (Bueno-Hernández and Llorente Bousquets 2006). Lyell believed that the number of living species was in equilibrium, so when some species became extinct, others had to be created in one region or another. Lyell's concept of "creation" means creation according to natural laws and processes (Nelson and Platnick 1981). Lyell's (1830–1833) *Principles of Geology* opposed rapid catastrophic changes, which he thought were based on interpretations of the Bible, proposing that gradual changes through time accounted for fossil remains of extinct species, and he emphasized environmental conditions to explain the creation and extinction of species (Bueno-Hernández and Llorente Bousquets 2006; Humphries and Ebach 2004). He connected fossils and contemporary patterns, bringing to biogeography a sense of history and evolution (Briggs 1995).

British naturalist Charles Darwin (1809–1882) made several contributions to biogeography. In fact, the key evidence that convinced him of evolution came from his study of geographic distribution (Mayr 1982). He was the first author to postulate that the facts of geographic distribution might be explained by a combination of a theory of evolution with the study of dispersal of plant and animal taxa (Bowler 1989). Darwin's interest in biogeography can be categorized into three loosely defined stages: the notebooks and unbound notes of the late 1830s, his unpublished essays on species from the 1840s, and the manuscripts and books of the 1850s, including *On the Origin of Species* (Camerini 1993). In his first sketch of the theory of natural selection, drafted in 1842, and in his *Essay* (1844), Darwin speculated on the changes in plant species ranges that were induced by climatic and altitudinal changes (Browne 1983). On the basis of Humboldt's account of mountain zonation, Darwin envisaged climatic belts moving up or down the sides of hills as climate changed or the land subsided or was elevated by geological activity.

In *The Origin of Species* Darwin (1859) wrote two chapters on biogeography. Contemporary creationists explained disjunct distributions as the result of multiple centers of creation. To counter this idea, Darwin invoked a theory of chance dispersal from single centers of origin. He wrote, "The simplicity of the view that each species was first produced within a single region captivates the mind. He who rejects it, rejects the *vera causa* of ordinary generation with subsequent migration, and calls in the agency of a miracle" (Darwin 1859:352). What is the essence of the Darwinian perspective? "The view of each species having been produced in one area alone, and having subsequently migrated from that area as far as its powers of migration and subsistence under past and present conditions permitted, is the most probable" (Darwin 1859:353).

Another British naturalist, Alfred Russel Wallace (1823–1913), is considered by several authors to be the father of evolutionary biogeography (Brown and Lomolino 1998; George 1964; Riddle 2005). Interestingly, his first four articles, published between 1855 and 1863, followed extensionist ideas, whereas in the fifth, published in 1864, he showed a radical change to permanentism (Bueno-Hernández and Llorente Bousquets 2003). His first important paper, *On the Law Which Has Regulated the Introduction of New Species* (Wallace 1855), constituted his first public announcement of his evolutionary hypothesis (Camerini 1993). His main thesis was that "every species has come into existence coincident both in space and time with a pre-existing closely allied species" (Wallace 1855:156), meaning that geographic distribution was not random but followed a simple law. Additionally, Wallace's law challenged species immutability: Species evolved. Wallace also used the term *creation*, but for him it meant creation from preexisting species. He clearly challenged special acts of creation, these being unnecessary to account for the existence of new species.

Wallace (1863a) published an article on the geography of the Malay archipelago, which was accompanied by a map depicting the boundary between the Asian and Australian biotas, known later as Wallace's line. Mapping this boundary served four functions: It was a method for communicating and organizing faunistic data, a potential device for predicting range limits of other species, a modern method of argumentation, and a method of analysis that tested evolutionary hypotheses positively (Camerini 1993). Another of Wallace's biogeographic projects was his attempt to map the boundary line between the Malays and the Papuans in Southeast Asia (Wallace 1863b). Although concerns with origins and explanations were very important to Wallace's biogeographic thinking, this work on human biogeography dealt not only with the search for evolutionary mechanisms but also with field mapping over geographic space (Vetter 2006). Although Wallace combined fact gathering with higher-level generalizing, on his return to England he found that the response to his theoretical interpretations was less than enthusiastic.

Wallace (1864) published *On Some Anomalies in Zoological and Botanical Geography*, in which he changed to a clear permanentist position (Bueno-Hernández and Llorente Bousquets 2003). The objective of this paper was to analyze some "anomalies" in the distribution of different taxa and to articulate a general system of regionalization for Earth. Following Sclater's ideas, Wallace's purpose was to look for natural relationships between the ontological regions. In contrast with his previous works, Wallace (1864) postulated that long-distance dispersal was the fundamental process that caused distributional patterns. Despite showing a firm permanentism, he still displayed a degree of extensionism:

> A great part of the southern portion of America is of more recent date than the central tropical mass, and must have had at some time a closer communication than at present with the Antarctic lands and Australia, the insects and plants of which finding a congenital climate, established themselves in the new country. (Wallace 1864:120)

Once converted to permanentism, Wallace (1876) wrote his monumental *The Geographical Distribution of Animals*, in which he summarized his previous findings and provided a general account of Sclater's system with numerous examples. According to Wallace, the purpose of a general regionalization of the earth was to allow comparative analyses of the geographic distribution of different taxa, communication between naturalists, and anomaly detection (Bueno-Hernández and Llorente Bousquets 2003). Regions should be useful and natural, so they must be based on the great geographic divisions of the earth, be rich in all taxonomic groups, and show some individuality (based on the presence or absence of taxa that are abundant in adjacent areas).

Patterson (1983) found in Wallace the beginnings of two approaches to biogeography. The first uses the distribution of life as factual evidence bearing on the history of the earth "may reveal to us, in a manner which no other evidence can, which are the oldest and most permanent features of the Earth's surface, and which are the newest" (Wallace 1876:8). The second uses theories of the history of the earth and life to explain distributions: "[They] will teach us to estimate the comparative importance of various groups of animals, and to avoid the common error of cutting the Gordian knot of each difficulty by vast hypothetical changes in existing continents and oceans—probably the most permanent features of our globe" (Wallace 1876:9).

Patterson (1983) considered Wallace's biogeographic approach original because of its commitment to evolution, its use of statistical comparisons of genera, its reliance on mammals, and its appeal to fossils. Michaux (1991) considered it an oversimplification to regard Wallace as a dispersalist. Instead, he attempted to interpret distributions in terms of the geological history. Bueno-Hernández and Llorente Bousquets (2003) analyzed Wallace's biogeographic ideas, finding a sharp contrast between his first papers and *On Some Anomalies in Zoological and Botanical Geography*. Originally, Wallace was an extensionist, interpreting faunistic affinities as indicating former connections and considering dispersal to be irrelevant, but after 1864 he believed that the earth's surface has been basically unchanged over its history and postulated that dispersals have had a fundamental role in shaping biogeographic patterns. Heads (2005a) noted that in the original edition of *The Malay Archipelago*, Wallace (1869) explained some "anomalies" in the fauna of Sulawesi with reference to past connections with Africa, but in the tenth edition (1890) he added a footnote stating that he had concluded that no such connecting land was needed to explain the facts.

Philip Lutley Sclater (1829–1913), a British ornithologist, published *On the General Geographical Distribution of the Members of the Class Aves* in 1858. He divided the world into six biogeographic regions that would reflect "the most natural primary ontological divisions of the Earth's surface" (1858:130), although he acknowledged that his system was rudimentary and that much work was left to be done. Sclater's biogeographic regions were modified by Wallace (1876) and other authors; however, they have stood the test of time, and many textbook accounts use them (Briggs and Humphries 2004; Brown and Lomolino 1998). In contrast to other biogeographers at that time, Sclater believed that organisms did not disperse to favorable habitats; rather, they changed over time in the same area. When explaining the presence of some mammal taxa in Madagascar, Sclater concluded,

> The anomalies of the mammal-fauna of Madagascar can be best explained by supposing that, anterior to the existence of Africa in its present shape, a large continent occupied parts of the Atlantic and Indian Oceans stretching out towards America on the west, and to India and its islands on the east; that this continent was broken up into islands, of which some have become amalgamated with the present continent of Africa, and some possibly with what is now Asia, and that [in] Madagascar and the Mascarene Islands we have existing relics of this great continent. (Sclater 1864:219)

This statement makes Sclater a precursor of vicariance biogeography.

Extensionists and Other Unorthodox Biogeographers

Darwin's and Wallace's dispersalist ideas were challenged by several authors. The extensionists, such as Forbes and Hooker, considered long-distance dispersal an unlikely process to explain disjunct distributions (Brown and Lomolino 1998). They chose instead to postulate ancient land bridges and continents now submerged in the oceans, which once linked the surviving continents (Bowler 1996). Extensionists initially had a profound influence, but at the end of the nineteenth century their ideas were abandoned.

British naturalist Joseph Dalton Hooker (1817–1911) brought with him hundreds of plant specimens from a voyage of exploration to Antarctica and the southern continents. After analyzing them (Hooker 1844–1860), he found that it was common for the same families and even genera to be present in widely separated areas as New Zealand, Australia, Tasmania, and Patagonia. When describing the flora of New Zealand, Hooker decided that species were surviving relicts of a once widespread flora that had grown on all the southern continents. He even suggested that these lands had been joined together, and geological events had caused their breakup. In modern terms, he was suggesting vicariance as a historical explanation (Briggs 1995; Brown and Lomolino 1998). The need to postulate former land bridges to explain disjunct distributions by Hooker and other so-called extensionists was overshadowed by Darwin and Wallace, and he had to abandon his ideas (Funk 2004). However, Hooker is another precursor of vicariance biogeography (Briggs and Humphries 2004).

Hermann von Ihering (1850–1930) was a German zoologist who lived in southern Brazil for more than four decades. On the basis of his observations of the freshwater fauna of southern South America, New Zealand, and Africa, as well as his geological knowledge, he came to be dissatisfied with Darwinian's dispersal over preexisting barriers. He postulated, instead, that biotic disjunctions could be explained by the existence of former land bridges that joined areas now widely separated. In *Die Geschichte des Atlantisches Ozeans* (von Ihering 1927), he delineated Archatlantis (North Atlantic), Archhelenis (South America), and Lemuria (Madagascar and India), which existed from the Cretaceous to the Eocene. Von Ihering was one of the earliest vicariance biogeographers, recognizing biotic affinities between southeastern Brazil and Chile, areas that he suggested were part of Archiplata, an ancient territory (Choudhury and Pérez-Ponce de León 2005). Von Ihering (1927) summarized his main biogeographic ideas while reconstructing the configuration of the continents and land bridges during the Cretaceous. Until recently, it was common to ridicule the ideas of von Ihering and others as "land bridge builders," but this is inaccurate because former lands were only a small part of their analyses (Heads 2005a).

Argentinean paleontologist Florentino Ameghino (1854–1911) was an opponent of Holarcticism, the approach that situates the origin of all taxa in the Northern Hemisphere. On the basis of paleontological findings made by him and his brother Carlos, he worked out the stratigraphy of South America. As a result of a systematic overestimation of the ages of the strata, Ameghino claimed that South American mammals predated and were ancestral to all others in the world. Many paleontologists engaged a debate with Ameghino, but finally the study of fossil invertebrates by Ortmann showed that Ameghino's correlations, based only on mammals, were wrong (Bowler 1996).

Daniele Rosa (1857–1944), an Italian lumbricologist (a specialist in earthworms), was an evolutionary biologist. In 1918 he published *Ologenesi*, where he developed a theory that emphasized evolution due to internal causes. Rosa's fundamental statements were the following: Evolution depends not only on external but also on internal factors to such an extent that it continues even in a uniform environment; the direction in which internally driven evolution acts is independent of the variation of external factors; and even so, evolution is not indefinitely linear; on the contrary, it is dichotomously ramified (Baroni-Urbani 1977). The theory of hologenesis assumes that cladogenesis proceeds dichotomously (Luzzatto et al. 2000). In biogeography, Rosa (1918) postulated cosmopolitan distribution of primitive species. This is relevant because instead of a "vacuum" theory, in which areas are devoid of taxa, this provides the possibility of a "primitive cosmopolitanism," a prerequisite for vicariance. Rosa's (1918) arguments against dispersal are similar to those of panbiogeographers and cladistic biogeographers in more recent times (Luzzatto et al. 2000).

In 1922 English botanist John Christopher Willis (1868–1958) published *Age and Area*. In this book, he postulated that age as an explanation of distribution was simpler than natural selection. This hypothesis was born when the author had the opportunity to study the flora of Ceylon (now Sri Lanka) (Willis 1915) and neighboring countries and observed that many endemic species were confined to small areas of the island, other species also distributed in India had larger areas, and those ranging beyond the peninsula had the largest areas. When defining very rare, rare, and rather rare species (fig. 3.1), Willis observed that very rare species occurred in one place only, rare species in areas about 10–30 miles across, and rather rare species in areas 30–60 miles across. He found that the idea that endemic species were local adaptations could not explain the observed facts. Some mechanical explanations were necessary, and Willis thought it was the age of the species. His hypothesis was formulated as follows:

> The area occupied at any given time, in any given country, by any
> group of allied species at least ten in number, depends chiefly, so long
> as conditions remain reasonably constant, upon the ages of the species
> of that group in that country, but may be enormously modified by the
> presence of barriers such as seas, rivers, mountains, changes of climate

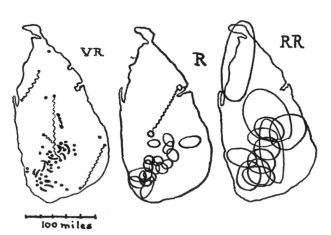

Figure 3.1 Geographic distribution of very rare (VR), rare (R), and rather rare (RR) species from Sri Lanka (modified from Willis 1922:56).

100 miles

from one region to the next, or other ecological boundaries, and the like, also by the action of man, and by other causes. (Willis 1922:63)

Croizat (1958a) commented on Willis's ideas, which he considered to represent another challenge to dispersalism. However, he criticized the idea that the widest-ranged plants are the oldest and the smallest-ranged the youngest. Despite this dismissal, Croizat considered that if his own work were "to be 'placed' somewhere in the stream of our times any how, it surely would not fall with Darwinism in the least (and even less with Matthewism, Mayrism, Simpsonism, Darlingtonism, Goodism, etc.) but find more or less distal place in the sphere of Willisism" (Croizat 1958a:116). Nelson (1994) considered Willis's curves to be artifacts derived from paraphyly and ancestor–descendant relationships.

French entomologist René Jeannel (1879–1965), a specialist in beetles (Coleoptera) and earlier supporter of Wegener's continental drift, published several monographs in which he discussed biogeographic distribution based on phylogenetic bases (Roig-Juñent 2005). In one of his most important works, *Les Migadopides (Coleoptera, Adephaga), une Lignée Subantarctique* (Jeannel 1938), he analyzed a tribe of the family Carabidae. He characterized Migadopini and identified within it different phyletic lineages. On the basis of its distribution in southern Chile, New Zealand, Australia, and Tasmania, Jeannel postulated a Mesozoic origin in the southern continents, when they integrated in a single landmass, but Africa and Madagascar were already separated. Darlington (1965) criticized Jeannel's (1938) conclusions, suggesting that the existence of fully winged genera, the presence of species in localities as far north as Uruguay in South America and New South Wales in Australia, the extreme morphological diversity of the tribe, and its close relationship with a tribe of the north temperate zone indicated that the ancestor of the tribe crossed the tropics a long time ago and that an Antarctic origin was not the only possibility. Although some of the taxa recognized by Jeannel seemed to be formed in order to fit continental drift, his work stands as an important pioneering effort to correlate distributional patterns with continental drift (Noonan 1979).

In 1942 Jeannel published his masterpiece, *La Genése de Faunes Terrestres*, in which he discussed the relevance of isolation to explain how different related lineages inhabited disjunct geographic areas. After defining distributional patterns, Jeannel postulated that it was possible to determine the origin, age of dispersal, and geological age of the taxa analyzed. In order to explain the main distributional patterns in South America, he postulated the existence of fragments of Gondwana named Paleantarctica (including Australia and Archiplata in South America) and Inabresia (comprising South Africa, Madagascar, India, and Archiguyana and Archibrasil in South America). The phyletic lineages from Paleantarctica are adapted to cold temperate conditions, whereas those from Inabresia are from tropical habitats.

The steps Jeannel followed in his analyses have been summarized as follows: (1) Undertake the taxonomic revision of the group, (2) establish the evolutionary significance of several characters, (3) define the main phyletic lineages, (4) determine the phylogenetic relationships between the phyletic lineages, (5) seek explanations for the distributional patterns based on paleogeographic and paleoclimatological data, and (6) support the conclusions with data from other taxa (Roig-Juñent 2005). Jeannel's emphasis on an accurate phylogenetic analysis of the taxa studied, his discussion of paleogeographic and paleoclimatological evidence, and the search for patterns among completely different groups place him as one of the outstanding precursors of evolutionary biogeography.

Other unorthodox biogeographic ideas came from authors who supported the idea of continental drift. Early in the twentieth century, Taylor (1910) postulated that the Atlantic Ocean originated with the separation of two continents that formerly constituted a single landmass. In 1912 Alfred Lothar Wegener (1880–1930) published *Die Entstehung der Kontinente und Ozeane*, where he postulated the existence of a former giant supercontinent, which in later editions he named Pangaea. His ideas were not accepted for several decades until the pioneering work of Hess (1962) provided geological evidence for continental drift.

The New York School of Zoogeography

In the United States, in the first decades of the twentieth century, a zoogeographic school originated with William Diller Matthew (1871–1930), founded on the dispersal of organisms over a static Earth. Simpson, Darlington, and Myers are its main authors (Nelson and Ladiges 2001; Williams 2007b). It has been named the New York school of zoogeography (Croizat 1958b, 1984b) and Holarcticism (Reig 1981).

Matthew was a Canadian paleontologist. His most widely known work is *Climate and Evolution* (Matthew 1915). He opposed land bridges across the oceans, advocating Holarctic centers of origin for all terrestrial vertebrates and successive dispersal events to Africa, to Australia through southeastern Asia, and to the Americas through the Bering Strait. Matthew summarized his theses as follows:

1. Secular climatic change has been an important factor in the evolution of land vertebrates and the principal known cause of their present distribution.

2. The principal lines of migration in later geological epochs have been radial from Holarctic centers of dispersal.

3. The geographic changes required to explain the present distribution of land vertebrates are not extensive and for the most part do not affect the permanence of the oceans as defined by the continental shelf.

4. The theories of alternations of moist and uniform with arid and zonal climates, as elaborated by Chamberlin, are in exact accord with the course of evolution of land vertebrates, when interpreted with due allowance for the probable gaps in the record.

5. The numerous hypothetical land bridges in temperate tropical and southern regions, connecting continents now separated by deep oceans, which have been advocated by various authors, are improbable and unnecessary to explain geographic distribution. On the contrary, the known facts point distinctly to a general permanency of continental outlines during the later epochs of geologic time, provided that due allowance be made for the known or probable gaps in our knowledge. (Matthews 1915:3)

The implications of Matthew's (1915) idea are rather simple. At any given period, the most advanced species are those inhabiting the original area where they evolved, and the most primitive species are those remote from the center of origin. This remoteness is not a matter of geographic distance but of inaccessibility to invasion, conditioned by the habitat and facilities for migration and dispersal. Classic dispersalism, especially that of the twentieth century, is a reflection of Matthew's (1915) work, which simplified and codified Darwin's and Wallace's ideas (Heads 2005a). Matthew reaffirmed the permanence of the great features of the earth, the origin of most taxa in the Northern Hemisphere, and the importance of climatic changes as the principal cause of the distribution of land vertebrates (Bowler 1996).

George Gaylord Simpson (1902–1984), a U.S. mammalogist and paleontologist, worked as one of Matthew's assistants and succeeded him in the position of curator of vertebrate paleontology at the American Museum of Natural History. In 1940, he published the essay *Mammals and Land Bridges*, in which he began developing his concepts of corridors, filter bridges, and sweepstake routes. He discussed the ideas about biotic expansions and contractions by Willis, Rosa, and Matthew. He dismissed Willis's "age and area," although he found this theory to involve "various interesting corollaries" (Simpson 1940:140) because conclusions reached were unsatisfactory for mammals, and he dismissed Rosa's hologenesis because it "seems at first sight so fantastic as hardly to warrant serious discussion" (Simpson 1940:141). With respect to Matthew (1915), he found that his theses were typical of those about mammals.

Simpson (1940) postulated that almost any taxon appears on a center of origin, then it spreads steadily in all directions until it encounters insuperable barriers, and after a time it begins to contract, often splitting into disjunct small areas (fig. 3.2). This model implies movement from place to place, even to some places where their immediate ancestor has never been. If no barrier exists between two areas,

Figure 3.2 Model postulated to explain the expansion and contraction of a taxon. Lines 1–11 represent distributional limits at different times; 1–5, primary expansion of the group on the landmass where it originally evolved; 5–8, expansion on a second landmass; 8–11, contraction. After 8 it is extinct in its homeland but survives abroad, and after 11 it is extinct everywhere (modified from Simpson 1940:145).

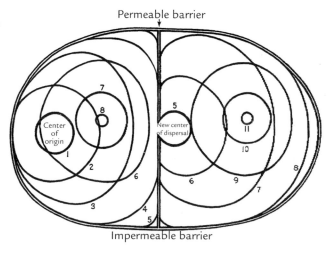

their faunas will be very similar. When two areas are separated by a strong barrier, they develop quite different faunas, their differences being roughly proportional to the lapse of time since the areas were connected. Simpson (1940) postulated that what is a barrier for one taxon may be an open route for another and that wide-open, nonselective connections or corridors are rare. Thus "filter bridges" are the best fit for the zoological evidence. For the instances of dispersal of single groups or unbalanced faunas that do not meet the criteria for filter bridge connections, Simpson postulated the existence of "sweepstake routes," which depending on chance may allow the dispersal of particular taxa.

Simpson's (1953) *Evolution and Geography* represents an essay on biogeography, where he formalized some key terms, such as *fauna*. For Simpson,

> The fauna of any one area is made up of numerous species which have evolved from diverse ancestral groups, which interact with each other in various ways, which (either as such or in ancestral forms) have been in the region for different lengths of time, which have come from different geographic sources, and which have at present different relationships to species existing in other regions. (Simpson 1953:31)

In order to analyze a fauna, Simpson considered five basic elements: (1) systematic classification of all groups present, implying their phylogenetic relationships; (2) ecological characteristics of each group; (3) antiquity of each group within the region; (4) geographic source of each group; and (5) current geographic affinities of each group.

Philip J. Darlington Jr. (1904–1983), a U.S. entomologist, was not a disciple of Matthew, but his work can be ascribed to the New York school of zoogeography. He was one of the last biogeographers to advocate dispersal over fixed continents as an explanation for current distributional patterns (Funk 2004). Darlington's (1957) *Zoogeography*, a classic biogeographic textbook, represents the climax at Harvard University of the New York school near the end of its development (Nelson and Ladiges 2001). It contains detailed accounts of the global distributional patterns of freshwater fishes, amphibians, reptiles, birds, and mammals, continental and island patterns, evolution of the patterns, transition zones, and zoogeographic principles. Darlington (1957:25–35) discussed seven working principles for zoogeography:

1. Formulate working principles before work is begun.
2. Work with facts rather than opinions, so far as possible, and understand the nature of the "facts" worked with.
3. Define and limit both the work to be done and the factual material to be worked with.
4. Present the selected material fully and fairly.
5. Remember that animals are living things, which are constantly evolving and multiplying in some place, and dying out in others, and thus forming new geographic patterns.
6 Understand and use fairly the clues to geographic histories of animals.

7. Try working hypotheses when facts fail, remembering always that the hypotheses are not facts.

Darlington (1957) characterized transition zones as areas where regional faunas meet or are separated by partial barriers, with overlapping of faunal elements, with progressive subtraction, in both directions. Transition zones (fig. 3.3) are complex, involving exclusive, transitional, and shared families and genera. He described the main transition zones of the world and their taxa, especially the Wallacea and the Central American–Mexican, currently known as the Mexican Transition Zone (Halffter 1987; Morrone 2006).

Darlington (1965) published *Biogeography of the Southern End of the World*, in which he held that many plants and invertebrates of southern South America, southern Australia, Tasmania, and New Zealand are phylogenetically related, whereas this pattern does not occur for vertebrates (except for some salt-tolerant fishes). Rainfall, cold, and other climatic and ecological factors affect the distribution of life in the far south now and have probably done so in the past. Southern distributions should be analyzed in relation to the whole world. Distribution of different taxa should be analyzed case by case to see whether they have diverse patterns (suggesting different histories) or common patterns (suggesting a common history). Darlington (1965) concluded that no great changes in the arrangement of the southern continents seem possible in the Late Cretaceous (except possibly the making and breaking of an isthmian link between Antarctica and southern South America). Several plant and animal taxa have been arriving continually from the north. Some groups of plants and invertebrates have been especially adapted and confined to the southern continents, where a characteristic far southern biota evolved, especially in wet forests and wet moorlands. Interestingly, in the conclusion of the book, he stated,

> Evidence from several independent sources, discussed and compared in preceding chapters, has forced me to conclude that the southern

Figure 3.3   Transition between two regional faunas, showing that each fauna consists of exclusive, transitional, and shared families; and transitional and shared families consist of exclusive, transitional, and shared genera (modified from Darlington 1957:453).

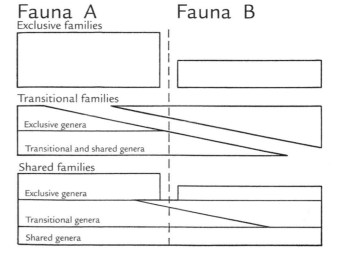

Fauna A   Fauna B

continents have drifted. I have therefore become a Wegenerian, but not an extreme one. I doubt the former existence of a Pangaea or Gondwanaland, and I think that movements of continents have been simpler and shorter than most Wegenerians support. Also, I am not absolutely sure of my conclusions. I shall therefore state the conclusions as probabilities rather than proven facts. (Darlington 1965:210)

Darlington's (1965) book best sums up two problems of classic dispersalism: a pre-plate tectonic version of geology (that since the Late Cretaceous, southern continents have not been in contact) and the assumption that a taxonomic group is as old as its earliest fossil representative (Humphries and Parenti 1999).

George Sprague Myers (1905–1985) was a U.S. ichthyologist. As a teenager, he was volunteer assistant at the American Museum of Natural History, where he came under the influence of Karl Schmidt, one of Matthew's disciples. Myers (1938) divided true freshwater fishes into primary and secondary classes; the former are physiologically incapable of surviving in seawater, whereas the latter may occasionally enter the sea. Myers's treatment of freshwater fishes was praised by several authors, and Darlington (1957) admitted that it has been a key contribution to the success of his treatment of the distributional patterns of freshwater fishes. By the 1960s, Myers (1963) abandoned "the Gospel according to St. Matthew" (Nelson and Ladiges 2001:400), rejecting notions of continental stability in favor of continental drift.

U.S. ornithologist Ernst Mayr (1904–2005), one of the most influential authors in evolutionary biology, played a key role in the Modern Evolutionary Synthesis, the unified theory of evolution developed in the 1940s. Although his biogeographic ideas share many similarities with those of the authors belonging to the New York school of zoogeography, they do not derive from Matthew but rather from Erwin Stresemann (1889–1972), a disciple of German ornithologist Otto Kleinschmidt (1870–1954). In fact, it is interesting to note that the biogeographic ideas of Mayr, Croizat, and Hennig might all lead back to Kleinschmidt (Williams 2007b).

In *Systematics and the Origin of Species*, Mayr (1942:155) analyzed in detail the process of geographic speciation: "A new species develops if a population which has become geographically isolated from its parental species acquires during this period of isolation characters which promote or guarantee reproductive isolation when the external barriers break down."

Mayr considered this "geographic speciation" to be the orthodox theory and disregarded the possibility of sympatric speciation. In order to detail the stages of geographic speciation, he provided an illustration (fig. 3.4) in which an ancestral species is differentiated into subspecies and the geographic barriers develop and isolate the subspecies, which by later range expansion establish a hybrid zone between the new species. The general approach followed by Mayr is the historical dynamic method, which stressed the need to examine entire faunas, analyzing the dispersal capabilities and distributional ranges of the species as well as the ecological and geological history of the area analyzed in order to understand the biogeographic history as a dynamic and continuing process (Haffer et al. 2000). Mayr

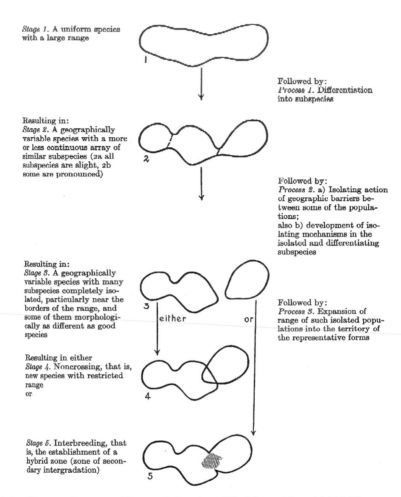

Stage 1. A uniform species
with a large range

Followed by:
Process 1. Differentiation
into subspecies

Resulting in:
Stage 2. A geographically
variable species with a more
or less continuous array of
similar subspecies (2a all
subspecies are slight, 2b
some are pronounced)

Followed by:
Process 2. a) Isolating action
of geographic barriers between some of the popula-
tions;
also b) development of isolating mechanisms in the
isolated and differentiating
subspecies

Resulting in:
Stage 3. A geographically
variable species with many
subspecies completely isolated, particularly near the
borders of the range, and
some of them morphologi-
cally as different as good
species

Followed by:
Process 3. Expansion of
range of such isolated populations into the territory of
the representative forms

Resulting in either
Stage 4. Noncrossing, that is,
new species with restricted
range
or

Stage 5. Interbreeding, that
is, the establishment of a
hybrid zone (zone of secondary intergradation)

either   or

Figure 3.4   Stages of geographic speciation (modified from Mayr 1942:160).

argued that knowledge of biogeography may offer critical insight into evolution;
however, he did not include it in the four disciplines dealing with evolution—systematics, genetics, paleontology, and ecology—when founding the Society for the
Study of Evolution in 1945 (Cain 1993).

Centers of Origin

The center of origin is the key concept of Darwinian dispersalism. For Darwin, ancestral species occupy restricted geographic areas, from which they are constantly
evolving and dispersing out to other parts of the earth by using different means of
dispersal. The dispersal abilities are then basic mechanisms that allow the organisms to reach distant areas. Barriers provide the isolation needed for evolutionary
differentiation. During the history of dispersalism, authors have proposed several
criteria for identifying centers of origin. Cain (1944) analyzed critically thirteen of

them, finding inconsistencies and contradictions. Some of these criteria are as follows (Cain 1944; Morrone, Espinosa Organista, et al. 1996):

- The area with the greatest taxonomic or ecological diversification of the taxon. It assumes that diversity increases through time, so the most ancient area—the center where the taxon appeared—harbors more species. Although the reasoning may seem to be correct, it lacks empirical support, and there are numerous examples of areas with great species richness that correspond to secondary centers of diversification, such as the Hawaiian islands for the genus *Drosophila* (Diptera: Drosophilidae).
- The area where the species has the greatest abundance of organisms. It assumes that because the center of origin has the optimal ecological conditions, it is the place where the organisms of the species will have the best development, whereas in the periphery, where the conditions are different, there will be fewer individuals. This criterion also lacks theoretical and empirical support, and there are numerous examples of species introduced to some area where they are so successful that they have become plagues.
- The area where the most recent or apomorphic species of the taxon analyzed is distributed. This criterion supposes a priori that new species arise in the centers of origin and push the most primitive to the periphery. This criterion assumes that no appreciable climatic changes or extinctions have occurred, which is erroneous. Furthermore, islands can harbor young species, and this does not imply that they constitute the centers of origin of the taxa to which they belong.
- The area where the oldest fossil of the taxon analyzed is found. It assumes that the fossil record is exhaustive, but we know that it is incomplete, so there is no way to know whether a fossil is the oldest of the group. In addition, the oldest known fossils of a group do not always represent primitive or ancestral forms, and there are even examples of recent species that show more plesiomorphic characters than fossil species of the same taxon.
- The area where the most primitive taxa of a monophyletic group are distributed. It is based on the progression rule of phylogenetic biogeography (Brundin 1966; Hennig 1950), which assumes a parallelism between the progression of morphological characters in the cladogram and the geographic distribution of the species, so that the ancestral area is identified with the most primitive species. This criterion is as aprioristic as the one that situates the center of origin in the area inhabited by the most recent species of the group. On the other hand, relictual distributions of ancient taxa would invalidate this criterion.
- The area where the largest organisms of the species are found. It assumes that organisms reach their largest size in the area where conditions are optimal, which occurs in the center of origin. It is similar to the criterion that locates the center of origin in the area with more abundance of organisms. Cases of gigantism on islands would invalidate this criterion.

Common to all these criteria is the aprioristic assumption that some areas are originally devoid of the taxa that eventually dispersed there (Andersen 1982). In the second half of the twentieth century, some authors (Briggs 1984; Müller 1973) reformulated this concept to refer to centers where an important part—not necessarily the initial—of the evolutionary history of a taxon has taken place. These "centers of evolutionary radiation," "secondary centers of evolution," or "centers of diversification" are identified based on the congruence of the distributional ranges of several species. It is interesting that the identification of these centers, at least from an operational viewpoint, is similar to the discovery of areas of endemism (Morrone 2002b). In fact, there are also parallels with the identification of Pleistocene refugia and panbiogeographic nodes, which suggests that the same patterns may be explained differently depending on the biogeographic approach (Contreras-Medina et al. 2001).

Briggs (1984, 1987) presented the most recent treatment of centers of origin under a dispersalist framework. His main conclusions are as follows:

- Information available suggests that centers of origin usually are placed in the tropics. They are large, are heterogeneous in topography, have warm temperatures, have maximum species diversity for the general part of the world where they are located, and harbor the most advanced species and genera of taxa that are well represented.

- On a worldwide basis, the study of major barriers separating biogeographic regions shows that species evolved in the centers of origin not only can spread out to occupy large portions of such regions but also can transgress the barriers and colonize adjacent regions. As a result, a given center of origin may eventually have a profound influence on the biotic composition of a large portion of the world. Because these centers harbor more advanced and more highly competitive species, they have very high resistance to invasion by species from other areas.

- The kind of evolution that goes on in the centers of origin probably is different from that which takes place in peripheral areas because they have larger populations and individuals have higher levels of genetic variation. The rate of evolutionary change is slower than in smaller, isolated populations, but in terms of producing phyletic lines, they are probably more successful.

- Data about the centers of origin and their probable mode of operation indicate that some parts of the earth, in terms of evolutionary progress, have been much more important than others. The complex community structure of the tropical areas is not well understood, yet many of their biotas are being destroyed for agricultural and other purposes. In an evolutionary sense, these areas represent the future of the living world and therefore should be preserved.

From a different perspective, Bremer (1992, 1995) formalized a cladistic method to estimate centers of origin or ancestral areas. Bremer reasoned that the search for the ancestral area of a taxon was a valid part of the study of its natural history. Ancestral area analysis assumes that areas inhabited by the most plesio-morphic species and areas represented on numerous branches of the cladogram are more likely to represent parts of the ancestral area of the group. These assumptions are similar to Hennig's (1950) progression rule, so ancestral area analysis may be considered a formalization of phylogenetic biogeography (Santos 2007). This new version of dispersalism assumes that the search for and identification of ancestral areas or centers of origin constitute valid biogeographic objectives and that they can found by determining which are the most basal areas in a cladogram and how many times they are represented in it. The deficiencies of this revived dispersalism, which ignores homology and uses paralogy to weigh areas and locate centers of origin, have been demonstrated eloquently by Ebach (1999). Simply by weighting replicated areas in basal clades, these methods overestimate the areas where there were fewer extinctions but fail to identify an area as ancestral. Thus they expose the main defect of the dispersalist approach: that the centers of origin are an artifact of paralogy (Morrone 2002b). Ancestral area analysis represents a common misapplication of a cladistic method to a biogeographic problem (Parenti and Humphries 2004). The optimization of areas on the internal nodes of an area cladogram to interpret the ancestral area or center of origin is inappropriate because it dismisses vicariance and favors implicitly the hypothesis of a center of origin. In fact, this misinterpretation becomes evident when authors refer to "basal" or "early divergent" extant taxa (Crisp and Cook 2005; Santos 2007). Area cladograms are biogeographic classifications, not explicit evolutionary hypotheses.

Phylogenetic Biogeography

Willi Hennig (1913–1976), a German entomologist, is best known as the father of phylogenetic systematics or cladistics, which had profound effects on the theory and practice of systematics (Nelson and Platnick 1981; Wiley 1981). Hennig (1950) described four criteria for determining character polarity: geological precedence, chorological precedence, ontogenetic precedence, and correlation of transformation series. The second combines the phylogeny of a monophyletic group with the distribution of its species to trace the progression in space.

Phylogenetic biogeography applies two rules. The chorological progression rule assumes a progression in the areas parallel to the progression in the characters in the cladogram, so that the areas inhabited by primitive species are deemed to be ancestral, whereas the areas inhabited by apomorphic species are situated far away from the center of origin. The deviation rule states that in every speciation event one sister species is more apomorphic ("deviated") than the other, which remains more similar to the ancestor. These rules are based on the assumption that the peripatric speciation model is common in nature. If this speciation model occurs again and again, we will have the impression of dispersal from one area to another.

Swedish entomologist Lars Brundin (1907–1993) published in 1966 *Transantarctic Relationships and Their Significance as Evidenced by Midges*. In this monograph he devoted a section to explaining Hennig's phylogenetic biogeographic methods and discussed the relevance of the sister group in the reconstruction of the biogeographic history of a taxon. Brundin (1966) was one of the first authors to accept the theory of continental drift and used it to support transantarctic disjunctions of the Chironomidae analyzed. Brundin's impact on biogeography has been arguably greater than Hennig's because of the clear exposition of his ideas and because he developed the method and convinced other authors of the relevance of phylogenetics in biogeography (Funk 2004). He saw the key to a deeper understanding of evolution through biogeography rather than paleontology (Williams and Ebach 2004).

Brundin (1981) characterized phylogenetic biogeography as the study of the history of monophyletic groups of nature's hierarchy in space and time. Methodologically, he stated that this approach studies the causal connections between phylogenesis (development of hierarchy in space and time), allopatry (vicariance), sympatry (dispersal), and paleogeographic events. In order to exemplify this approach, one may examine Brundin's (1981) analysis of one of the taxa analyzed in 1966, the tribe Podonomini. Its area cladogram (fig. 3.5) shows that the genera *Podonomus* and *Podochlus* exhibit a three-area pattern including Australia, Patagonia, and New Zealand, whereas *Rheochlus* and *Podonomopsis* have developed a two-area pattern comprising Australia and Patagonia. Together these genera are involved in multiple sympatric vicariance between Patagonia and Australia, where the Australian species of each genus stands out as a young and more apomorphic offshoot of an older Patagonian group. On the basis of this cladogram, Brundin concluded that the occurrence in Australia of species of *Podonomus, Podochlus, Rheochlus,* and *Podonomopsis* is due to dispersal from South America or

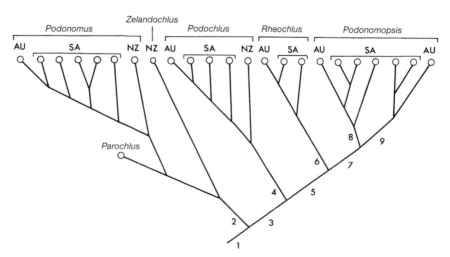

Figure 3.5   Part of the cladogram of the tribe Podonomini, with indication of the areas inhabited by the taxa: AU, Australia; NZ, New Zealand; SA, South America (modified from Brundin 1981:116).

East Antarctica of the common ancestor with their sister species in Patagonia. He postulated that dispersal occurred before the break between East Antarctica and Australia between the Paleocene and Eocene. Within the apomorphic genus *Podonomopsis*, however, there are two cases of two-area vicariance. The New Zealand taxa differ from the Australian ones by their older relative age. *Rheochlus* and *Podonomopsis* are not represented in New Zealand, and if they ever occurred in West Antarctica, they were probably too young to reach New Zealand before its separation from West Antarctica in the upper Cretaceous. They might have originated in Patagonia or in Patagonia + East Antarctica. The tribe as a whole originated in New Zealand–West Antarctica–southern South America.

Despite its aprioristic dispersal approach, phylogenetic biogeography was the first approach to use an explicit phylogenetic hypothesis (Ball 1976; Crisci et al. 2000; Humphries and Parenti 1999; Wiley 1981). Wiley (1981) found it problematic that phylogenetic biogeography followed a single speciation model, peripheral isolation, and that analyses usually are concerned with the taxon of interest, freeing the researcher from examining other groups.

Panbiogeography

Italian American botanist Léon Croizat (1894–1982) was one of the most controversial figures in the history of biogeography of the twentieth century. He challenged not only the dispersalist explanations of the geographic distribution but also the relevance of natural selection as the preponderant agent of evolutionary change. Croizat's ideas suggest the possibility of a new, more integral evolutionary synthesis (Colacino and Grehan 2003; Grehan 2001c; Llorente Bousquets et al. 2000; Morrone 2000c, 2000g). Through panbiogeography (Croizat 1952, 1958b, 1964, 1976)—and especially after its "hybridization" with Hennigian phylogenetic systematics to originate cladistic biogeography (Nelson and Platnick 1981)—he contributed significantly to the development of a new biogeographic approach.

Croizat developed his panbiogeographic approach to test classic dispersalism by comparing hundreds of plant and animal distributions, trying to determine whether they agree with the expectation of chance dispersal according to their particular dispersal abilities. Croizat found a limited number of repetitive distributional patterns or generalized tracks, which had little relation to present-day geography and comprised taxa with very different means of dispersal. This finding is contrary to what one might expect—a chaos of conflicting distributions—from the dispersal model. Panbiogeography assumes that geographic barriers evolve along with biotas, or "life and earth evolve together" (Croizat 1964:iv). When Croizat (1958b) tried to associate the distributional patterns with geological events, he found that Wegener's theory was insufficient with respect to the sequence of geological events, especially for explaining complex biogeographic patterns in the areas bordering the Pacific Ocean. For this reason, he thought that Wegener's theory might explain old (pre-Permian) patterns satisfactorily, but it was incongruous with most modern patterns, so he rejected it, choosing to arrive at his own conclusions in light of the observation of the distribution patterns.

In *Space, Time, Form: The Biological Synthesis,* Croizat provided some precision for his panbiogeographic method, which consists of "finding out what nature herself tells of it all when speaking through strictly factual scores" (Croizat 1964:iii). The analytical tool Croizat chose was the construction of tracks, the "graphs of geographic distribution" or the "primary coordinates" that "open the way to an enquiry into factors of *time* and *form* material to our considerations on *space*" (Croizat 1964:7). By comparing these tracks it is possible to discover patterns, which elucidate both morphological differentiation ("form making") and translation in space. In a diagram (fig. 3.6), Croizat showed how an ancestral species (A) is initially widely distributed in an area although differentiated in subordinated taxa, such as subspecies or races (a and b), and how mountains, lakes, and volcanoes arise and fragment the original distribution, inducing the differentiation of new species. This process involves two different stages: immobilism, which is responsible for vicariance, and mobilism, which allows the expansion of distributions.

Craw and Heads (1988) wrote an interesting interpretation of *Space, Time, Form,* postulating that it represents the deconstruction of Darwin's (1859) *Origin of Species.* In the latter, Darwin followed the sequence variation under domestication and natural selection (chapter 1), the problem of species (chapter 2), struggle for existence (chapter 3), natural selection (chapter 4), laws of variation (chapter 5), difficulties on the theory (chapter 6), instinct (chapter 7), objections to the theory (chapter 8), hybridization (chapter 9), the geological record (chapters 10 and 11), biogeography (chapters 11 and 12), the natural system, morphology, and embryology (chapter 13), and conclusions (chapter 14). Croizat's book reversed Darwin's order: biogeography and its relationship to geology (chapters 1–3), evolution in relation to biogeography (chapter 4), laws of growth (chapters 5 and 6), the problem of species (chapter 7), and natural selection (chapter 8). This shows clearly the preeminence Croizat gave to space, in contrast with Darwin's emphasis on form.

Figure 3.6   Diagram showing the origin of two species after the development of a barrier (modified from Croizat 1964:188).

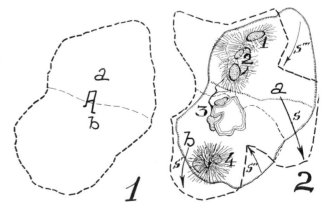

In the early times of panbiogeography few scientists seem to have read seriously or to have considered Croizat's contributions. Corner (1959), Cranwell (1962), Brundin (1966), Aubréville (1969, 1970, 1974a, 1974b, 1975), and Ball (1976) are some of the few authors who commented on these ideas positively. Taylor (1960) analyzed the geographic distribution of a species of mollusks with a panbiogeographic approach. Löve (1967) presented a positive general impression of *Panbiogeography* and *Space, Time, Form*. Later European references to the works of Croizat and panbiogeographic analyses appeared in France (Cusset and Cusset 1988a, 1988b; Janvier 1984; Lourenço 1998) and Italy (Zunino 1992). In the United States, Simpson, Darlington, and Mayr knew Croizat's works but did not mention them. It is interesting that during a brief period, Simpson corresponded with Croizat (correspondence that Simpson discontinued). Ernst Mayr wrote,

> Neither Simpson nor anyone else has affected my treatment of
> Croizat, but only his totally unscientific style and methodology. Time
> is too short to argue with such authors and one cannot simply refer to
> Croizat without a detailed analysis. I am prepared to be criticized for
> this, but any scientist has to make the decision where to draw the line.
> (Nelson 1977:452)

Usually, authors criticized Croizat's writing style. For example, it was considered "verbose, meandering and tangential, archaic in style, loaded with seeming internal contradictions, and with subjects being picked and dropped at will" (Keast 1991:468).

In New Zealand Croizat's ideas were welcomed initially. In the 1950s, Lucy Cranwell, a botanist from the Auckland Museum, corresponded with Croizat, and after moving to the United States they kept contact (Grehan 1989). Robin Craw, Michael Heads, John Grehan, Ian Henderson, and Rod Page discussed panbiogeographic principles and applied them to the analysis of the biogeography of the country (Craw 1979, 1982, 1984a, 1984b, 1985; Craw and Gibbs 1984; Craw and Sermonti 1988; Craw et al. 1999; Grehan 1988b, 1989, 2001a; Heads 1984, 1985a, 1985b, 1986; Henderson 1991). Supporters of panbiogeography and cladistic biogeography initiated a lively debate about the relative values of both approaches (Craw 1982, 1983, 1988a, 1988b; Craw and Page 1988; Craw and Weston 1984; Humphries and Parenti 1999; Humphries and Seberg 1989; Page 1987; Platnick and Nelson 1988; Seberg 1986). In 1984, Craw and Gibbs edited a special volume of the journal *Tuatara*, dedicated to analyze the influence of *Panbiogeography* and *Principia Botanica* in biogeography and systematic botany, respectively. In addition to contributions by Craw, Grehan, and Heads, it included a posthumous contribution by Croizat (1984b), a translation to English of a paper on the Darwinian theory (Croizat 1984a) originally published in Spanish in 1977, biographical data, discussions on panbiogeography, orthogenesis and Darwinism, and a list of Croizat's publications. A special volume of *Rivista di Biologia—Biology Forum* (Italy), edited by Craw and Sermonti (1988), included contributions by Craw, Grehan, and Heads on theoretical issues and empirical applications by Climo and Chiba on mollusks and butterflies, respectively. A special issue of the *New Zealand Journal of Zoology* (1989), more ambitious than the two previous, included theoretical, methodological, and empirical contributions, as well as discussions on tectonics and the application of panbiogeography to biodiversity conservation (Holloway 1992; Keast 1991; Nelson and Ladiges 1990). The book by Craw et al. (1999) represents the maturity of New Zealand panbiogeographers, and anyone with a serious interest in panbiogeography should read it.

Despite the important development of panbiogeography in New Zealand, in the 1980s important political changes took place in the country, with consequences for the economic support of scientific research. More orthodox biogeographers took advantage of this circumstance, and the expansion of panbiogeography in the country was limited (Grehan 2001c). Finally, New Zealand panbiogeographers either lost their jobs, kept them but stopped writing on panbiogeography, or had to find employment overseas (Heads 2005a). New Zealand panbiogeographers made a major contribution to the field. In fact, many developments in panbiogeography, especially in Latin America, owe more to Craw, Heads, and Grehan than to Croizat himself.

In Latin America, the primary interest in panbiogeography and cladistic biogeography developed in the 1990s. In Argentina, Mexico, Brazil, Colombia, Chile, and Venezuela, panbiogeographic and cladistic biogeographic ideas were discussed and applied to the resolution of biogeographic problems (Heads 2005a; Morrone 2004a). Morrone and Crisci (1990, 1995) argued that the panbiogeographic and cladistic biogeographic methods can be applied as successive stages of the same analysis. Initially, a panbiogeographic analysis allows ordering of the studied taxa in different sets, according to their biotic origin. Then, through cladistic biogeography, it is possible to determine the relationships between the areas that integrate each of generalized tracks formerly identified. Morrone (2001c, 2004a) identified the first stage with primary biogeographic homology and the second with secondary biogeographic homology.

Refuge Theory

German ornithologist Jürgen Haffer (1969, 1974, 1978, 1981) analyzed the biogeography of birds inhabiting the Amazon forest. He found six areas of higher endemism, which contained about 25% of the bird forest species of the area, and between them areas where related species hybridized (Cox and Moore 1998). Haffer hypothesized that during the glacial periods of the Northern Hemisphere, these centers of endemism were islands of rain forest surrounded by a sea of grasslands (Lomolino et al. 2006). These refugia allowed many forest species to survive therein and eventually evolve in isolation as separate species. When the conditions allowed the forests to expand, these species could spread over them. As a result, these species could meet with their related species and hybridize, whereas others that had developed reproductive isolation mechanisms did not.

Refuge theory initially provided an interesting explanation for the high diversity and endemism of tropical rain forests. In recent decades, however, many authors have questioned its general validity (Lomolino et al. 2006). Colinvaux (1997, 1998) argued that ice age climate was not sufficiently arid to fragment the Amazonian forest and that vicariance occurred because climatic change created islands in the elevated areas. Thus, on one hand, the refuge hypothesis would have the facts reversed: Ice age climatic change raised an archipelago of islands while the forest sea remained intact (Colinvaux 1997). On the other hand, molecular and morphological data indicate that many endemic species are much older than the hypothesized refugia (Cracraft and Prum 1988; Marshall and Lundberg 1996). Finally, Haffer's hypothesis is not the most parsimonious explanation for the diversity of the tropical forests, which may derive, at least in part, from the fact that they are much more spatially heterogeneous than once assumed (Lomolino et al. 2006).

Cladistic Biogeography

At the moment of the formulation of panbiogeography, most of the scientific community reacted negatively or ignored it. Nevertheless, U.S. ichthyologists Gareth Nelson and Donn Eric Rosen and U.S. arachnologist Norman Platnick, from the American Museum of Natural History of New York, knew and appreciated it. In 1973, when Nelson was the editor of *Systematic Zoology*, he invited Croizat to send a manuscript on panbiogeography. Several of the experts whom Nelson contacted as reviewers of the manuscript refused to comment on it, and of those who accepted, only one recommended publication, whereas fifteen asked for substantial modifications and four suggested rejection (Hull 1988). The negative commentaries referred to Croizat's style of writing or his emphasis on criticizing his opponents. However, most of the reviewers were curious about Croizat's ideas, although they would have preferred a more concise presentation. Among the commentaries of those who declined to review the work, one irritated Nelson: "Study of Croizat's voluminous work has convinced me that he is a member of the lunatic fringe" (Nelson 1977:451). Nelson sent Croizat his manuscript with the suggestions, but when he received the corrected version, it still did not satisfy the reviewers, so he wrote to Croizat, suggesting that he and Rosen review the manuscript and become coauthors. Nelson found it very unlikely that a work coauthored with two researchers of a prestigious scientific institution, one the editor of *Systematic Zoology* and the other the president of the society that published the journal, would be rejected. Croizat accepted, and Nelson and Rosen added an introductory section, footnotes, and a discussion of the principles of phylogenetic systematics, and the work was eventually published (Croizat et al. 1974). Croizat found it unacceptable, suggesting that his ideas had been distorted by Nelson and Rosen. It is interesting that this work, rejected by Croizat and his followers from New Zealand, has been one of his most cited and influential papers (Zunino 1992). Some years later, Craw and Heads tried to have the original manuscript of Croizat published in *Systematic Zoology*, with the intention of clarifying this apparently confusing situation, but their request was denied (Grehan 2001c).

A remarkable article by Nelson (1974) shows a break with a previous article (Nelson 1969) in which he attempted to formalize phylogenetic biogeography (see also Parenti 2007). Nelson (1974:555) wrote, "Having considered the arguments of Croizat (1964 and other papers), I now believe that dispersal is not realistically resolvable by that formalization." He provided an example (figs. 3.7a–3.7c) in which an ancestral species inhabiting what was to become South America and Africa split by a vicariance event into a species in southern South America and another widespread in northern South America and Africa. A further vicariance event then split the latter into a species in South America and another in Africa. If the phylogenetic relationships of these species are correctly inferred, two alternative explanations are possible: a dispersal event from South America to Africa (fig. 3.7d) or a vicariance explanation (fig. 3.7e). The first explanation, which resolves dispersal when no dispersal occurred, is wrong. Nelson (1974) concluded by rejecting as aprioristic all rules to resolve centers of origin and dispersal without reference to general patterns of vicariance.

Nelson and Platnick's (1981) *Systematics and Biogeography: Cladistics and Vicariance* marks the culmination of the dispute between Croizat and New York bioge-

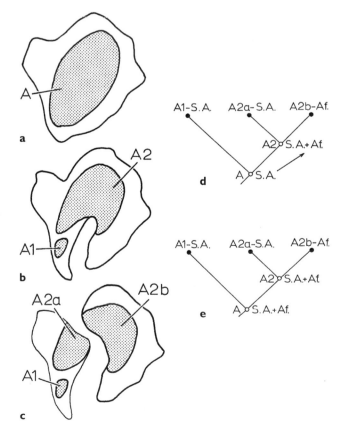

Figure 3.7   Vicariance of an ancestral species widespread in South America and Africa: (a) ancestral species A; (b) vicariance of species A into descendant species A1 and A2; (c) subsequent splitting of A2 into descendant species A2a and A2b; (d) dispersal interpretation; (e) vicariance interpretation (modified from Nelson 1974:556; reproduced with permission of *Systematic Biology*).

ographers. Simultaneously, it marks the birth of cladistic biogeography as an independent approach of panbiogeography. According to these authors,

> The views presented in this volume have their source largely in the work of two biologists, the late Willi Hennig, author of a 1966 book called "Phylogenetic systematics," and Leon Croizat, author of a 1964 book called "Space, time, form: The biological synthesis," and in the writings of a philosopher of science, Sir Karl Popper. Hennig and Croizat have not found their work particularly compatible (Hennig never cited Croizat, and Croizat [1976] has published negative comments on Hennig), and neither one has indicated any interest in Popper's views or cited them as being compatible with his own. Yet both Hennig and Croizat have made substantial, and substantially similar, contributions in (1) pointing out major inadequacies in some conventional methods in systematics and biogeography, respectively, and (2) suggesting significantly improved methods for those fields. We believe that the contributions of both Hennig and Croizat can be readily (and fruitfully) understood within the context of Popper's view of the nature and growth of scientific knowledge, and that the ideas of all three men are largely compatible. (Nelson and Platnick 1981:ix)

Rosen, Nelson, and Platnick associated Croizat's panbiogeography with Hennig's phylogenetic systematics, creating vicariance or cladistic biogeography (Nelson 1969, 1973, 1974, 1978a, 1983; Nelson and Platnick 1980, 1981; Platnick and Nelson 1978; Rosen 1974, 1976, 1978, 1979, 1981). Cladistic biogeography assumes a correspondence between taxonomic relationships and area relationships (fig. 3.8). If we compare area cladograms derived from taxonomic cladograms of different groups of plant and animals inhabiting a certain region, we may recognize the

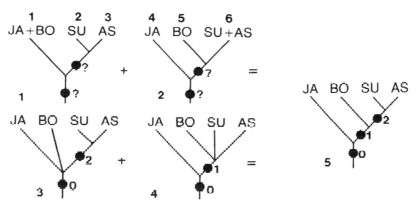

Figure 3.8 Cladistic biogeography replaces terminal taxa in the taxonomic cladograms with the areas inhabited by them in order to obtain area cladograms and then combines them into general area cladograms. AS, Asia; BO, Borneo; JA, Java; SU, Sumatra (modified from Nelson and Platnick 1981:426).

general pattern of fragmentation of the areas analyzed (Morrone and Crisci 1995). Croizat (1982) found this combination of panbiogeography with phylogenetic systematics to be completely unacceptable because he had already denounced Hennig, claiming that he had plagiarized the ideas of Rosa (Croizat 1978).

Despite the important development of cladistic biogeography in the United States, the book that summarizes the evolution of the discipline in the late twentieth century was published in Great Britain (Humphries and Parenti 1999). It represents the second edition of a 1986 book, although it incorporates several new methodological developments. Anyone interested in cladistic biogeography should read it.

Panbiogeographers Versus Cladistic Biogeographers

Craw and Weston (1984) discussed the methods of scientific research programs developed by Lakatos (1970, 1978) and applied it to the dispersalist, panbiogeographic, and cladistic biogeographic approaches. According to Lakatos (1978) a research program consists of three parts: the hard core, the positive heuristic, and the protective belt. The hard core is the set of assumptions that define the program and are irrefutable; scientists protect them as part of their research tradition. The hard core of a research program develops gradually over years. The positive heuristic is a set of methodological rules that specify the research policy of the adherents of the research program, trying to anticipate anomalies and how to address them. The protective belt is a flexible set of auxiliary hypotheses that are constructed, readjusted, or discarded as directed by the positive heuristic.

Research programs are either progressive or degenerating. Programs are progressive if each modification leads to novel predictions, which are confirmed by subsequent research within the confines of the program or by the chance discovery of new facts. A degenerating program is characterized by ad hoc modifications or by the persistent failure of its predictions to be corroborated empirically. Craw and Weston (1984) considered classic dispersalism not a unified research program but several programs with different assumptions (for similar viewpoints, see Savage 1982 and Morrone 2002b). Additionally, the dispersalist approach is not intended to provide predictions but instead to produce unique, narrative explanations of the biogeographic distributions of particular taxa. Craw and Weston (1984) restricted their analysis to panbiogeography and cladistic biogeography. They suggested that a fundamental distinction between them—their hard cores—concerns accepting the possibility of random distributional patterns (cladistic biogeography) and rejecting it (panbiogeography). The aim of cladistic biogeography is to determine whether such patterns exist, whereas panbiogeography assumes a priori that patterns are congruent. It follows that cladistic biogeography involves no preference for particular biogeographic processes, whereas panbiogeography emphasizes vicariance. After comparing some predictions formulated by panbiogeographic and cladistic biogeographic analyses, as Lakatos (1970, 1978) considered for judging competing research programs, Craw and Weston (1984) concluded that panbiogeography was more successful than cladistic biogeography. However, they pointed out that cladistic biogeography was still in its infancy, whereas panbiogeography was twenty-five years old, so it was not fair to expect the same progress from both.

Seberg (1986) offered a critique of panbiogeography. After acknowledging the relevance of Croizat's contributions as inspiration for cladistic biogeography, he analyzed some panbiogeographic concepts. He maintained that despite Croizat's dismissal of dispersal and centers of origin, he used both concepts extensively. Seberg (1986) believed that Croizat's massive compilation of data was worth mentioning as a relevant contribution but that his only original contribution was the concept of track. Then, he criticized Craw and Weston's (1984) analysis, suggesting that it has been futile because Lakatos's (1978) descriptions were not intended to analyze contemporary research programs. He found it difficult to view panbiogeography as a progressive research program, being more like an adjunct to historical geology.

Some cladistic biogeographers also criticized panbiogeography from the methodological point of view (Platnick and Nelson 1988; Seberg 1986), indicating that the procedure for orienting tracks is ambiguous because it can take into account the minimum distance criterion or phylogenetic relationships. Tracks can also be oriented in agreement with main massings (centers of numerical, genetic, or morphological richness), which resemble the dispersalist criteria to determine centers of origin. Platnick and Nelson (1988) distinguished Croizat's original ideas from those of his followers from New Zealand, which they preferred to call "minimum-spanning tree biogeography." Its main characteristic is that it allows one to establish biogeographic relationships with little or no phylogenetic information on the taxa analyzed. It would allow a solution of biogeographic relationships based only on geographic proximity, which would represent a shortcut that would avoid the large amount of time cladistic analyses may take. Page (1987) justified this procedure, arguing that considering the minimum distance between the present localities does not imply assuming that evolution is parsimonious; it is simply a methodological resource that can be applied in the absence of other information for track construction. If there were cladistic information that the closest geographic neighbor is not the next phylogenetic relative, then it would be necessary to consider the phylogenetic criterion. Page (1987) also argued in favor of the use of minimum-spanning trees, indicating that they can be calculated exactly and efficiently. Platnick and Nelson (1988) indicated that the ease of calculating them in itself is not a reason to consider them a more appropriate tool. Joining the geographically nearest localities could lead to erroneous conclusions when they belong to taxa that were not also the nearest relatives or when the present geography is not the same as that when they evolved.

Cladistic biogeographers have also criticized panbiogeography for using phylogenetic information only to orient the individual tracks, which would represent a partial application of Hennig's phylogenetic biogeography, which supposes a priori that the most ancestral forms are situated in the center of origin. Cladistic information is not used to reveal the relationships between areas, which can be masked by present geography, so the construction of tracks may reflect only present geographic proximity (Platnick and Nelson 1988). Craw (1982:305) had previously argued that there are "ample theoretical statements in Croizat's work demonstrating a keen appreciation of the relationship between phylogenetics and biogeography." In fact, Croizat (1958b) clearly distinguished between ancestor–descendant relationships and common ancestry, something that even today may not be clear to some practicing systematists.

Craw (1988a) justified the panbiogeographic approach, applying the Hennigian concept of reciprocal illumination to systematics and biogeography. Panbiogeography assumes that spatial relationships can suggest genealogical relationships. A classic example is the redefinition of the taxonomic relationships between carnivorous plants by Croizat, based on their spatial relationships. Nevertheless, Platnick and Nelson (1988) noticed that Croizat did not base his conclusions exclusively on geographic proximity but took morphological characters into consideration to regroup them. Cladists accept that the incongruence of a taxon with a recognized biogeographic pattern may suggest an incorrect systematic hypothesis, but the incongruence by itself does not constitute evidence. Any congruence revealed by panbiogeography will be uncertain because it is possible that it is produced by geographic proximity, and even assuming that the construction of tracks may reveal causal connections between areas, they are unable to reveal their exact relationship (Platnick and Nelson 1988).

Cenogenesis, Cenocrons, and Horofaunas

Argentinean paleontologist and evolutionary biologist Osvaldo Reig (1929–1992) made some important contributions to biogeography. Reig (1962:132–133) published a brief article in which he introduced some precisions on biogeographic concepts:

> All modern and scientific biogeographic research is born from a common theory to all biology, the theory of evolution, where the historical, genetic and dynamic conceptualization is the starting point. At the present time there cannot be any branch of biogeography that does not bear these attributes, because the mere configuration of distribution data presupposes an historical theory with whose aid facts are interpreted.

Reig thought that referring to "historical biogeography" disregarded the fact that the objects of inquiry are communities. He introduced the term *cenogenesis* (originally coined by Sukachev 1958) to refer to "the development of associations through geological time" (Reig 1962:133). Reig analyzed Simpson's (1950) contribution, which, closely following Matthew's (1915) theses, postulated the origin of all South American mammals in the Holarctic realm. Reig did not invalidate it totally, but he restricted its application to Cenozoic mammals, allowing other tetrapods to show relationships with the austral continents. As part of his critique, Reig questioned the concept of faunistic strata, which Simpson used to explain the fact that different taxa have occupied some areas during determined periods of time of different duration. He replaced it with the term *cenocron*, synthesizing in its etymology the notions of community and time.

Reig (1981) published a more extensive work that represents the maturity of his conceptions (Morrone 2003a), in which his differences with Simpson's approach are more evident. He took an "equidistant position between the Holarcticism of the Matthewian school and the *ad hoc* Australism of other biogeographers, trying to formulate an integrative theory of the biogeographic history of South American vertebrates" (Reig 1981:13). Holarcticism, formulated by Matthew, Simpson, and Darlington, emphasizes the origin of all South American vertebrates in the Northern Hemisphere. Australism, represented by Ameghino, postulates that all mammals that inhabited the world in the Cretaceous originated in Patagonia. Reig's (1981) synthetic theory, which he considered analogous to the theory developed by Halffter (1964, 1965) for the Scarabaeidae (Coleoptera) of the Mexican Transition Zone, rescued some postulates of Simpson but incorporated alternative interpretations to explain pre-Cenozoic faunistic connections, based on vicariance due to continental drift. Thus Reig combined both dispersal and vicariance into a single model. Reig (1981) concluded his proposal by postulating the reconstruction of "horofaunas," or main patterns of change in the fauna of South American mammals. Horofaunas imply successive events of adaptive evolution of different lineages that interact, maximize the exploitation of available resources, and reach certain stability, covering all possible ecological niches.

Taxon Pulses

Erwin (1979, 1981) expanded the "taxon cycle" concept, introduced by Darlington (1943) and Wilson (1961), into a model that accounts for habitat shifts along certain pathways (fig. 3.9). It postulates the existence of a unidirectional progression along a sequence of habitats from tropical wetlands toward forest canopies, high latitudes, grasslands, deserts, caves, and mountains. Taxon pulses take place when generalists (plesiotypes) radiate into new habitats and become specialists (apotypes) and even superspecialists (superapotypes). Plesiotypes generate several pulses, which overtake previous pulses and replace them along the pathways. Taxa may be categorized into sets that reflect how far they have departed from the generalist basal stage.

Liebherr and Hajek (1990) tested the taxon pulse model using eight taxa of New World Carabidae (Insecta: Coleoptera) for which both phylogenetic hypotheses and habitat data were available. They compared cladistic transformations of habitat preference with patterns generated randomly under a null hypothesis, finding that only one of the groups analyzed exhibited a statistically significant pattern supporting the taxon pulse. They suggested that the failure of the model as a predictive hypothesis may result from historical changes in climate that allow

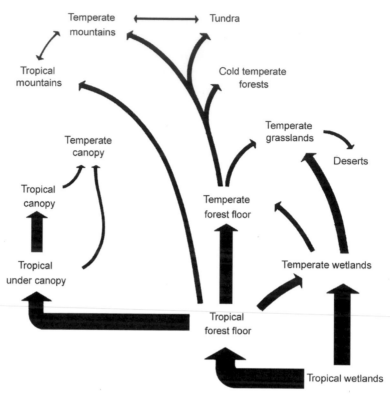

Figure 3.9 · Pathways of lifestyles in carabid beetles, according to the taxon pulse model identified by Erwin (1981).

range expansions for species restricted previously to small habitats, and habitat shifts do not progress in linear transformation series.

Brooks (2005) stated that the taxon pulse model was more appropriate for cladistic biogeography than the vicariance model. According to him, taxon pulse–driven biotic diversification differs from vicariance-driven biotic diversification in three ways. First, as diversification is driven by biotic expansions, one expects to find general patterns associated with dispersal, not just with vicariance. Second, episodes of biotic expansion, even those involving large areas, will lead to reticulated historical relationships between areas, with biotas comprising species of different ages derived from different sources. Finally, the absence of particular clades in particular areas is more parsimoniously explained as a lack of participation in that particular expansion episode by a particular clade rather than dispersal with extinction.

Phylogeography

Advances in molecular biology in the late twentieth century allowed the development of phylogeography, whose aim is to analyze the geographic distribution of genealogical lineages (Avise 2000; Avise et al. 1987). This approach has been designed explicitly to combine the power of molecular genetic methods and analyses (Riddle and Hafner 2004).

Despite the potential conceptual links between phylogeography and biogeography, some panbiogeographers (Heads 2005b) and cladistic biogeographers (Ebach and Humphries 2002; Humphries 2000; Parenti 2007) have criticized it. Ebach and Humphries (2002), Nelson and Ladiges (2003), and Heads (2005a, 2005b) held that phylogeography has reinvented dispersal biogeography. Heads (2005b:679) noted, "Unfortunately, the vast majority of molecular results are interpreted using dispersalist theory and dubious clock calibrations." Parenti (2007:62) stated,

> Application of these techniques to systematics and biogeography has been, on the one hand, invigorating, as it has generated many novel hypotheses of phylogenetic relationships, yet at the same time, detrimental, as it has revived untestable hypotheses of centers of origin, recognition of ancestors, and dismissal of the importance of Earth history at all levels, not just plate tectonics, in biogeography.

I concur with Riddle and Hafner (2004) that a synthesis that uses the reciprocal strengths of intraspecific phylogeography and cladistic biogeography represents an interesting prospect for evolutionary biogeography.

Conclusions

Evolutionary biogeography represents a research tradition in the sense of Laudan (1977), namely, a family of theories related by shared goals and methods. Its origin may be traced to the nineteenth century, when Buffon's law was formulated and tested. As a result of its influence, the idea of a single center of origin for all species was replaced by the idea of several "centers of creation," where creative activity was especially intense. Darwin and Wallace explained Buffon's law by chance dispersal, followed in the twentieth century by Matthew, Simpson, Myers, Darlington, and Mayr. Some unorthodox authors such as Willis, Rosa, Jeannel, von Ihering, and Croizat emphasized vicariance as a more appropriate general explanation. Panbiogeographers and some cladistic biogeographers, such as Nelson, Platnick, Rosen, Humphries, Parenti, and Ebach, emphasized the idea that Earth and life evolve together, looking for patterns due to vicariance. Other cladistic biogeographers, such as Brooks and Lieberman, assumed more complex scenarios, invoking dispersal in addition to vicariance. The methodological developments of panbiogeography, cladistic biogeography, and phylogeography illustrate well how evolutionary biogeography has been invigorated by new techniques and approaches, mainly from cladistics and molecular biology. However, tensions and contradictions continue to exist because none of the approaches could absorb the others.

The notions of centers of origin and dispersal have been recurrent, although it is debatable whether Linnaeus's Garden of Eden, Darwin and Wallace's centers of origin, and phylogeographers' ancestral areas really represent the same concept.

The notion of vicariance is also recurrent. Is there necessarily a conflict between dispersal and vicariance? Some believe that these processes should be considered only after patterns are revealed, taking a classificatory approach (Ebach and Goujet 2006; Williams 2007b). The evolutionary approach followed herein is based on the dispersal–vicariance model that incorporates both processes and fulfills the classificatory objective as a step in the analysis.

Most biogeographers from the twentieth century have considered the relationships between space, time, and form in one way or another (Humphries 2004). Panbiogeographers followed Croizat and his space–time–form sequence, stating that biogeography should be given precedence over systematics. Dispersalists and cladistic biogeographers used the sequence form–time–space or form–space–time, giving preeminence to systematics over biogeography. Evolutionary biogeography, as conceived herein, follows the sequence space–form–time: The panbiogeographic analysis (space) constitutes the first step of the analysis, form is incorporated through the phylogenetic hypotheses used in the cladistic biogeographic analysis of the second step, and time is finally introduced through fossils, intraspecific phylogeography, and molecular clocks.

For Further Reading

Bowler, P. S. 1996. *Life's splendid drama.* Chicago: University of Chicago Press.
Browne, J. 1983. *The secular ark: Studies in the history of biogeography.* New Haven, Conn.: Yale University Press.
Lomolino, M. V., D. F. Sax, and J. H. Brown, eds. 2004. *Foundations of biogeography: Classic papers with commentaries.* Chicago: University of Chicago Press.

For Discussion

1. Most practicing biogeographers are not very interested in the history of biogeography. Did the historical development of evolutionary biogeography, reviewed in this chapter, help you understand its principles better? Do you think it is important for practicing biogeographers to study the history of their discipline? If so, why?

2. Dispersal and vicariance have been portrayed as alternative processes during most of the history of biogeography, although some authors have considered both of them relevant processes. Search for biogeographic papers published in recent decades and classify them as "dispersalist," "vicariancist," or "dispersal–vicariancist." Are they related to any particular approach or method?

3. Carefully read the following article:

   Nelson, G. 1978. From Candolle to Croizat: Comments on the history of biogeography. *Journal of the History of Biology* 11:269–305.

   a. List the basic ideas identified by the author.

   b. State each idea as a question.

4. Do you think that Darwinian dispersalism, the New York school of zoogeography, and phylogeography constitute a single research tradition? Provide arguments supporting your view.

CHAPTER 4

Identification of Biotic Components

Biotic components are sets of spatiotemporally integrated taxa that coexist in given areas. Their unity is due to their common history, although they may not represent monophyletic entities because of reticulation due to geodispersal and biogeographic convergence. Each biotic component usually consists of a particular set of cenocrons that have been integrated at different times. If taxa studied have a wide distribution in the fossil record or a molecular clock can be calibrated, it would be possible to recognize these cenocrons according to their geological age. The identification of biotic components, the basic biogeographic units, is the first stage of an evolutionary biogeographic analysis. In this chapter I present the two basic approaches for studying biotic components: panbiogeography and identification of areas of endemism. I also introduce some basic methods and provide case studies.

Biotic Components

There are two basic ways to represent biotic components: generalized tracks and areas of endemism. The former are studied by panbiogeography, whereas the latter are the units of cladistic biogeography. We may distinguish generalized tracks and areas of endemism by their scales (larger or smaller, respectively), although they both represent biotic components (Morrone 2001c, 2004a). The aim of panbiogeography is to recognize generalized tracks, whereas cladistic biogeography emphasizes the recognition of areas of endemism and their relationships as fundamental issues (Morrone and Crisci 1995; Nelson and Platnick 1981; Szumik et al. 2002, 2006).

Panbiogeography

Any attempt to define panbiogeography—the prefix *pan-* ("all") tries to go beyond the traditional distinction between phytogeography and zoogeography—must contend with a great variety of opinions. Patterson (1983) considered it "phenetic" because it is based on global similarity. Mayr (1982) described it as "eccentric." Nelson (1989) considered panbiogeography an "evolutionary metatheory." Stace (1989) defined it as "vicariance in global distributional patterns." Grene (1990) considered it "faddish." For some authors, panbiogeography is simply a precursor of cladistic biogeography (Cox 1998; Grande 1990; Nelson and Platnick 1981; Warren and Crother 2001), whereas for others it is an alternative research program (Colacino 1997; Craw and Weston 1984; Espinosa Organista and Llorente Bousquets 1993; Grehan 2001d; Humphries and Seberg 1989; Morrone et al. 1996; Zunino and Zullini 1995).

Panbiogeography emphasizes the spatial or geographic dimension of biodiversity to allow a better understanding of evolutionary patterns and processes (Craw et al. 1999). Its main objective is to highlight the relevance of geographic distributions as direct objects of analysis. The spatial dimension of organisms is a prerequisite for any evolutionary study because geography is the substrate where life occurs. Crisci et al. (2000:48) stated, "Panbiogeography is like real estate: the most important thing is location."

Croizat (1958b, 1964) formulated his approach in terms of three metaphors: "Earth and life evolve together," "space + time + form = the biological synthesis," and "life is the uppermost geological layer." His main objectives were to establish biogeography as an independent science free from prior commitments to geological or geophysical theories and to ensure a synthesis between biology and geology (Craw and Page 1988). Until the development of panbiogeography, biogeographers followed the fashionable geological ideas of their times. The lack of a method by which biogeography and geology could be integrated is illustrated by the cases in which former opponents of continental drift rapidly transferred their models to mobile models of Earth history.

Panbiogeography is based on four assumptions (Craw et al. 1999):

- Distributional patterns constitute an empirical database for biogeographic analyses.
- Distributional patterns provide information about where, when, and how plants and animals evolved.
- The spatial and temporal component of these distributional patterns can be represented graphically.
- Testable hypotheses about historical relationships between the evolution of distributions and Earth history can be derived from geographic correlations between distribution graphs and geological or geomorphic features.

With regard to the relevance of panbiogeography for evolutionary biology, Grehan (1988b) postulated that the fusion of space, time, and form implies the recognition of a single evolutionary process, where "organism" and "environment" are not isolated entities. Classically, the absolute separation of organisms and environments has been reflected in the search for mechanisms or forces acting on passive organisms (natural selection) or passive environments (orthogenesis). Panbiogeography assumes that taxa exist as part of their environments and vice versa; they are coconstructed, so mechanisms should be focused on the relationship itself rather than on the organisms or the environment. This "constructivist interaction" (Oyama 2000) provides a conceptual framework for the development of a new synthesis in evolutionary biology, where the disciplines that contribute to it can derive insight from the others.

A panbiogeographic analysis comprises three basic steps (fig. 4.1):

1. Construct individual tracks for two or more different taxa (fig. 4.1a–4.1f).
2. Obtain generalized tracks based on the comparison of the individual tracks (fig. 4.1g).
3. Identify nodes in the areas where two or more generalized tracks intersect (fig. 4.1h).

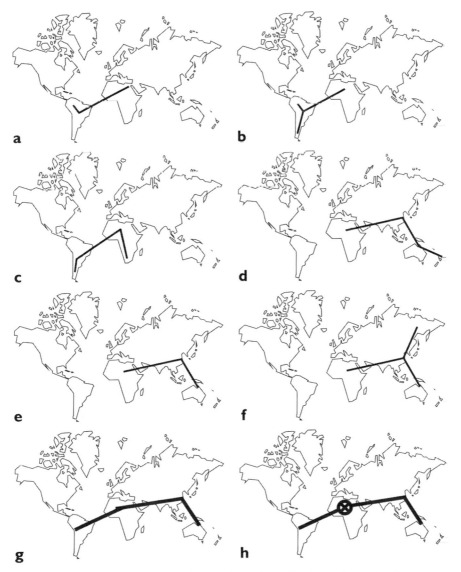

Figure 4.1   Steps of a panbiogeographic analysis. (a–f) Obtaining individual tracks; (g) identifying generalized tracks; (h) identifying nodes.

Individual Tracks

An individual track is the basic unit of panbiogeography. It can be defined as the primary spatial coordinates of a species or supraspecific taxon (Crisci et al. 2000). Operationally, an individual track is a line graph drawn on a map that connects the different localities or distributional areas of a taxon according to their geographic proximity. According to Craw (1988a), the concept of individual tracks is not original to Croizat but comes from van Steenis (1934–1935).

From the topological viewpoint, an individual track is a minimum-spanning tree that for n localities contains $n - 1$ connections (Page 1987). When a track is drawn, the criterion for connecting the different localities of a species is simple. When any locality is chosen, the nearest locality to it is found, and they are connected by a line; then, this pair of localities is connected with the nearest locality to any of them; the nearest locality to any of the three is united, and so on (fig. 4.2). The result is an unrooted cladogram, where the sum of the segments connecting the localities is minimal, following a sort of geographic parsimony. An alternative formalization, based on minimal Steiner trees (where extra localities are added in order to reduce the length of the tree), is provided by Zunino et al. (1996).

How can we interpret individual tracks? Each taxon has a distributional area or range, namely, the area where it is distributed. In order to study geographic distributions, biogeographers need some sort of representation or abstraction (Gaston 2003; Rapoport 1975). Dot maps (fig. 4.3a) plot points in the localities where the taxon has been recorded, and for some biogeographers they convey accurately the known records. Traditionally, biogeographers have enclosed the points with a free-form line around the peripheral localities, obtaining an outline map (fig. 4.3b). However, outline maps may be so generalized that important localities are not sufficiently highlighted (Heads 1994). Rapoport (1975) developed the mean propinquity method, which consists of connecting the points on a map by means of arcs, then establishing their mean distance, and finally surrounding every point with a circle whose ratio equals the obtained mean distances (fig. 4.3c). Individual tracks are another representation of the geographic range of a taxon (fig. 4.3d) in which space geometry is interpreted as an explicit component (Craw et al. 1999; Grehan 2001c).

Individual tracks can be oriented. Orienting an individual track consists of formulating a hypothesis on the sequence of the disjunctions implied in it. The most common way to orient a track is designating a baseline (fig. 4.4a), which represents a spatial correlation between the individual track and a geographic or geological feature. For identifying a baseline, we have to analyze the geological characteristics of greatest relevance. On a global scale, ocean or marine basins can be used as baselines, whereas on smaller scales, we may use some evident geological features such as rivers and mountain chains. If we are orienting a strictly terrestrial track, the situation is more complex because we have to decide whether a mountain range is more relevant than a river or any other geological feature. I find it difficult to understand why baselines have been treated as equivalent to "centers of origin" (Craw et al. 1999; Grehan 1988a, 1994). Although it is not the same as dispersalism, this may give rise to confusion.

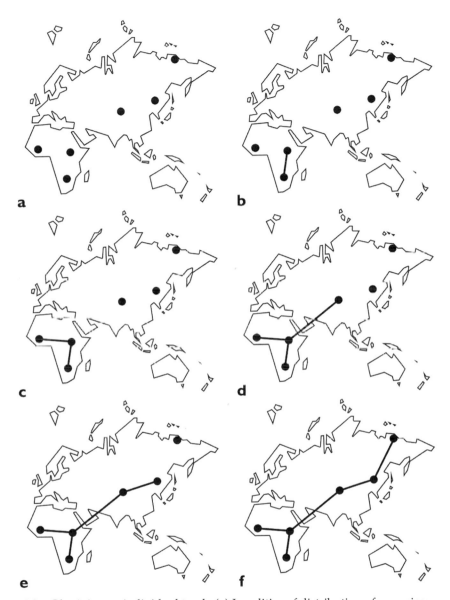

Figure 4.2 Obtaining an individual track. (a) Localities of distribution of a species;
(b) choosing a locality and joining it to its nearest locality; (c–f) joining the remaining
localities based on their proximity.

A phylogenetic criterion can be used for orienting the track of a supraspecific
taxon. The interpretation of the phylogenetic criterion in Croizat's work was confined to the fact that tracks connected areas or localities of the same taxonomic
group (family, genus, or species), thus implying monophyly. Page (1987) suggested the possibility of using cladistic information for orienting individual tracks
(fig. 4.4b), and his suggestion was accepted by Craw (1988a), Henderson (1989),

Figure 4.3 Alternative representation of a species distribution. (a) Dot map; (b) outline map, where a free-form line encloses the peripheral localities; (c) application of Rapoport's (1975) mean propinquity method; (d) individual track.

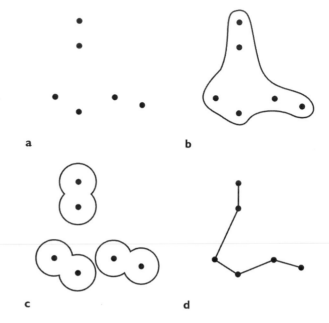

and Grehan (1991). McDowall (1978) discussed the possibility of testing generalized tracks, noting the problem of using phylogenetic criteria to construct the individual tracks on which they are based. Platnick and Nelson (1988) noticed that the application of this criterion can be analogous to that of Hennig's progression rule. If the results of panbiogeographic analyses represent hypotheses of primary biogeographic homology, which we will falsify in a cladistic biogeographic analysis (Morrone 2001c), it seems problematic to orient the tracks by means of phylogenetic information. This implies that the phylogenetic hypotheses are part of both the panbiogeographic and cladistic biogeographic analyses, falling in a circular sequence of reasoning. Therefore, I find it inappropriate to use this criterion.

Another criterion for orienting individual tracks is the location of main massings, which are defined as the greatest concentration of biological diversity in the range of the taxon, such as number of species or genetic diversity (fig. 4.4c). In general, main massings represent areas of numerical, genetic, or morphological diversity of a group (Page 1987), which may be identified by a grid analysis (Craw et al. 1999). Platnick and Nelson (1988) and Humphries and Seberg (1989) complained that Croizat referred to main massings as "dispersal centers," "places of origin," "centers of emergence," "ancestral centers of radiation," and "centers of origin," and even Page (1990c) admitted that the concept of main massing was "horribly vague." If a track is oriented from the main massing toward the periphery, the inference involved would be similar to that from dispersal biogeographers (Crisci et al. 2000), so this criterion might also be inappropriate.

Of the available criteria for orienting tracks, the less problematic is the baseline. When the analyses are undertaken on continental scale, however, the use of

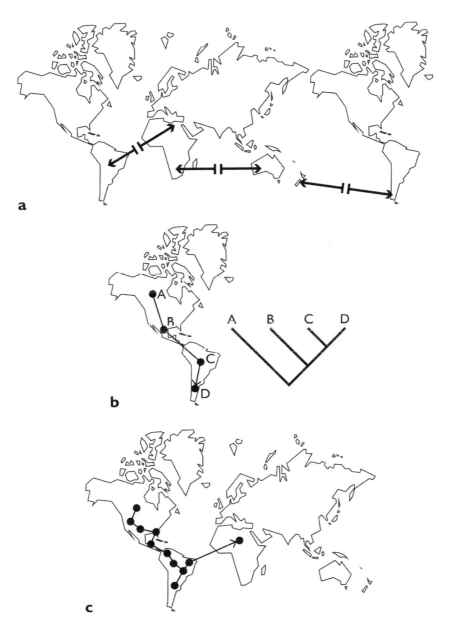

Figure 4.4   Criteria for orienting individual tracks. (a) Baselines; (b) phylogenetic information; (c) main massing.

geological or tectonic characteristics is somewhat more difficult to carry out (Morrone 2004a). For this reason, most of the published panbiogeographic analyses do not orient the individual tracks.

In order to draw individual tracks on maps, authors have used different symbols to indicate localities (e.g., circles, triangles, squares, stars) and different types

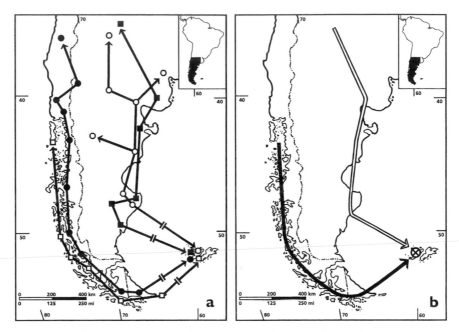

Figure 4.5   Graphic representation of tracks and nodes. (a) Individual tracks; (b) generalized tracks and node.

of lines to connect them (e.g., dotted, solid). The representation of two or more individual tracks on the same map is problematic. Fortino and Morrone (1997) suggested that a solid line provides the best visualization and will not be confused with lines representing borders on the map (fig. 4.5a). In order to avoid confusion between taxa and to improve legibility, localities should be represented with different symbols, such as circles and squares. Triangles should not be used because they can be confused with the arrows that indicate the direction of oriented tracks. If more symbols are needed, squares and circles can be black, white, or, if possible, in different colors. In order to represent a baseline, two lines parallel to each other and perpendicular to the track (fig. 4.5a) may be used to emphasize the idea of rupture and separation (Fortino and Morrone 1997).

Generalized Tracks

Generalized or standard tracks result from the significant superposition of different individual tracks (Zunino and Zullini 1995). They indicate the preexistence of ancestral biotic components that became fragmented by geological or tectonic events (Craw 1988a). Generalized tracks are the result of a comparative analysis of the individual tracks (fig. 4.5b). When we compare oriented individual tracks, we can determine that they belong to the same generalized track when they agree in both their structure and their direction (Craw 1988a). Nihei and de Carvalho (2005) suggested that generalized tracks could be recognized only when there is phylogenetic evidence supporting them (e.g., they consist of sister clades). I consider generalized tracks and areas of endemism to be different representations of biotic components (fig. 4.6).

The generalized track is the most important concept in panbiogeography. However, McDowall (1978) noted that there are still some questions research should address: How many individual tracks must coincide in order for a generalized track to be identified? How good must the coincidence be for an individual track to be considered part of a generalized track? How does one interpret and weigh the patterns indicated by noncongruent or conflicting generalized tracks? Graphic representation of generalized tracks has been analyzed by Fortino and Morrone (1997), who suggested the use of lines twice as thick as those used for individual tracks (fig. 4.5b). When two or more generalized tracks have to be represented on the same map, double or triple lines could be used.

From an epistemological perspective, identifying generalized tracks is an inductive process. Although inductivism seems to be a crude and primitive way to conceive scientific knowledge (Ball 1976; Reig 1981), it can play a basic part of pattern seeking, as recognized by Rosen (1988a, 1988c). Generalized tracks and areas

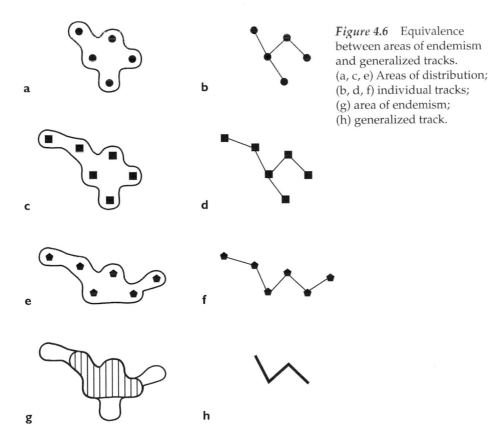

Figure 4.6 Equivalence between areas of endemism and generalized tracks. (a, c, e) Areas of distribution; (b, d, f) individual tracks; (g) area of endemism; (h) generalized track.

of endemism represent valid conjectures, the most basic hypothetical statements, which may be falsified by cladistic biogeographic analyses (Morrone 2001c).

Nodes

Nodes are complex areas where two or more generalized tracks superimpose (fig. 4.5b). They are usually interpreted as tectonic and biotic convergence zones, areas of ancient geography around which evolution has taken place (Heads 2004). The recognition of nodes is one of the most important contributions of panbiogeography. Croizat and his followers have based many of their critiques to cladistic biogeography on its inability to distinguish these types of complex areas because of their attachment to the implicit hierarchy in cladograms. This would not be totally correct because compound areas would behave as species of hybrid origin, showing conflicting relationships with different "paternal" areas.

Nodes are particularly interesting from the evolutionary biogeographic viewpoint because they allow us to speculate on the existence of compound or complex areas. On a global scale, some major nodes include Mesoamerica, Chocó (Colombia), Fouta Djallon plateau (Guinea), Madagascar, New Caledonia, and the area around Wallace's line in Indonesia (Heads 2004). Nodes may represent the location of endemism, high diversity, distributional boundaries, disjunction, anomalous absence of taxa, incongruence and convergence of characters, and unusual hybrids, among other features (Heads 2004).

In order to provide an objective procedure to identify nodes, Henderson (1989) suggested that they may correspond to points with high density of terminal track vertices, such as 1° vertices, which are endpoint vertices that only have one connecting link to another point. A higher number of 1° vertices is found at the periphery of a minimum-spanning tree, and the highest density of 1° vertices occurs where different individual tracks come into contact (Grehan 1991). Where 1° vertices of two individual tracks are in close proximity, their highest density will be close to the boundary between the two taxa. In cases of complex, overlapping distributions, this approach allows an objective nodal analysis by testing a null hypothesis of random distribution (Craw et al. 1999). In order to represent nodes graphically, Fortino and Morrone (1997) suggested using an "x" enclosed by a circle (fig. 4.5b).

Some authors (Craw et al. 1999; Grehan 2001a; Henderson 1989) also considered as nodes the localities of intersection of two or more individual tracks. These represent the geometric center of "form making" or the boundary between two sister species (Henderson 1989). Nodes found at the intersection of generalized tracks therefore should be called generalized nodes.

Areas of Endemism

The definitions and criteria for areas of endemism are complex issues (Linder 2001; Morrone 1994b; Platnick 1991; Szumik et al. 2002; Viloria 2005). There are several definitions of areas of endemism:

- Fairly small areas that have a significant number of species that occur nowhere else; areas delimited by the coincident distributions of taxa that occur nowhere else (Nelson and Platnick 1981:390, 468).
- An area of endemism can be defined by the congruent distributional limits of two or more species (Platnick 1991:xi).
- Geographic region comprising the distributions of two or more monophyletic taxa that exhibit a phylogenetic and distributional congruence and have their respective relatives occurring in other such defined regions (Harold and Mooi 1994:262).
- Areas of nonrandom distributional congruence between different taxa (Morrone 1994b:438).
- Areas defined by the distributions of endemic taxa occurring in those areas (Humphries and Parenti 1999:6).
- Areas delimited by the congruent distribution of at least two species of restricted range (Linder 2001:893).
- Areas delimited by barriers, the appearance of which entails the formation of species restricted by these barriers (Hausdorf 2002:648).
- Areas that have many different groups found there and nowhere else (Szumik et al. 2002:806).

Some of these definitions entail extensive sympatry (Platnick 1991), although this congruence does not demand complete agreement on those limits at all possible scales of mapping (Hausdorf 2002; Linder 2001; Morrone 1994b; Morrone and Crisci 1995; Wiley 1981). Some definitions derive explicitly from a vicariance model (Harold and Mooi 1994; Hausdorf 2002), whereas others are neutral (Morrone 1994b; Szumik et al. 2002). Some refer to species (Hausdorf 2002; Linder 2001; Platnick 1991) and others to taxa (Harold and Mooi 1994; Humphries and Parenti 1999; Morrone 1994b).

How do we recognize an area of endemism? Müller (1973) suggested a protocol for working out "dispersal centers," which has been applied to identify areas of endemism (Morrone 1994b; Morrone et al. 1994). It consists basically of plotting the ranges of species on a map and finding the areas of congruence between several species. This approach assumes that the species' ranges are small compared with the region itself, that the limits of the ranges are known with certainty, and that the validity of the species is not in dispute. According to Linder (2001), areas of endemism should meet four criteria: They must have at least two endemic species; the ranges of the species endemic to them should be maximally congruent; they should be narrower than the whole study area, so that several areas are located; and they should be mutually exclusive.

Several issues concerning areas of endemism should be addressed. Crisp et al. (1995) suggested that the alternative procedures for identifying areas of endemism were controversial, especially questioning whether the hierarchical model of parsimony analysis of endemicity (PAE) was adequate for that purpose. Humphries and Parenti (1999) argued that including species that are ecologically very different can argue for a historical rather than ecological explanation for the areas of endemism identified. Linder (2001) proposed three optimality criteria to help choose the best estimate of the areas of endemism: the number of areas identified, the proportion of the species restricted to the areas of endemism, and the congruence of the distributions of the species restricted to the areas of endemism. Roig-Juñent et al. (2002) enumerated some problems with the identification of areas of endemism: lack of distributional data, bias toward locality data, and subjectivity in drawing the exact limits of the areas of endemism.

Methods

Six methods can be applied in panbiogeography and the identification of areas of endemism (Morrone 2004a; Morrone and Crisci 1995). I will deal herein with the minimum spanning-tree method (Croizat 1958b, 1964), track compatibility (Craw 1988a), PAE (Rosen 1988b), and endemicity analysis (Szumik et al. 2002). Connectivity and incidence matrices (Page 1987), which have never been applied empirically, and nested areas of endemism analysis (Deo and DeSalle 2006), an adaptation of the nested clade analysis (Templeton 1998), are not dealt with herein.

Minimum-Spanning Tree Method

Croizat did not focus on explicit methods (Grehan 1991; Humphries and Parenti 1999), but one may consider the minimum-spanning tree method as the first formalization of panbiogeography (Craw 1988a; Page 1987). It consists of delineating on maps the individual tracks of different taxa and then superimposing them in order to find the generalized tracks. Nodes are identified in the areas where two or more generalized tracks superimpose. With few individual tracks, the minimum-spanning tree method is easy to apply.

Algorithm It consists of the following steps (Morrone 2004b):

1. Construct individual tracks for different taxa, connecting the localities where they are distributed by a minimum-spanning tree.
2. If possible, orient the individual tracks with baselines.
3. Recognize similar tracks (in oriented tracks, they should have the same direction), which will be considered as part of the same generalized track.
4. Recognize nodes in the areas where two or more generalized tracks superimpose.
5. Indicate on a map the generalized tracks, baselines, and nodes.

Software Trazos2004 (Rojas Parra 2007).

Empirical Applications Abrahamovich et al. (2004), Aguilar-Aguilar and Contreras-Medina (2001), Álvarez Mondragón and Morrone (2004), Andersen (1982), Candela and Morrone (2003), Carvalho et al. (2003), Christiansen and Culver

(1987), Contreras-Medina and Eliosa León (2001), Contreras-Medina et al. (1999), Corona and Morrone (2005), Croizat (1958b, 1964, 1976), De Marmels (2000), Escalante et al. (2004), Fontenla (2005), Franco Rosselli and Berg (1997), González-Zamora et al. (2007); Grant et al. (2006), Grehan (2001a, 2001b, 2001c, 2007), Grehan and Rawlings (2003), Heads (1986, 1989, 1996, 2001, 2005b), Katinas et al. (1999), Kolibác (1998), López Almirall (2005), López-Ruf et al. (2006), Lopretto and Morrone (1998), Luna-Vega and Alcántara (2002), Luna-Vega and Contreras-Medina (2000), Márquez and Morrone (2003), Menu-Marque et al. (2000), Morrone (1993c, 1994c, 1996b, 1996c, 2000a, 2000b, 2000f, 2000h, 2001a, 2001b, 2001d, 2001f), Morrone and Gutiérrez (2005), Morrone and Pereira (1999), Morrone et al. (2002, 2004), Nihei and de Carvalho (2005), Ochoa et al. (2003), Roig-Juñent et al. (2003), Rosas-Valdez and Pérez-Ponce de León (2005), Rosen (1976), Soares and de Carvalho (2005), and Torres-Miranda and Luna-Vega (2006).

CASE STUDY 4.1 Biogeography and Evolution of North American Cave Collembola

Plant and animal taxa inhabiting caves are interesting from both biogeographic and evolutionary viewpoints. Collembola (Hexapoda) possess many cave-inhabiting genera in the New World, some of them showing different levels of adaptation to life in caves, known as troglomorphy. Christiansen and Culver (1987) analyzed some cave Collembola from the United States and Mexico in order to determine whether significant evolutionary novelty occurs centrally and moves out or starts in peripheral associates; whether the present distribution of taxa is better explained by dispersal, vicariance, or a combination of both processes; and whether a panbiogeographic analysis is applicable to cave Collembola.

Christiansen and Culver (1987) analyzed the geographic distribution of ninety species. They found two generalized tracks in the United States. One runs in straight line east–west along the 37° and 38° parallels, generally bending northward in West Virginia and Virginia (fig. 4.7a). Another generalized track starts in southeastern Missouri and curves first south and then north, ending in south central Pennsylvania (fig. 4.7b). According to the authors, both tracks are very similar and may be variants of a single generalized track, although they treated

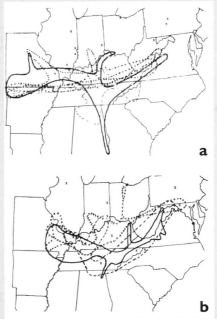

Figure 4.7 Biogeographic analysis of U.S. cave Collembola by Christiansen and Culver (1987). (a) Generalized track I; (b) generalized track II; (c) geographic distribution of troglomorphy of cave Entomobryinae (solid black: most troglomorphic species; dots: least troglomorphic species).

(continued)

CASE STUDY 4.1 Biogeography and Evolution of North American Cave Collembola *(continued)*

them as distinct based on the different taxa involved in each of them. For Mexico, the Caribbean, and South America the patterns were much less clear, and no generalized tracks could be identified.

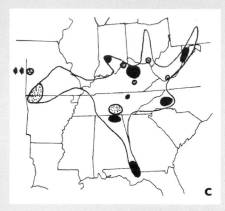

Christiansen and Culver (1987) also analyzed the levels of troglomorphy (characters that change progressively with increasing time in caves) exhibited by some of the taxa analyzed and correlated them with the geographic range. They found that smaller geographic ranges were associated with increased troglomorphy, and they considered this evidence of dispersal from a center of novelty and a series of more troglomorphic species spreading out. Distribution of single lineages showed peripheral or central evolutionary novelties, but examination at a larger scale (fig. 4.7c) did not reflect a clear trend for the most advanced or most primitive taxa to be peripheral or central. The authors concluded that the panbiogeographic method was useful to analyze cave Collembola in the restricted sense of providing a simple technique to look for evidence of past dispersal events.

References

Christiansen, K., and D. Culver. 1987. Biogeography and the distribution of cave Collembola. *Journal of Biogeography* 14:459–477.

CASE STUDY 4.2 Distributional Patterns of Mexican Marine Mammals

Aguilar-Aguilar and Contreras-Medina (2001) undertook a panbiogeographic analysis of the marine mammals (Cetartiodactyla, Sirenia, and Carnivora) represented in Mexico, with the aim of identifying their distributional patterns. The authors selected forty-five species from published lists (Aurioles 1993; Salinas and Ladrón de Guevara 1993), excluding those that were cosmopolitan or with a very restricted distribution. Some individual tracks are represented in figs. 4.8a–4.8f.

Three generalized tracks were identified (fig. 4.8g). One generalized track is located in the northern Pacific Ocean, ranging from the eastern coasts of Asia to the western coasts of North America. Another generalized track in the American Pacific is located in the central and southern Pacific Ocean, including the Gulf of California. The third generalized track, located in the central Atlantic Ocean, ranges from equatorial Africa to South America, the Caribbean, and the Gulf of Mexico, including parts of eastern North America.

The two Pacific Ocean generalized tracks overlap in a node in the Gulf of California (fig. 4.8g). Aguilar-Aguilar and Contreras-Medina (2001) found that, on one hand, the northern Pacific Ocean generalized track is congruent with the distributions of the beetle genus *Diaulota*, the chaetognath *Sagitta euneritica,* and the algal genus *Macrocystis.*

(continued)

CASE STUDY 4.2 Distributional Patterns of Mexican Marine Mammals *(continued)*

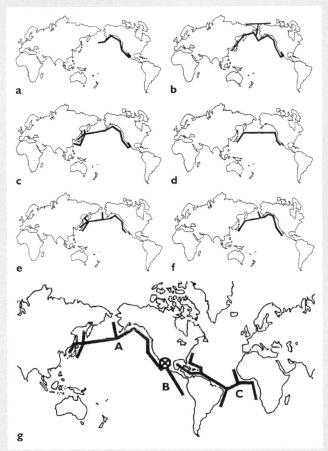

Figure 4.8 Biogeographic analysis of marine mammals by Aguilar-Aguilar and Contreras-Medina (2001). (a–f) Individual tracks: (a) *Mirounga angustirostris;* (b) *Eschrichtius robustus;* (c) *Lagenorhynchus obliquidens;* (d) *Lissodelphis borealis;* (e) *Phocoenoides dulli;* (f) *Berardius bairdi;* (g) generalized tracks and node. A, northern Pacific Ocean track; B, American Pacific track; C, central Atlantic Ocean track.

On the other hand, despite not inhabiting the Mexican coasts, several species of marine mammals display a congruent distribution with the northern Pacific Ocean track, such as the marine wolves *Eumetopias jubatus* and *Callorhinus ursinus* and the whales *Mesoplodon carlhubbsi* and *M. stejnegeri*. This generalized track is compatible with the reconstruction of the northern part of the Pacifica paleocontinent (Nur and Ben-Avraham 1980). The American Pacific generalized track is congruent with a generalized track proposed by Croizat (1958b) for terrestrial organisms and with the distribution of the chaetognath *Sagitta bierii,* the jellyfish *Phyalopsis diegensis,* and the algae *Codium picturatum*. The Atlantic Ocean generalized track is congruent with Wegener's (1929) proposal and is similar to the transoceanic Atlantic track found by Croizat (1958b) for terrestrial organisms.

From the geological viewpoint, it is difficult to interpret the complex history of the Gulf of California and its relationship with the Baja California Peninsula (Ferrusquía-Villafranca 1998). The coastal area of California and the Baja California Peninsula do not form part of the North American plate but belong to the Pacific plate, where the San Andreas Fault marks the boundary between it and the North American plate.

CASE STUDY 4.2 Distributional Patterns of Mexican Marine Mammals   *(continued)*

References

Aguilar-Aguilar, R. and R. Contreras-Medina. 2001. La distribución de los mamíferos marinos de México: Un enfoque panbiogeográfico. In *Introducción a la biogeografía en Latinoamérica: Teorías, conceptos, métodos y aplicaciones,* ed. J. Llorente Bousquets and J. J. Morrone, 213–219. Mexico, D.F.: Las Prensas de Ciencias, UNAM.

Aurio\les, G. D. 1993. Biodiversidad y estado actual de los mamíferos marinos de México. *Revista de la Sociedad Mexicana de Historia Natural* 44:397–412.

Croizat, L. 1958. *Panbiogeography,* Vols. 1 and 2. Caracas, Venezuela: Author.

Ferrusquía-Villafranca, I. 1998. Geología de México: Una sinopsis. In *Diversidad biológica de México,* ed. T. P. Ramamoorthy, R. Bye, A. Lot, and J. Fa, 3–108. Mexico, D.F.: Instituto de Biología, UNAM.

Nur, A. and Z. Ben-Avraham. 1980. Lost Pacifica continent: A mobilistic speculation. In *Vicariance biogeography: A critique,* ed. D. E. Rosen and G. Nelson, 341–358. New York: Columbia University Press.

Salinas, M. and P. Ladrón de Guevara. 1993. Riqueza y diversidad de los mamíferos marinos. *Ciencias* 7:85–93.

Wegener, A. 1929. *The origin of continents and oceans.* Dover: Dover Publications.

Track Compatibility

Craw (1988a, 1989a) formalized a quantitative method based on character compatibility (Lequesne 1982). Individual tracks are coded in an area × track matrix that is analyzed for track compatibility, where two individual tracks are compatible or congruent with each other if one is a subset of the other or they are the same in a pairwise comparison. The largest set of compatible tracks is called a clique, and it is used to construct the generalized track.

The individual tracks (figs. 4.9a–4.9d) are coded and input into an area × track matrix, where the presence of a track in an area is indicated with "1" and its absence with "0" (fig. 4.9e). The matrix is analyzed in search of compatibility between the tracks. The set of compatible tracks represents a clique that is used to construct the generalized track connecting the areas, which is drawn on a map (fig. 4.9f). In order to allow the statistical evaluation of the generalized tracks, matrices of the same size as the real matrix are generated randomly. If the number of randomly generated matrices that display cliques equal to or greater than that obtained from the real data is large, the statistical meaning of the generalized track will be small.

Algorithm The algorithm consists of the following steps (Craw 1989a; Espinosa Organista et al. 2002; Grehan 2001c; Morrone 2004b; Morrone et al. 1996):

1. Construct an $r \times c$ matrix, where r (rows) represent the localities or distributional areas and c (columns) represent the individual tracks. Each entry is "1" or "0," depending on whether the track is present or absent.

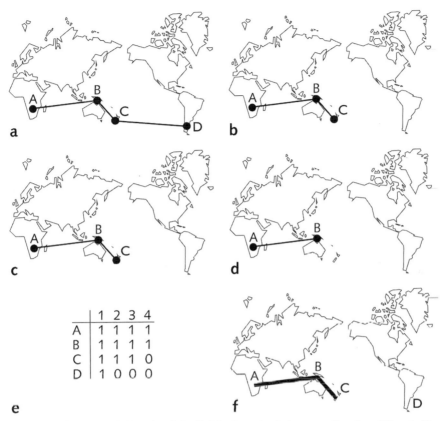

Figure 4.9   Track compatibility. (a–d) Individual tracks 1–4 connecting localities A–D; (e) matrix of localities × individual tracks; (f) generalized track obtained.

2. Use a compatibility program to find the largest clique of compatible tracks.

3. Map out the largest cliques as generalized tracks connecting the localities or areas.

4. Use a statistical test to evaluate the percentage of randomly generated matrices where the largest clique size is as large as or larger than the largest clique in the real data in order to provide a statistical test of the level at which the largest clique attains significance.

5. Identify baselines (if possible) for the generalized tracks.

6. Represent the generalized tracks, nodes, baselines, and main massings on a map.

Software CLIQUE of package PHYLIP (Felsenstein 1993), CLINCH (Fiala 1984), SECANT 2.2 (Salisbury 1999), and TNT (Goloboff et al., retrieved May 25, 2008, from http://www.zmuc.dk/public/phylogeny/TNT/).

Empirical Applications Craw (1988a), Crisci et al. (2001), Katinas et al. (2004), Morrone (1992), Morrone and Lopretto (1994), Posadas et al. (1997), and Quijano-Abril et al. (2006).

CASE STUDY 4.3 Biogeography of the Subantarctic Islands

The Subantarctic islands surround Antarctica in the Atlantic, Indian, and Pacific oceans. The biogeographic relationships of their biota were analyzed by means of a panbiogeographic approach by Morrone (1992). The biogeographic units of the analysis were the Falklands, Tierra del Fuego, the Patagonian steppes, the Magellan area, and the Juan

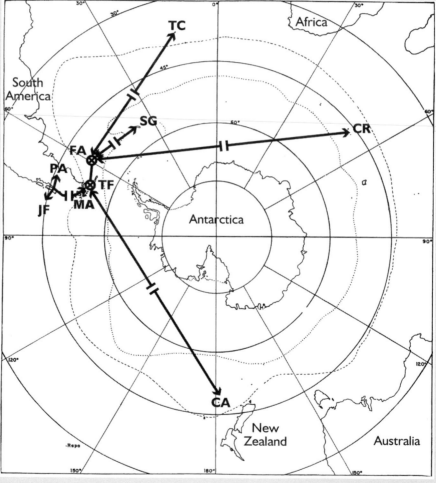

Figure 4.10   Biogeographic analysis of the Subantarctic islands by Morrone (1992): generalized tracks obtained. CA, Campbell Islands and other Subantarctic islands of New Zealand; CR, Crozet Island and other Subantarctic islands of the Indian Ocean; FA, Falkland Islands; JF, Juan Fernandez Islands; MA, Magellan; PA, Patagonian steppes; SG, South Georgia Island; TC, Tristan da Cunha-Gough Islands; TF, Tierra del Fuego.

(continued)

CASE STUDY 4.3 Biogeography of the Subantarctic Islands (continued)

Fernández and South Georgia islands in South America; Campbell Island and other New Zealand Subantarctic islands; the Tristan da Cunha-Gough Islands; and Crozet, Marion, and Prince Edward islands in the Indian Ocean. The analysis was based on fifty taxa, which included ferns, angiosperms, Collembola, Coleoptera (Carabidae, Staphylinidae, Tenebrionidae, and Curculionidae), Diptera (Ephydridae), Hemiptera (Delphacidae), Mallophaga, Crustacea (Cladocera, Isopoda, and Copepoda), Oligochaeta, and Mollusca (Endodontidae). The track compatibility analysis was undertaken with PHYLIP (Felsenstein 1986).

From the compatibility analysis of the matrix, twenty-one cliques were obtained. The longest cliques delineated five generalized tracks (fig. 4.10), which are divided into two main groups having baselines in the Atlantic and Pacific oceans, respectively. Three generalized tracks with baselines in the Atlantic Ocean connect the Falklands with Tristan da Cunha-Gough, with South Georgia Island, and with Crozet Island, respectively, and then continue from the Falklands to Tierra del Fuego and the Magellan area. A generalized track with a baseline in the Pacific Ocean connects Campbell Island with Tierra del Fuego, and then it continues to the Falklands and Magellan. The fifth generalized track, with a baseline in the Pacific Ocean, joins the Juan Fernandez Islands with the Magellan area and the Falklands. The five generalized tracks share two segments: one connects the Falklands and Tierra del Fuego, and the other connects the latter with the Magellan area. The Falklands and Tierra del Fuego were identified as nodes.

This analysis highlighted the complex biotic relationships of most of the studied areas, especially Tierra del Fuego and the Falklands. Other analyses based on insect taxa (Carvalho and Couri 2002; Morrone 1993c, 1994a; Morrone et al. 1994) have found similar patterns.

References

Carvalho, D. J. B. de and M. S. Couri. 2002. A cladistic and biogeographic analysis of *Apsil* Malloch and *Reynoldsia* Malloch (Diptera, Muscidae) of southern South America. *Proceedings of the Entomological Society of Washington* 104:309–317.

Felsenstein, J. 1986. *Phylogenetic inference package (PHYLIP)*. Seattle: University of Washington.

Morrone, J. J. 1992. Revisión sistemática, análisis cladístico y biogeografía histórica de los géneros *Falklandius* Enderlein y *Lanteriella* gen. nov. (Coleoptera: Curculionidae). *Acta Entomológica Chilena* 17:157–174.

Morrone, J. J. 1993. Revisión sistemática de un nuevo género de Rhytirrhinini (Coleoptera: Curculionidae), con un análisis biogeográfico del dominio Subantártico. *Boletín de la Sociedad de Biología de Concepción* 64:121–145.

Morrone, J. J. 1994. Distributional patterns of species of Rhytirrhinini (Coleoptera: Curculionidae) and the historical relationships of the Andean provinces. *Global Ecology and Biogeography Letters* 4:188–194.

Morrone, J. J., S. Roig-Juñent, and J. V. Crisci. 1994. Cladistic biogeography of terrestrial subantarctic beetles (Insecta: Coleoptera) from South America. *National Geographic Research and Exploration* 10:104–115.

CASE STUDY 4.4 Biogeography of the Sierra de Chiribiquete (Colombia)

Cortés and Franco (1997) analyzed the relationships of the flora of the Sierra de Chiribiquete, in the Guyanan shield in Colombia. They used the track compatibility technique, with the routine clique of the PHYLIP package (Felsenstein 1986), to analyze the species × locality data matrix.

Fifteen of the 102 species analyzed led to the recognition of the longest generalized track (fig. 4.11b), which connects the Sierra de Chiribiquete with other areas of the Guyanan shield. Smaller generalized tracks connect the Sierra de Chiribiquete with the Chocó area, the Magdalena River basin, the Brazilian shield, the southern Andes, and

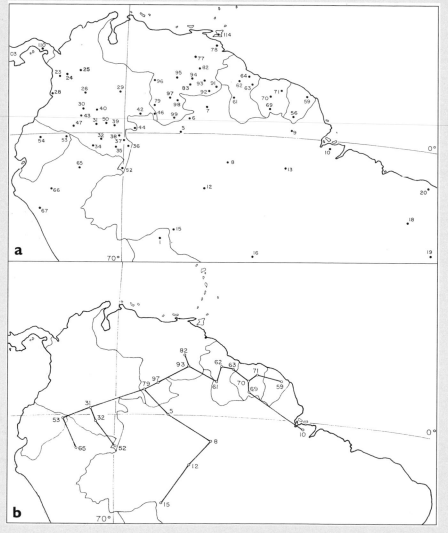

Figure 4.11   Biogeographic analysis of the Sierra de Chiribiquete by Cortés and Franco (1997). (a) Localities analyzed; (b) longest generalized track obtained, connecting Chiribiquete with some Amazonian localities.

(continued)

CASE STUDY 4.4 Biogeography of the Sierra de Chiribiquete (Colombia) *(continued)*

Central America, apparently indicating older relationships between areas that were fractured with the uplift of the Andes. Croizat (1976) gave examples of similar generalized tracks based on the distribution of the genera *Platypsaris* (Cotingidae) and *Rapatea* (Rapateaceae) and suggested an ancestral, pre-Oligocene distribution (Croizat 1976).

A node was identified by Cortés and Franco (1997) in northern Antioquia that corresponds to the node of Santander, previously recognized by Croizat (1976). This node is important because it is based on generalized tracks directed toward South and Central America. Croizat defined Colombia as a whole as a node; additionally, he found sectors in the country that constituted smaller nodes (e.g., Santander, San Andrés, Providencia, and Macarena). According to Croizat, these tracks are the result of the collision of a large piece of geological coast of Central and South America in the Caribbean Sea. In the marine region between San Andrés, Providence, and central Panama, there is a concentration of generalized tracks that connect the Antilles, islands of coastal Central and South America, the Galápagos Islands, and areas of the northeastern Pacific.

References

Cortés, B. R. and P. Franco. 1997. Análisis panbiogeográfico de la flora de Chiribiquete, Colombia. *Caldasia* 19:465–478.

Croizat, L. 1976. *Biogeografía analítica y sintética ("panbiogeografía") de las Américas.* Caracas, Venezuela: Biblioteca de la Academia de Ciencias Físicas, Matemáticas y Naturales.

Felsenstein, J. 1986. *Phylogenetic inference package (PHYLIP).* Seattle: University of Washington.

Parsimony Analysis of Endemicity

PAE is also known as parsimony analysis of shared presences (Rosen and Smith 1988), simplicity analysis of endemicity (Crisci et al. 2000), parsimony analysis of distributions (Trejo-Torres and Ackerman 2001), parsimony analysis of species sets (Trejo-Torres 2003), cladistic analysis of distributions and endemism (Porzecanski and Cracraft 2005), and parsimony analysis of community assemblages (Ribichich 2005). It was formulated originally by Rosen (1985) and fully developed by Rosen (1988b) and Rosen and Smith (1988). PAE (fig. 4.12) constructs cladograms based on the parsimony analysis of a presence–absence data matrix of species and supraspecific taxa (Cecca 2002; Cracraft 1991; Escalante and Morrone 2003; Morrone 1994a, 1994b, 1998; Myers 1991; Nihei 2006; Porzecanski and Cracraft 2005; Posadas and Miranda-Esquivel 1999; Rosen 1988b; Rosen and Smith 1988; Trejo-Torres 2003). PAE cladograms may allow one to infer the three biogeographic processes: Synapomorphies are interpreted as vicariance events, parallelisms as dispersal events, and reversals as extinction events.

Crisci et al. (2000) distinguished three variants of PAE according to the units analyzed: localities, areas of endemism, and grid cells. There are several other

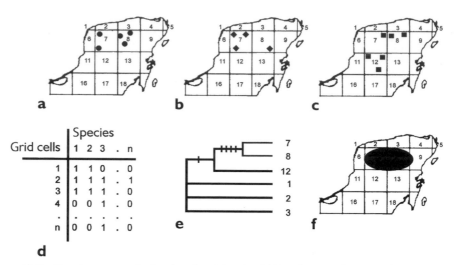

Figure 4.12   Parsimony analysis of endemicity. (a–c) Distributional maps of three species, with grid cells superimposed; (d) data matrix of grid cells × species; (e) cladogram of grid cells; (f) area of endemism obtained.

units, such as hydrological basins (Aguilar-Aguilar et al. 2003), real and virtual islands (Luna-Vega et al. 1999, 2001; Maldonado and Uriz 1995; Trejo-Torres and Ackerman 2001), transects (García-Trejo and Navarro 2004; León-Paniagua et al. 2004; Navarro et al. 2004; Trejo-Torres and Ackerman 2002), communities (Ribichich 2005), and political entities (Cué-Bär et al. 2006). García-Barros (2003) proposed a more appropriate classification based on the objectives of the analysis: to infer historical relationships between areas, to identify areas of endemism, and to classify areas (as a phenetic association method).

In order to root the PAE cladograms, a hypothetical area with all "0" is added to the matrix. However, some authors (Cano and Gurrea 2003; Ribichich 2005) have used an area coded with all "1." This alternative rooting groups areas according to shared absences, which would imply depletion through time starting from a cosmopolitan biota (Cecca 2002).

PAE may be used for panbiogeographic analyses, where the clades obtained are considered generalized tracks (Craw et al. 1999; Luna-Vega et al. 2000; Morrone and Márquez 2001). With the aim that the "1" appears only once and does not revert to "0" (as in a compatibility analysis), Luna-Vega et al. (1999) undertook the parsimony analysis with PAUP 4.0.1 (Swofford 1999), setting Goloboff concavity $k = 0$. Luna-Vega et al. (2000) and García-Barros et al. (2002) proposed that when the most parsimonious cladograms have been obtained, it is possible to remove or exclude the taxa supporting the different clades and analyze the reduced matrix to search for alternative clades supported by other taxa. This procedure has been named parsimony analysis of endemicity with progressive character elimination (PAE-PCE) (García-Barros 2003; García-Barros et al. 2002).

Equal (nondifferential) weighting is usually used for PAE. However, Linder (2001) suggested a protocol to weight species inversely to their distribution areas so that widespread species do not obscure the analyses by introducing homoplasy into the data. It consists of four steps:

1. Weight each species in each grid cell by the inverse of its distribution range so that a species restricted to a grid cell would be scored as 1, a species restricted to two grid cells as 0.5, three grid cells as 0.33, and so forth.

2. Transform the values to a scale of 0–20, multiplying each by 20 and rounding the product. This results in a matrix with values of 20 (single-grid endemics), 10 (two-grid endemics), 7 (three-grid endemics), and so forth.

3. Simplify the matrix by changing the 20s and 10s to 9 (because single-grid species carry no grouping information in a parsimony analysis) so that each species will be represented by a single-digit value in each grid cell score.

4. Input the matrix into a parsimony analysis, treating the characters' states (0 to 9) as additive.

PAE has received some criticism. Linder and Mann (1998) criticized Morrone's (1994b) approach for identifying areas of endemism with PAE because grid cells can be used only as presence–absence data, and undercollecting may result in grid cells being omitted. Some authors suggested that PAE is not a valid historical method because it does not take into account the phylogenetic relationships of the taxa analyzed (García-Barros et al. 2002; Humphries 1989, 2000; Santos 2005). According to Rosen (1988b; see also Nihei 2006; Trejo-Torres 2003; Trejo-Torres and Ackerman 2002), there are two possible interpretations for PAE cladograms: static and dynamic. The former assumes that cladograms constitute an alternative to phenetic classification methods, whereas according to the latter, cladograms are hypotheses on the historical or ecological relationships of the areas analyzed. If we interpret the external area with all "0" as an area lacking suitable conditions for the taxa to survive therein (ecological interpretation), relationships will indicate ecological affinities. If we interpret the external area as a geologically ancient area, where none of the taxa has yet evolved (historical interpretation), relationships will indicate biotic interchanges or vicariance events. Most of the authors who have used PAE explored historical interpretations of the detected patterns, usually from a vicariance viewpoint; for ecological interpretations, see Trejo-Torres and Ackerman (2002), Trejo-Torres (2003), and Ribichich (2005).

Enghoff (2000) considered PAE an extreme "assumption 0" approach because only the widespread taxa provide evidence of area relationships. Morrone and Márquez (2001) and Brooks (2005) considered PAE an incomplete implementation of Brooks parsimony analysis (BPA). Szumik et al. (2002) criticized the use of PAE for identifying areas of endemism because an explicit optimality criterion is used a posteriori to select areas of endemism found by what they considered less appropriate means. Brooks and Van Veller (2003) criticized the use of PAE as a cladistic biogeographic method, which is erroneous because it has a different objective.

Parenti and Humphries (2004) suggested that PAE adopts protocols directly from phylogenetic systematics and violates some of the basic assumptions of cladistic biogeography.

Nihei (2006) presented a revision of PAE, including a discussion of its history and applications. He suggested that most of the criticisms dealt with its method rather than its theory and that they usually resulted from the confusion between the dynamic and static approaches. Nihei warned biogeographers applying PAE to be aware of the problems and limitations of both dynamic and static PAE and to evaluate new variations of PAE.

Algorithm   PAE-PCE consists of the following steps (Craw 1989b; Crisci et al. 2000; Grehan 2001c; Lomolino et al. 2006; Morrone 1994b, 2004b; Posadas and Miranda-Esquivel 1999; Vargas 2002):

1. Construct an $r \times c$ matrix, where r (rows) represents the units analyzed (e.g., localities, distributional areas, grid cells) and c (columns) represents the taxa. Each entry is "1" or "0," depending on whether the taxon is present or absent in the locality. A hypothetical area coded with all "0" is added to the matrix in order to root the resulting cladograms.
2. Analyze the matrix with a parsimony algorithm. If more than one cladogram is found, calculate the strict consensus cladogram.
3. Connect on a map the area relationships supported by two or more taxa as generalized tracks or areas of endemism.
4. Remove the taxa supporting the previous generalized tracks or areas of endemism.
5. Repeat steps 2–4 until no more taxa support any clade.

Software   Hennig86 (Farris 1988), PHYLIP (Felsenstein 1993), NONA (Goloboff 1998), PAUP (Swofford 2003), Pee-Wee (Goloboff, available at http://www.zmuc. dk/public/phylogeny/Nona-PeeWee/), and TNT (Goloboff et al., retrieved May 25, 2008, from http://www.zmuc.dk/public/phylogeny/TNT/). For reading and editing data files and cladograms: Winclada (Nixon 1999), compatible with NONA, Pee-Wee, and Hennig86; and MacClade (Maddison and Maddison, retrieved May 25, 2008, from http://macclade.org/macclade.html), compatible with PAUP.

Empirical Applications   Aguilar-Aguilar et al. (2003, 2005), Ahrens (2004), Andrés Hernández et al. (2006), Bates and Demos (2001), Bates et al. (1998), Bellan and Bellan Santini (1997), Biondi and D'Alessandro (2006), Bisconti et al. (2001), Cano and Gurrea (2003), Carrillo-Ruiz and Morón (2003), Carvalho et al. (2003), Cavieres et al. (2001, 2002), Chen and Bi (2007), Conran (1995), Contreras-Medina et al. (2007a), Corona et al. (2005, 2007), Costa et al. (2000), Cracraft (1991, 1994), Craw (1988a, 1989b), Crisci et al. (2001), Cué-Bär et al. (2006), Dapporto et al. (2007), Da Silva and Oren (1996), Da Silva et al. (2004), Davis et al. (2002), De Grave (2001), Deleporte and Colyn (1999), Escalante et al. (2003, 2005, 2007b, 2007c), Espadas-

Manrique et al. (2003), Espinosa Organista et al. (2000, 2006), Espinosa Pérez and Huidobro Campos (2005), Fattorini (2002), Fattorini and Fowles (2005), Fernandes et al. (1995), Fontenla (2003, 2005), García-Barros (2003), García-Barros et al. (2002), García-Trejo and Navarro (2004), Garraffoni et al. (2006), Geraads (1998), Glasby and Álvarez (1999), Goldani and Carvalho (2003), Goldani et al. (2002), Huidobro et al. (2006), Ippi and Flores (2001), Katinas et al. (2004), Koehler (2000), León-Paniagua et al. (2004), Linder (2001), Linder and Mann (1998), Löwenberg-Neto and de Carvalho (2004), Luis Martínez et al. (2005), Luna-Vega et al. (1999, 2001), Maldonado and Uriz (1995), Marino et al. (2001), Marks et al. (2002), Martínez Gordillo and Morrone (2005), Melo Santos et al. (2007), Méndez-Larios et al. (2005), Michaux and Leschen (2005), Mihoc et al. (2006), Moline and Linder (2006), Moreno et al. (2006), Morrone (1994a, 1998), Morrone and Coscarón (1996), Morrone and Escalante (2002), Morrone and Lopretto (1995), Morrone et al. (1997, 1999, 2002), Mota et al. (2002), Myers (1991), Navarro et al. (2004), Pizarro Araya and Jerez (2004), Porzecanski and Cracraft (2005), Posadas (1996), Posadas et al. (1997), Quijano-Abril et al. (2006), Racheli and Racheli (2003, 2004), Raherilalao and Goodman (2005), Reyes-Castillo et al. (2005), Ribichich (2005), Riddle and Hafner (2006), Roig-Juñent et al. (2002), Rojas Soto et al. (2003), Ron (2000), Rosen (1988a), Rosen and Smith (1988), Rovito et al. (2004), Rundle et al. (2000), Seeling and Fauth (2004), Sfenthourakis and Giokas (1998), Silva and Gallo (2007), Smith (1988, 1992), Smith and Xu (1988), Soares and de Carvalho (2005), Trejo-Torres and Ackerman (2001, 2002), Tribsch (2004), Unmack (2001), Vargas et al. (1998, 2003), Vergara et al. (2006), Waggoner (1999, 2003), Watanabe (1998), Winfield et al. (2006), Xu (2005), Yeates et al. (2001), and Zhang (2002).

CASE STUDY 4.5 Biogeography of the Mexican Cloud Forests

Mexican mountain cloud forests, located in humid and temperate zones between 600 and 3,000 m height, exhibit a high biotic diversity and a fragmented distribution, similar to an archipelago, where each island has a particular biotic composition. Some authors (Luna et al. 1989; Puig 1989; Rzedowski 1978) suggested that these forests are made up of three different cenocrons: one comprising Nearctic taxa, basically corresponding to temperate trees; another with Neotropical taxa, basically grasses, epiphytes, and shrubs; and another with endemic taxa, not very important at the generic level but more significant at the specific level. Luna-Vega et al. (1999) undertook a PAE with the purpose of postulating a preliminary hypothesis on the relationships between different fragments of Mexican cloud forests.

The authors analyzed twenty-four localities of Mexican cloud forests belonging to the Mountain Mesoamerican phytogeographic region (Rzedowski 1978) or the Mexican Transition Zone (Halffter 1987; Morrone 2004c). They selected species of vascular plants, using distributional data from several floristic studies and databases from the Comisión Nacional para el Conocimiento y Uso de la Biodiversidad (Conabio). They built a data matrix of 1,267 species × 24 localities, including also a hypothetical area coded with all "0" to root the cladogram. The data matrix was analyzed with PAUP 4.0.1 (Swofford 1999).   *(continued)*

CASE STUDY 4.5 Biogeography of the Mexican Cloud Forests *(continued)*

A single cladogram was obtained (fig. 4.13a), with 3,581 steps, a consistency index of 0.354, and a retention index of 0.291. It shows five main clades (fig. 4.13b): the Serranías Transístmicas and part of the coast of the Gulf of Mexico, Sierra Madre Oriental, and Serranías Meridionales (El Triunfo and Montebello in Chiapas, San Martín Volcano and Teocelo in Veracruz, and Chinantla in Oaxaca); the northern part of the

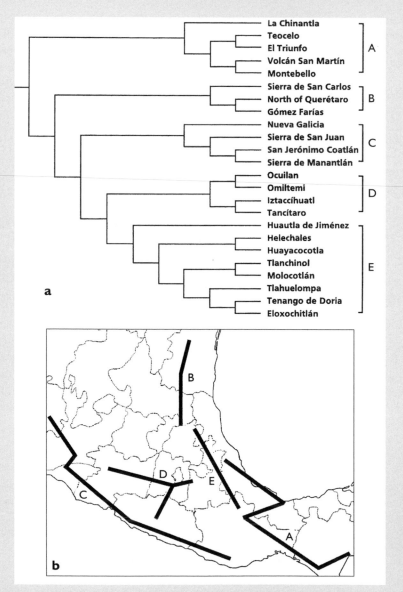

Figure 4.13   Biogeographic analysis of the Mexican cloud forests by Luna-Vega et al. (1999). (a) PAE cladogram obtained; (b) generalized tracks identified.

(continued)

CASE STUDY 4.5 Biogeography of the Mexican Cloud Forests *(continued)*

Sierra Madre Oriental (Sierra de San Carlos and Gómez Farías in Tamaulipas, and northern Querétaro); the Pacific portion of the Serranías Meridionales (Serranía de San Juan and Nueva Galicia in Nayarit, Jalisco, and Colima; Manantlán in Jalisco; and San Jerónimo Coatlán in Oaxaca); the central portion of the Serranías Meridionales (Tancítaro in Michoacán, Omiltemi in Guerrero, Ocuilan in Morelos and the state of Mexico, and Iztaccíhuatl in Mexico City and the state of Mexico); and the southern part of the Sierra Madre Oriental plus a part of the Serranías Meridionales (Tlanchinol, Eloxochitlán, Molocotlán, Tlahuelompa, and Tenango de Doria in Hidalgo; Huayaco-cotla and Helechales in Veracruz; and Huautla de Jiménez in Oaxaca).

On the basis of these results, Luna-Vega et al. (1999) concluded that five major Mexican forest units diverged sequentially from a former continuous forest as a consequence of past geological and climatic events. They also concluded that some of the phytogeographic provinces recognized traditionally, such as the Sierra Madre Oriental, the Sierra Madre del Sur, and the Serranías Meridionales, did not represent natural units because their cloud forests exhibited different, complex relationships.

References

Halffter, G. 1987. Biogeography of the montane entomofauna of Mexico and Central America. *Annual Review of Entomology* 32:95–114.

Luna, I., L. Almeida, and J. Llorente. 1989. Florística y aspectos fitogeográficos del bosque mesófilo de montaña de las cañadas de Ocuilan, estados de Morelos y México. *Anales del Instituto de Biología de la UNAM, Serie Botánica* 59(1):63–87.

Luna-Vega, I., O. Alcántara, D. Espinosa Organista, and J. J. Morrone. 1999. Historical relationships of the Mexican cloud forests: A preliminary vicariance model applying parsimony analysis of endemicity to vascular plant taxa. *Journal of Biogeography* 26:1299–1305.

Luna-Vega, I., O. Alcántara, J. J. Morrone, and D. Espinosa Organista. 2000. Track analysis and conservation priorities in the cloud forests of Hidalgo, Mexico. *Diversity and Distributions* 6:137–143.

Morrone, J. J. 2004. La Zona de Transición Sudamericana: Caracterización y relevancia evolutiva. *Acta Entomológica Chilena* 28:41–50.

Puig, H. 1989. Análisis fitogeográfico del bosque mesófilo de montaña de Gomez Farías. *Biotam* 1(2): 34–53.

Rzedowski, J. 1978. *Vegetación de México*. México D.F.: Limusa.

Swofford, D. L. 1999. *PAUP*: Phylogenetic analysis using parsimony (*and other methods)*. *Version 4.0 beta*. Sunderland, Mass.: Sinauer.

CASE STUDY 4.6 Distribution of Butterflies in the Western Palearctic

García-Barros (2003) analyzed the geographic distribution of species of butterflies of the families Papilionidae, Pieridae, Lycaenidae, and Nymphalidae of the western Palearctic. The author analyzed 196 species and subspecies endemic to the study area, in 245 grid cells of 1° × 1°. García-Barros used Winclada (Nixon 1999) and NONA (Goloboff 1993) to undertake a PAE-PCE. He also used PAUP (Swofford 2003), with Goloboff constant $k = 0$ to obtain a cladogram, as suggested by Luna-Vega et al. (2000) and García-Barros et al. (2002), and additionally obtained a phenogram applying the Unweighted Pair Group Method with Arithmetic Mean (UPGMA), using the Jaccard coefficient, with program SPSS (SPSS Inc. 2001).

The first PAE resulted in more than 1,000 cladograms (616 steps, consistency index of 0.31, and retention index of 0.80). In the strict consensus cladogram, ten areas of endemism were identified (fig. 4.14a). The best-supported areas using bootstrap were areas II and III. PAE-PCE did not allow identification of other areas of endemism. The analysis with PAUP with $k = 0$ led to more than 1,000 cladograms (673 steps, consistency index of 0.29, and retention index of 0.77), which contained areas I–X and XII. Areas XI and XIII were identified in the second and third runs of PAE-PCE, respectively. The UPGMA phenogram allowed recognition of seventeen areas (fig. 4.14b). For the Iberian Peninsula, in particular, García-Barros (2003) illustrated the corresponding parts of the

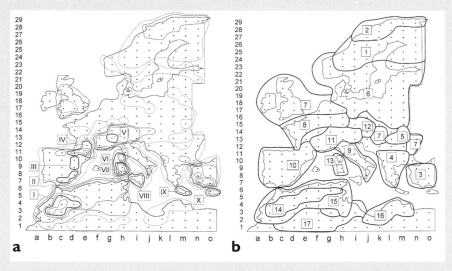

Figure 4.14 Biogeographic analysis of butterflies from the western Palearctic. (a) Areas of endemism identified by PAE-PCE; (b) areas of endemism identified by the Unweighted Pair Group Method with Arithmetic Mean (UPGMA) phenogram; (c) portion of the phenogram corresponding to the Iberian Peninsula; (d) portion of the cladogram corresponding to the Iberian Peninsula; (e) nearest neighbors for each Iberian grid cell; (f) nearest non-Iberian neighbor for each Iberian grid cell (modified from García-Barros 2003:240–243; reproduced with permission of *Graellsia*).

(continued)

CASE STUDY 4.6 Distribution of Butterflies in the Western Palearctic *(continued)*

phenogram (fig. 4.14c) and PAE cladogram (fig. 4.14d), as well as the nearest neighbor of each grid cell in the peninsula (fig. 4.14e) and the nearest non-Iberian neighbor (fig. 4.14f). These diagrams illustrate the absence of a clear North African cenocron in the Iberian Peninsula, in contrast to other taxa (García-Barros et al. 2002).

García-Barros (2003) concluded that the results based on the PAE cladogram were better than those based on the phenogram because the latter included some areas based on absent species. In addition, the procedure of setting $k = 0$ was more effective than PAE-PCE. He suggested that compatibility algorithms (e.g., CLIQUE from PHYLIP; Felsenstein 1993) may be more appropriate than those based on parsimony for identifying areas of endemism.

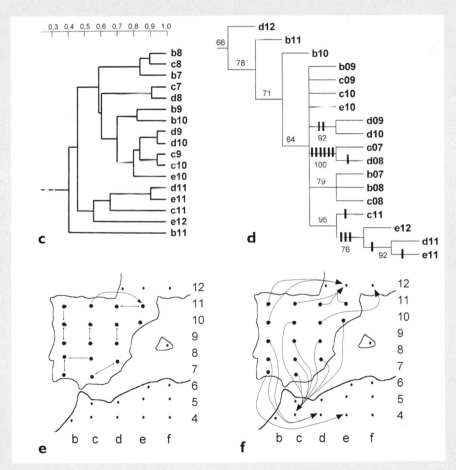

(continued)

CASE STUDY 4.6 Distribution of Butterflies in the Western Palearctic *(continued)*

References

García-Barros, E. 2003. Mariposas diurnas endémicas de la región Paleártica Occidental: Patrones de distribución y su análisis mediante parsimonia (Lepidoptera, Papilionoidea). *Graellsia* 59:233–258.

García-Barros, E., P. Gurrea, M. J. Luciáñez, J. M. Cano, M. L. Munguira, J. C. Moreno, H. Sainz, et al. 2002. Parsimony analysis of endemicity and its application to animal and plant geographical distributions in the Ibero-Balearic region (western Mediterranean). *Journal of Biogeography* 29:109–124.

Goloboff, P. 1998. *NONA ver. 2.0.* Retrieved May 25, 2008, from http://www. cladistics.com/about_nona.htm.

Nixon, K. C. 1999. *WinClada ver. 1.0000.* Ithaca, N.Y.: Author. Retrieved May 25, 2008, from http://www.cladistics.com/about_winc.htm.

Swofford, D. L. 2003. *PAUP*: Phylogenetic analysis using parsimony (*and other methods). Version 4.* Sunderland, Mass.: Sinauer. Retrieved May 25, 2008, from http://paup. csit.fsu.edu/.

Endemicity Analysis

Szumik et al. (2002) proposed a method that takes into consideration the spatial position of the species in order to identify the set of grid cells that represent an optimal area of endemism according to a score based on the number of species endemic to it (Szumik and Roig-Juñent 2005). In order to assign the values of endemicity to the sets of grid cells evaluated, Szumik et al. (2002) suggested four criteria (fig. 4.15a–4.15d):

1. First criterion (fig. 4.15a): The distribution of a species must adjust perfectly to the area to contribute to the score, so it must be present in all the grid cells of the set.

2. Second criterion (fig. 4.15b): A species can contribute to the score if it is present in some grid cells outside the area as long as the cell is adjacent to the area. This criterion does not require that all species contributing to the score have identical distributions.

3. Third criterion (fig. 4.15c): It is not required that all the grid cells of the set have identical species composition, but only species occurring in each one of the cells contribute to the score.

4. Fourth criterion (fig. 4.15d): A species may be absent from a given cell but still contribute to the score. Only a species that is evenly distributed in the area satisfies this criterion.

Szumik and Goloboff (2004) developed an endemicity value that gives weight to each species, considering its adjustment to the evaluated area. The degree of adjustment between the distributional area of each species and the area of endemism

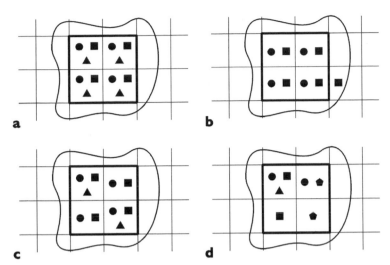

Figure 4.15   Assignment of scores under different criteria in endemicity analysis. (a) Area with score 3 under the first criterion (contributed by the three species); (b) area with score 2 under the second criterion (in addition to species "circle," species "square" contributes to the score); (c) area with score 2 under the third criterion (species "triangle" does not contribute to the score because it is found only in some cells); (d) area with score 4 under the fourth criterion (the four species contribute to the score, although none of them is found in all the cells).

under evaluation depends on the relationship between the number of grid cells where it is found and the total number of grid cells. Additionally, the endemicity value increases with the number of grid cells where the presence of the species is assumed or inferred and decreases with the number of grid cells outside the area of endemism where the species is observed or assumed to be present (Szumik and Goloboff 2004).

Algorithm   It consists of the following steps (Szumik et al. 2002, 2006; Szumik and Goloboff 2004; Szumik and Roig-Juñent 2005):

1. Plot species localities on a map with a grid.
2. Assign values of endemicity to all possible sets of grid cells, counting the species that may be considered endemic to them according to the four criteria defined by Szumik et al. (2002).
3. Choose the sets of grid cells with the highest endemicity scores.
4. Draw the sets of grid cells on a map as areas of endemism.

Software   NDM and VNDM (Goloboff 2004; Szumik et al. 2006).

Empirical Applications   Domínguez et al. (2006), Moline and Linder (2006), Szumik and Roig-Juñent (2005).

CASE STUDY 4.7 Areas of Endemism in Southern South America

Southern South America, basically the area south of 30° south parallel (Crisci et al. 1991b), harbors a very interesting biota, with high endemicity and strong biotic relationships with other austral continents. Szumik and Roig-Juñent (2005) analyzed data from 191 species of Carabidae (Coleoptera), distributed in 208 1° × 1° grid cells with software NDM (Goloboff 2004).

Fifty-three (28%) species were present in a single grid cell. A first analysis of endemicity was undertaken in order to determine whether these species were defining single grid cell areas of endemism or were the result of insufficient data or artifacts due to small grid cells. Four small areas of endemism were found, one of them corresponding to the Juan Fernandez Islands. The second analysis of endemicity was conducted adding or deleting two grid cells each time and defining areas of endemism by four or more endemic species, with an endemicity value corresponding to two species with perfect sympatry. Fifteen sets of grid cells were found, but some of these sets were nested within others or differed in a single grid cell, so six areas of endemism were recognized (figs. 4.16a and 4.16b). Another analysis was undertaken filling empty grid cells (assumed presences) that show disjunct distributions, using option Alt-F from VNDM, which assumes a continuous distribution of a species within the maximum polygon of distribution. Ten sets of grid cells were found, some of them similar to those obtained previously (figs. 4.16c and 4.16d).

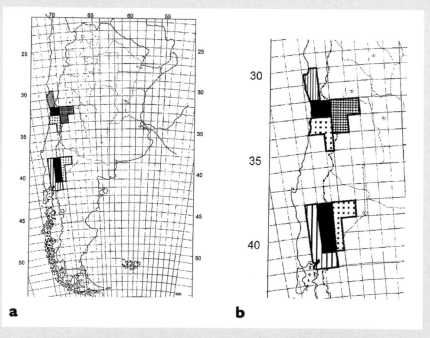

Figure 4.16   Biogeographic analysis of Carabidae from southern South America by Szumik and Roig-Juñent (2005). (a–b) Areas of endemism recognized initially; (c–d) areas of endemism recognized filling the empty grid cells.

(continued)

CASE STUDY 4.7 Areas of Endemism in Southern South America
(continued)

The analysis of endemicity allowed the authors to recognize several biogeographic provinces previously identified (Morrone 2001a): Western Patagonia, Maule, Coquimbo, and Valdivian Forest. Szumik and Roig-Juñent (2005) found that the analysis filling the empty grid cells gave better results. Because their analysis involved only species of Carabidae, they concluded that a more complete analysis of the biota may allow one to recognize other areas.

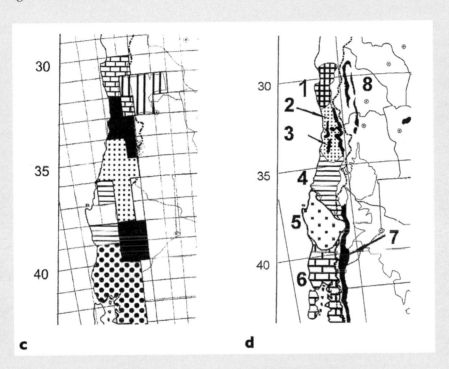

References

Crisci, J. V., M. M. Cigliano, J. J. Morrone, and S. Roig-Juñent. 1991. Historical biogeography of southern South America. *Systematic Zoology* 40:152–171.

Goloboff, P. 2004. NDM/VNDM programs ver. 1.5. Retrieved May 25, 2008, from http://www.zmuc.dk/public/phylogeny/Endemism/.

Morrone, J. J. 2001. *Biogeografía de América Latina y el Caribe.* Saragossa, Spain: Manuales y Tesis SEA, no. 3.

Szumik, C. and S. Roig-Juñent. 2005. Criterio de optimación para áreas de ende-mismo: El caso de América del Sur austral. In *Regionalización biogeográfica en Iberoamérica y tópicos afines: Primeras Jornadas Biogeográficas de la Red Iberoamericana de Biogeografía y Entomología Sistemática (RIBES XII.I–CYTED),* ed. J. Llorente Bousquets and J. J. Morrone, 495–508. Mexico, D.F.: Las Prensas de Ciencias, UNAM.

Evaluation of the Methods

There are no studies evaluating the performance of all available methods. Linder (2001) compared PAE using equally weighted data, PAE using his weighting procedure, and UPGMA using the Jaccard coefficient. He found that PAE with equally weighted data performed worst under the criterion of optimality and that weighted PAE outperformed the other two methods in the quality of the results. In addition, the majority of the areas of endemism identified in the consensus cladogram from the latter analysis had at least two endemic species, thus making it unnecessary to check each area of endemism.

Moline and Linder (2006) compared PAE and endemicity analysis. They found that both identified adequate areas of endemism, differing in the number of areas identified, the proportion of species endemic to them, the congruence of the distributions of species restricted to the areas, and the performance score calculated.

For Further Reading

Craw, R. C., J. R. Grehan, and M. J. Heads. 1999. *Panbiogeography: Tracking the history of life*. New York: Oxford University Press.

Croizat, L. 1958. *Panbiogeography*, Vols. 1 and 2. Caracas, Venezuela: Author.

Croizat, L. 1964. *Space, time, form: The biological synthesis*. Caracas, Venezuela: Author.

Grehan, J. R. 2001. Panbiogeography from tracks to ocean basins: Evolving perspectives. *Journal of Biogeography* 28:413–429.

Harold, A. S. and R. D. Mooi. 1994. Areas of endemism: Definition and recognition criteria. *Systematic Biology* 43:261–266.

Hausdorf, B. and C. Hennig. 2003. Biotic element analysis in biogeography. *Systematic Biology* 52(5):717–723.

Nihei, S. S. 2006. Misconceptions about parsimony analysis of endemicity. *Journal of Biogeography* 33:2099–2106.

Szumik, C. A., F. Cuezzo, P. A. Goloboff, and A. E. Chalup. 2002. An optimality criterion to determine areas of endemism. *Systematic Biology* 51(5):806–816.

Problems

Problem 4.1

Boeckella is a genus of freshwater copepods, with thirty-eight species distributed in South America, Australia, New Zealand, and Antarctica (Menu-Marque et al. 2000). On the basis of the localities of the South American species *B. gracilipes* (fig. 4.17a) and *B. gracilis* (fig. 4.17b) obtain their individual tracks by connecting the localities of each species according to their geographic proximity.

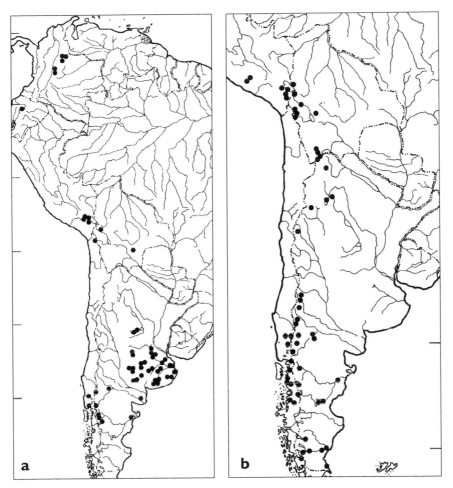

Figure 4.17   (a) Localities of distribution of *Boeckella gracilipes*; (b) localities of distribution of *B. gracilis*.

Problem 4.2

Mexico harbors a complex biota, especially in what is known as the Mexican Transition Zone (Halffter 1987; Morrone 2006). A panbiogeographic analysis may be used to detect the different biotic components contributing to this biota. On the basis of the seventy-two individual tracks of figs. 4.18–4.26 (Contreras-Medina and Eliosa-León 2001):

   a. Obtain the generalized tracks, by overlapping the individual tracks, and draw them on the map in fig. 4.27.
   b. Obtain the nodes, in the areas where different generalized tracks overlap, and draw them on the map in fig. 4.27.

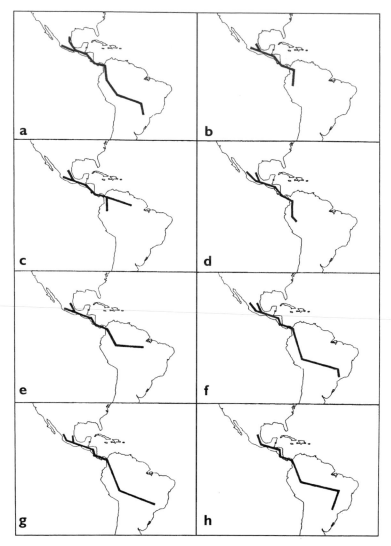

Figure 4.18   Individual tracks. (a) *Bufo marinus;* (b) *Eleutherodactylus rugulosus;* (c) *Leptodactylus labialis;* (d) *Phrynohias venulosa;* (e) *Boa constrictor;* (f) *Iguana iguana;* (g) *Veniliornis fumigatus;* (h) *Xenops minutus.*

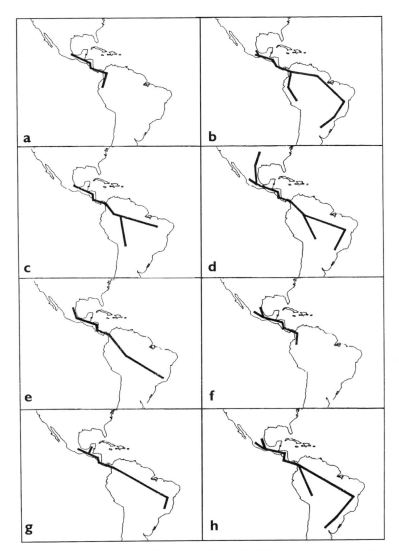

Figure 4.19   Individual tracks. (a) *Alouatta palliata;* (b) *Chironectes minimus;* (c) *Cyclopes didactylus;* (d) *Dasypus novemcinctus;* (e) *Didelphis marsupialis;* (f) *Felis yagouaroundi;* (g) *Galictis vittata;* (h) *Lutra longicaudis.*

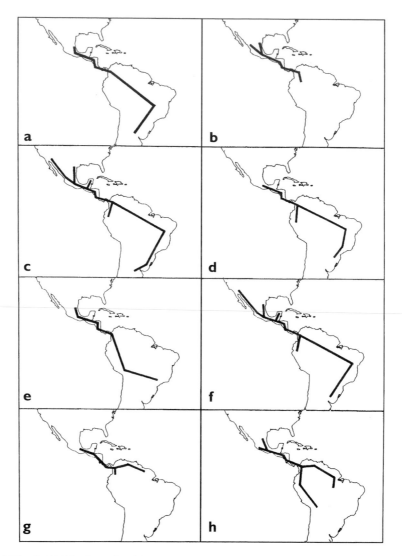

Figure 4.20   Individual tracks. (a) *Mazama americana*; (b) *Molossus ater*; (c) *Nasua narica*; (d) *Potos flavus*; (e) *Sylvilagus brasiliensis*; (f) *Tayassu tajacu*; (g) *Amblytropidia trinitatis*; (h) *Arsenura armida*.

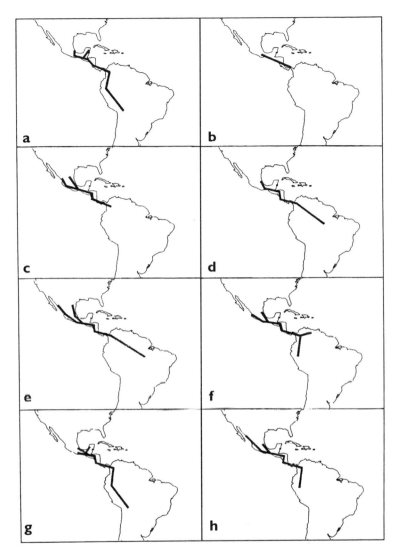

Figure 4.21   Individual tracks. (a) *Battus lycidas*; (b) *Carinisphindus isthmensis*; (c) *Catasticta flisa*; (d) *Dismorphia theucharila*; (e) *Leptophobia aripa*; (f) *Lienix nemesis*; (g) *Parides sesostris*; (h) *Pereute charops*.

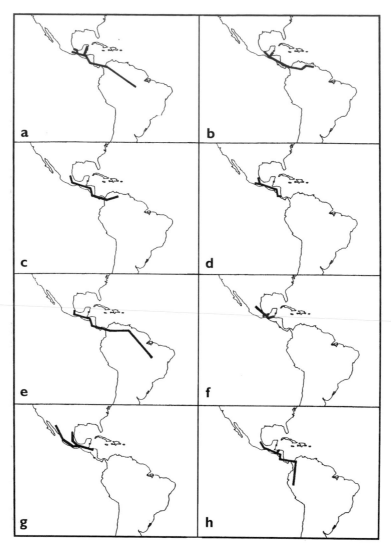

Figure 4.22   Individual tracks. (a) *Silvittetix* spp.; (b) *Simulium sanboni;* (c) *Deppea* spp.; (d) *Drymis granadensis;* (e) *Roupala montana;* (f) *Ceratozamia* spp.; (g) *Dioon* spp.; (h) *Podocarpus oleifolius.*

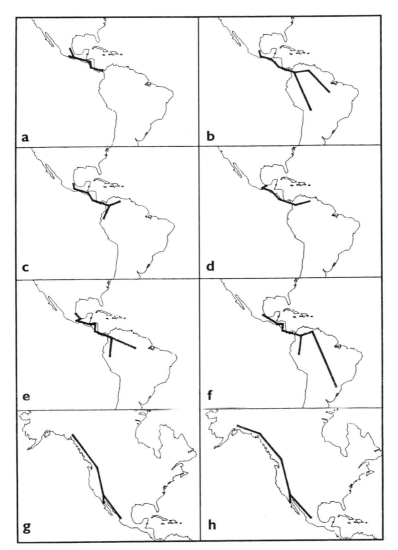

Figure 4.23   Individual tracks. (a) *Antrophytum ensiforme;* (b) *Hymenophyllum fendlerianum;* (c) *H. myriocarpum;* (d) *Lindsaea klotzschiana;* (e) *Trichomanes collariatum;* (f) *T. diaphanum;* (g) *Falco mexicanus;* (h) *Otus kennicottii.*

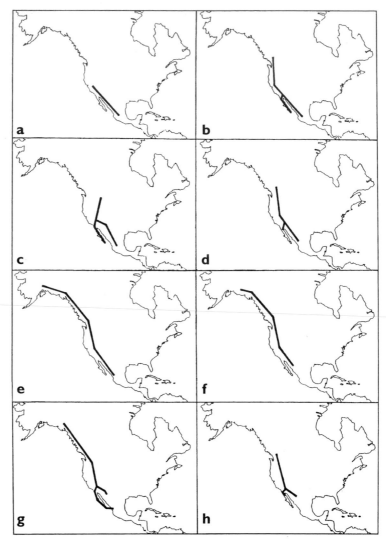

Figure 4.24   Individual tracks. (a) *Pseudacris regilla;* (b) *Myotis californicus;* (c) *Odocoileus hemionus;* (d) *Ovis canadensis;* (e) *Sorex vagrans;* (f) *Spermophilus beecheyi;* (g) *Anthocharis sara;* (h) *Euchloe hyantis.*

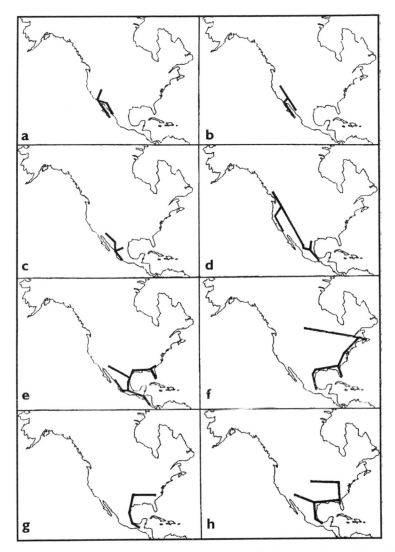

Figure 4.25   Individual tracks. (a) *Horesidotes cinereus*; (b) *Ligurotettix coquilletti*; (c) *Neophasia terlooti*; (d) *Pseudotsuga menziesii*; (e) *Achurum* spp.; (f) *Actias luna–A. truncatipennis*; (g) *Amblyscirtes celia*; (h) *Anaea andrea*.

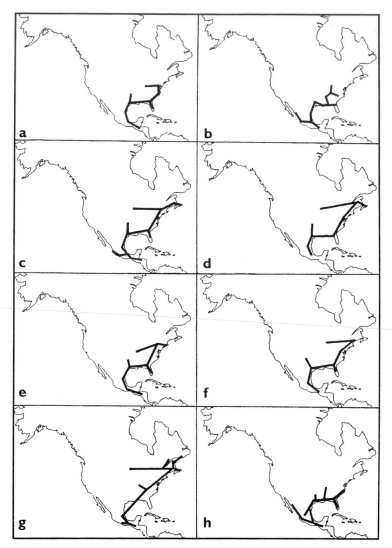

Figure 4.26   Individual tracks. (a) *Cyllopsis gemma;* (b) *Cymatoderella collaris;* (c) *Carpinus caroliniana;* (d) *Fagus grandifolia;* (e) *Liquidambar styraciflua;* (f) *Nyssa sylvatica;* (g) *Pinus strobus;* (h) *Taxodium* spp.

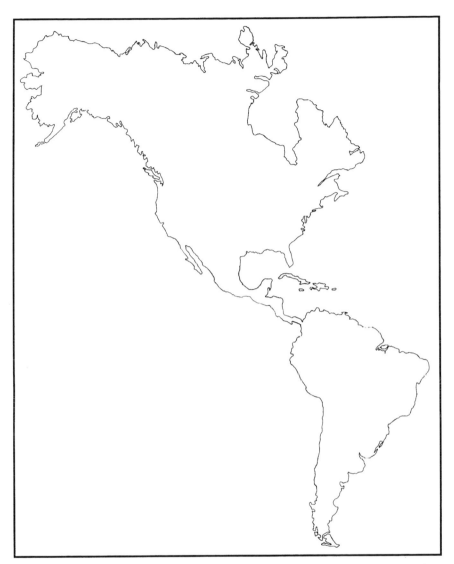

Figure 4.27   Map of the Americas to represent the generalized tracks and nodes obtained in the analysis.

Problem 4.3

The Neotropical plant genus *Bursera* (Burseraceae) is particularly well diversified in Mexico. On the basis of the maps with localities of Mexican species of *Bursera* (figs. 4.28–4.33) (Morrone 2001g), obtain the areas of endemism and represent them on the map in fig. 4.33b. The analysis may be done by hand or using any of the methods available.

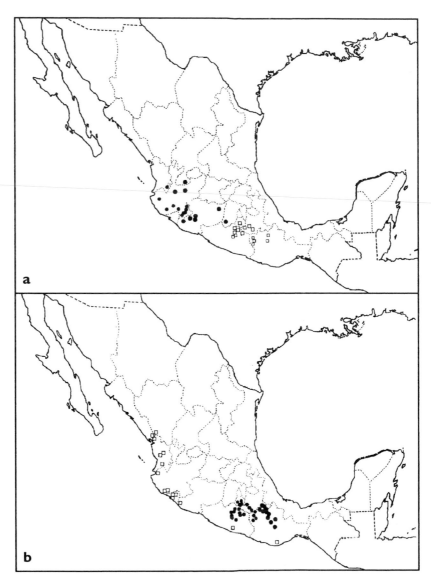

Figure 4.28   Localities of Mexican species of *Bursera*. (a) *B. acuminata* (black circles), *B. aloexylon* (white squares); (b) *B. aptera* (black circles), *B. arborea* (white squares).

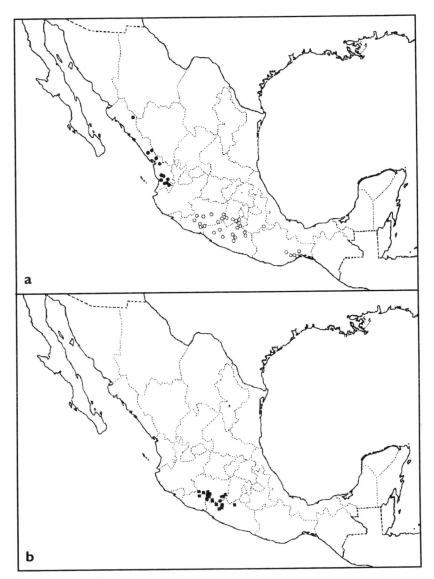

Figure 4.29   Localities of Mexican species of *Bursera*. (a) *B. attenuata* (black circles), *B. bicolor* (white circles); (b) *B. coyucensis*.

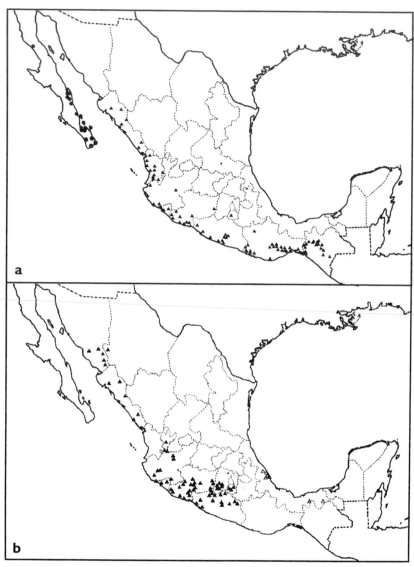

Figure 4.30   Localities of Mexican species of *Bursera*. (a) *B. epinnata* (black circles), *B. excelsa* (black triangles); (b) *B. grandifolia* (black triangles), *B. graveolens* (white triangles).

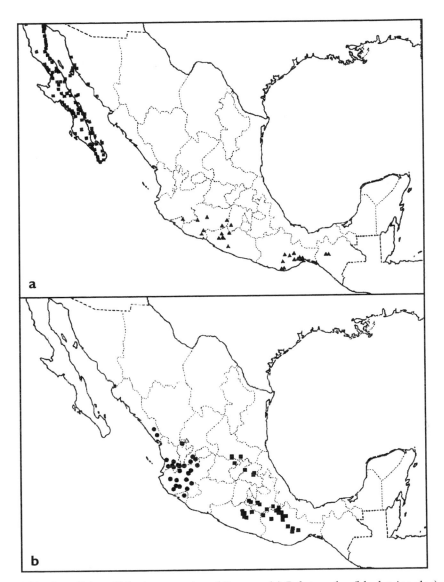

Figure 4.31   Localities of Mexican species of *Bursera*. (a) *B. heteresthes* (black triangles), *B. hindsiana* (black squares); (b) *B. morelensis* (black squares), *B. mutifolia* (white triangles), *B. multijuga* (black circles).

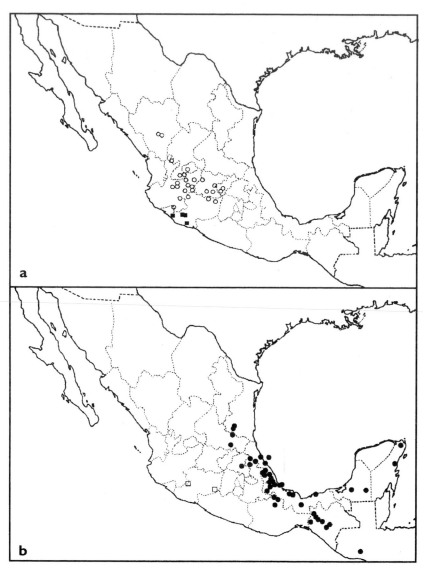

Figure 4.32   Localities of Mexican species of *Bursera*. (a) *B. occulta* (black squares),
B. palmeri (white circles); (b) *B. simaruba* (black circles), *B. staphyleoides* (white squares).

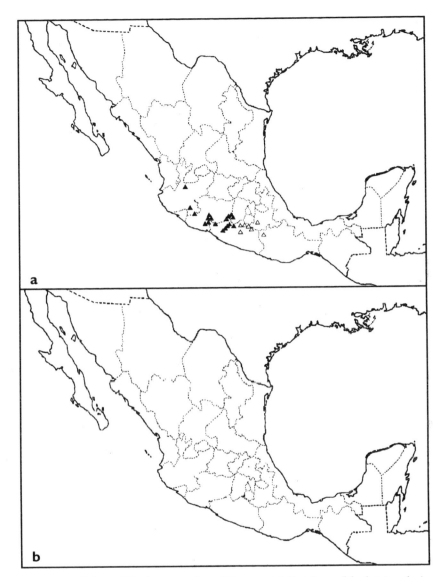

Figure 4.33   Localities of Mexican species of *Bursera*. (a) *B. trimera* (black triangles), *B. vejar-vazquezii* (white triangles); (b) map of Mexico to represent the areas of endemism obtained.

Problem 4.4

PAE based on grid cells has proven to be an adequate method for identifying areas of endemism. On the basis of the distributions of forty-one hypothetical species from southern South America (figs. 4.34–4.37; Morrone, Espinosa Organista, et al. 1996):

a. Number the grid cells, excluding those where no species is distributed.

b. Build a data matrix of grid cells (files) × species (columns).

c. Apply PAE with an appropriate software (Hennig86, Farris 1988; PHYLIP, Felsenstein 1993; NONA, Goloboff 1998; PAUP, Swofford 2003; Pee-Wee, Goloboff, retrieved May 25, 2008, from http://www.zmuc.dk/public/phylogeny/Nona-PeeWee/; TNT, Goloboff et al., retrieved May 25, 2008, from http://www.zmuc.dk/public/phylogeny/TNT/) in order to find the cladograms and, if multiple equally parsimonious cladograms are obtained, calculate the strict consensus cladogram.

d. On the basis of the clades supported by two or more species, draw the areas of endemism in the map of fig. 4.37d.

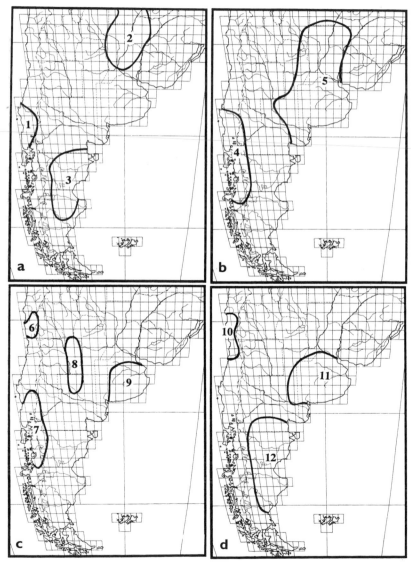

Figure 4.34   Maps of southern South America with distributions of hypothetical species 1–12.

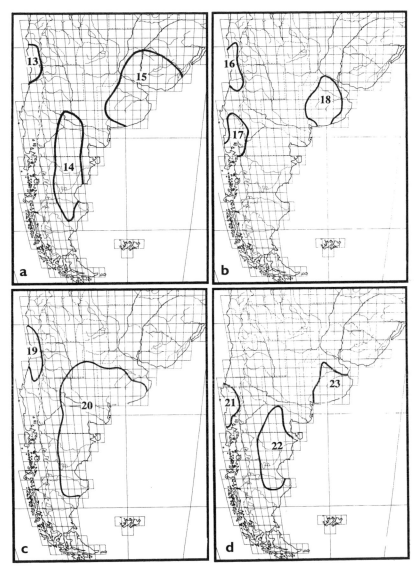

Figure 4.35   Maps of southern South America with distributions of hypothetical species 13–23.

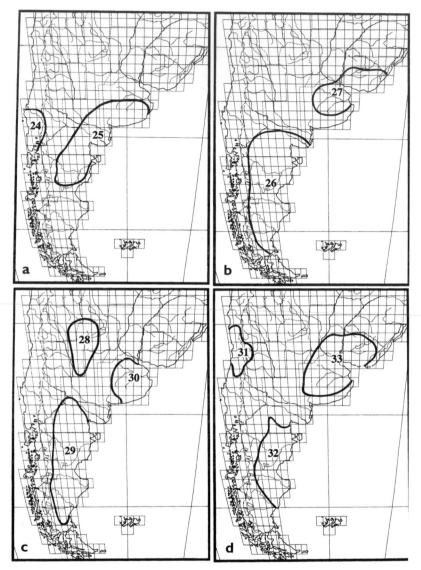

Figure 4.36   Maps of southern South America with distributions of hypothetical species 24–33.

For Discussion

1. One way of evaluating a biogeographic article is to postulate alternative explanations that the authors might have overlooked. Try to think of other explanations for the data presented in the following articles from a panbiogeographic viewpoint:

   Costa, L. P. 2003. The historical bridge between the Amazon and the Atlantic forest of Brazil: A study of molecular phylogeography with small mammals. *Journal of Biogeography* 30:71–86.

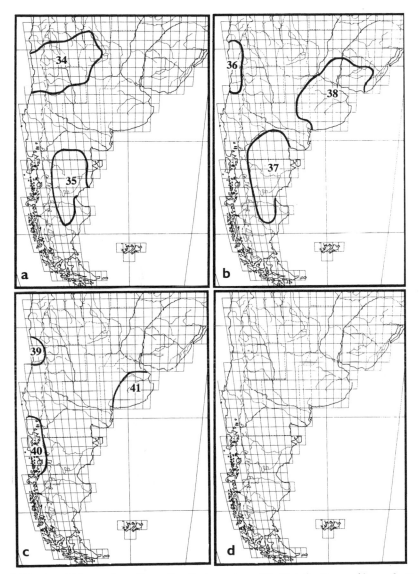

Figure 4.37   (a–c) Maps of southern South America with distributions of hypothetical species 34–41; (d) map of southern South America to represent the areas of endemism obtained.

Sanmartín, I. 2003. Dispersal vs. vicariance in the Mediterranean: Historical biogeography of the Palearctic Pachydeminae (Coleoptera, Scarabaeoidea). *Journal of Biogeography* 30:1883–1897.

Swenson, U. and K. Bremer. 1996. Pacific biogeography of the Asteraceae genus *Abrotanella* (Senecioneae, Blemnospermatinae). *Systematic Botany* 22:493–508.

2. Read the following article on the biogeography of the Amazonian rain forest. What weaknesses can you identify in the arguments presented? What alternative methods or additional data would you use?

Ron, S. R. 2000. Biogeographic area relationships of lowland Neotropical rainforest based on raw distributions of vertebrate groups. *Biological Journal of the Linnean Society* 71:379–402.

3. List similarities and differences between the following pairs of concepts:

    a. Area of distribution and individual track.

    b. Area of endemism and generalized track.

    c. Area of endemism and node.

4. Carefully read the following article:

    Heads, M. J. 2004. What is a node? *Journal of Biogeography* 31:1883–1891.

    a. Identify and extract the main ideas.

    b. Transform each main idea into a question.

5. Carefully read the following articles:

    Brooks, D. R. and M. G. P. Van Veller. 2003. Critique of parsimony analysis of endemicity as a method of historical biogeography. *Journal of Biogeography* 30:819–825.

    Nihei, S. S. 2006. Misconceptions about parsimony analysis of endemicity. *Journal of Biogeography* 33:2099–2106.

    Santos, C. M. D. 2005. Parsimony analysis of endemicity: Time for an epitaph? *Journal of Biogeography* 32:1284–1286.

    a. Establish the arguments favoring PAE and those criticizing it.

    b. What are your conclusions?

CHAPTER 5

Testing Relationships Between Biotic Components

Because the generalized tracks that result from panbiogeographic analyses are unrooted, they connect geographic areas but do not specify a precise sequence of fragmentation. For example, given a generalized track joining Australia, New Zealand, and Chile, which of the three areas first separated from the others? In order to determine this sequence, phylogenetic data must be incorporated. In this chapter I present the basic approach of cladistic biogeography, introduce some methods, and provide case studies.

Cladistic Biogeography

This approach assumes a correspondence between the phylogenetic relationships of the taxa and the relationships between the areas they inhabit (Nelson and Platnick 1980, 1981; Platnick and Nelson 1978). Cladistic biogeography uses information on the cladistic relationships between the taxa and their geographic distribution to form hypotheses on the relationships between areas. If several taxa show the same pattern, such congruence is evidence of common history (Crisci and Morrone 1992; Crisci et al. 2000; Ebach and Humphries 2002; Enghoff 1996; Humphries and Parenti 1999; Humphries et al. 1988; Morrone 1997; Morrone and Crisci 1995; Sanmartín and Ronquist 2002; Wiley 1988a; Zunino and Zullini 1995). We may characterize cladistic biogeography considering that it originated from the joining of three largely independent research programs: Hennig's (1950) phylogenetic systematics, Croizat's (1958b, 1964) panbiogeography, and Wegener's (1929) continental drift. To them, Nelson and Platnick (1981) added the deductive–hypothetical method of Popper (1959, 1963).

We can raise an analogy between systematics and biogeography. In systematics we study taxa, classifying them by their shared characters, whereas in biogeography we study areas, classifying them by their shared taxa. This implies equivalence between taxa (systematics) and areas (biogeography). However, this correspondence has been questioned by Hovenkamp (1997, 2001), who suggested that what we should reconstruct is not the sequence of area fragmentation but the sequence of vicariance events.

Cladistic biogeography poses three questions (Nelson and Platnick 1978):

- Is endemism geographically nonrandom, and if so, what are the areas of endemism?
- Given a list of areas of endemism, are the interrelationships of their endemic taxa geographically nonrandom, and if so, what patterns are formed by their interrelationships?
- Given one or more patterns of interrelationships, as represented by one or more general area cladograms, does the pattern correlate with the geological history?

Cladistic biogeography is based on geographic congruence, or the finding of identical patterns between unrelated taxa that are generally interpreted as having a common cause. For example, the breakup of the supercontinent Pangaea 250 mya produced a general pattern of vicariance between different groups of American continental taxa. The uplift of the Isthmus of Panama produced general patterns of biogeographic convergence in the north–south exchange of different groups of continental organisms, as well as vicariance of different groups of marine organisms. Congruence is detected when an initial pattern has been established. In cladistic biogeography, homology (geographic congruence) is usually considered to be the result of vicariance, although there may be instances of congruence due to dispersal (Lieberman 2000; van Welzen et al. 2001). This is why this approach, although originally known as vicariance biogeography (e.g., Nelson and Platnick 1981; Rosen and Nelson 1980), is now widely known as cladistic biogeography (Humphries and Parenti 1986, 1999; Page 1988; Parenti 1981, 2007).

A cladistic biogeographic analysis consists of three basic steps (fig. 5.1):

1. Construction of taxon–area cladograms, from the taxonomic cladograms of two or more different taxa (figs. 5.1a–5.1c), by replacing their terminal taxa with the areas they inhabit (figs. 5.1d–5.1i).
2. Obtaining resolved area cladograms from the taxon–area cladograms (when demanded by the method applied) (figs. 5.1j–5.1l).
3. Obtaining a general area cladogram, based on the information contained in the resolved area cladograms (fig. 5.1m).

Taxon–Area Cladograms

Taxon–area cladograms are obtained by replacing the name of each terminal taxon in the cladograms of the taxa analyzed with the area where it is distributed. For example, if a taxon (1 (2 (3, 4))) has species 1 distributed in North America, species 2 in Africa, species 3 in South America, and species 4 in Australia, by replacing the four species in the cladogram with the areas where they are distributed, we obtain the taxon–area cladogram (North America (Africa (South America, Australia))).

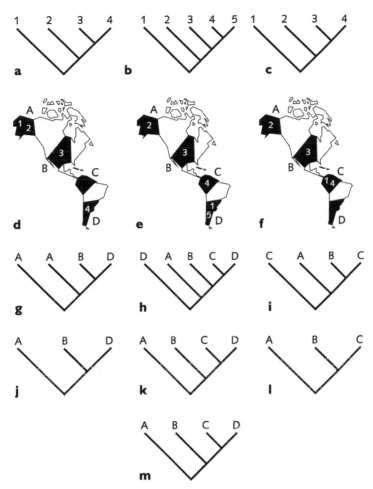

Figure 5.1 Steps of a cladistic biogeographic analysis. (a–c) Three taxonomic cladograms; (d–f) maps showing the distribution of the species of the three taxa analyzed; (g–i) taxon–area cladograms; (j–l) resolved area cladograms; (m) general area cladogram.

Resolved Area Cladograms

The construction of taxon–area cladograms is simple when each taxon is endemic to a single area and each area has only one taxon, but it is more complex when taxonomic cladograms include widespread taxa, redundant distributions, and missing areas. In these cases, some methods require that taxon–area cladograms be turned into resolved area cladograms (Crisci et al. 2000; Morrone 1997; Morrone and Carpenter 1994; Morrone and Crisci 1995; Nelson 1984; Nelson and Platnick 1981; Page 1988; Sanmartín and Ronquist 2002; Warren and Crother 2001). These are also known as fundamental area cladograms (Nelson and Platnick 1981), area cladograms (Page 1990a), and areagrams (Ebach et al. 2005a; Swenson et al. 2001; van Welzen 1992).

Widespread Taxa When any of the terminal taxa of a taxon–area cladogram inhabits two or more of the studied areas, it is a widespread taxon (Nelson and Platnick 1981) or a mast (for "multiple areas on a single terminal") (Ebach et al. 2005a). For example, if a taxon (1 (2, 3)) has species 1 distributed in both North America and Africa, species 2 in South America, and species 3 in Australia, by replacing the three species in the cladogram by the areas where they are distributed, we obtain the taxon–area cladogram (North America–Africa (South America, Australia)). As a result of the widespread taxon, North America and Africa appear together in the taxon–area cladogram.

Under assumption 0 (Zandee and Roos 1987), the areas inhabited by a widespread taxon are considered as a monophyletic group in the resolved area cladogram, meaning that the taxon is treated as a synapomorphy of the areas. Under assumption 1 (Nelson and Platnick 1981), the widespread taxon is not considered as a synapomorphy in constructing the resolved area cladograms, and the areas inhabited by it can constitute monophyletic or paraphyletic groups in the resolved area cladograms. Under assumption 2 (Nelson and Platnick 1981), only one of the occurrences is considered as evidence, whereas the other can "float" in the resolved area cladograms, therefore constituting the areas involved monophyletic, paraphyletic, or polyphyletic groups. The three assumptions show an inclusion relationship because topologies obtained under assumption 0 are included within those obtained under assumption 1, and those obtained under assumption 1 are included within those from assumption 2 (fig. 5.2).

Are there any biogeographic processes implicit in the assumptions? Under assumption 0, we assume that the pattern is due exclusively to vicariance. When we apply assumption 1, in addition to vicariance, we recognize the possibility that some species have not responded to it with speciation or that some species have became extinct. Finally, assumption 2 adds the possibility of dispersal to the processes implicit in the previous assumptions.

Some authors prefer assumption 2, especially to deal with widespread taxa (Humphries 1989, 1992; Humphries et al. 1988; Morrone and Carpenter 1994; Nelson and Platnick 1981), considering that widespread taxa are a source of ambiguity, because a future analysis can show that a widespread taxon really represents two or more different taxa, not necessarily related, and inhabiting different areas; a taxon may have a widespread distribution due to dispersal; and a taxon may have a wide distribution because it did not respond with speciation to a vicariance event. Other authors accept the informative value of widespread taxa, thus preferring assumption 0 (Brooks 1990; Enghoff 1996; Wiley 1988a; Zandee and Roos 1987). Enghoff (1995) and van Veller et al. (1999) considered assumption 2 to be less informative because it offers more solutions than assumptions 0 or 1. Zandee and Roos (1987), Wiley (1988a), Enghoff (1996), and van Veller et al. (1999, 2000) argued that assumptions 1 and 2 distort the phylogenetic relationships between the terminal taxa of the taxon–area cladogram. However, Page (1989a, 1990a) indicated clearly that assumptions 1 and 2 are interpretations about relationships between areas, not between taxa. The main criticism of assumption 0 is that it is too

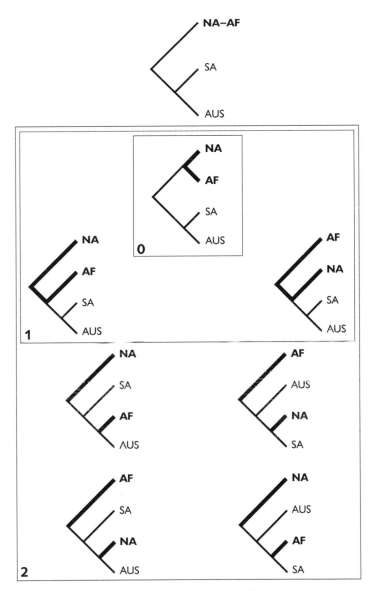

Figure 5.2   Possible resolved area cladograms obtained for a taxon–area cladogram with a widespread taxon, as monophyletic (assumption 0), monophyletic and paraphyletic (assumption 1), and monophyletic, paraphyletic, and polyphyletic groups of areas (assumption 2). AF, Africa; AUS, Australia; NA, North America; SA, South America.

restrictive, not considering the possibility of dispersal to explain the distributions of widespread taxa (Page 1989a, 1990a).

Van Soest (1996) and van Soest and Hajdu (1997) proposed an alternative treatment for widespread taxa called "no-assumption coding." These authors reasoned that because taxa have different means of dispersal and are affected differentially by the environment, when comparing different taxon–area cladograms, usually there will be areas of greater and smaller size partially superposed. If the smaller areas are chosen as units of the analysis, then the greater ones will generate widespread taxa, and later manipulation with assumptions 0, 1, and 2 will imply the assumption of some of the processes previously indicated. Additionally, current biogeographic methods allow areas to occur on a single position in the general area cladogram, whereas the history of the biota may indicate various positions. To remedy these limitations, van Soest (1996) proposed a different coding of widespread taxa. How does it work? If two or more taxon–area cladograms have taxa widespread in the same areas, this combination can be treated as a single area (figs. 5.3a–5.3d). The comparison between the different area cladograms will determine whether the large area is united in a general area cladogram with its constituent areas (demonstrating that in fact it is a single area) or it constitutes a paraphyletic or polyphyletic group (indicating that the individual areas are independent). In the example, both general area cladograms obtained (figs. 5.3e and 5.3f) show that area E has two different relationships: It is the sister area of A (fig. 5.3e), and the sister area of D (fig. 5.3f).

Sanmartín and Ronquist (2002) proposed another treatment for widespread taxa, within the approach of the event-based methods, because assumptions 0, 1, and 2 do not specify the costs these methods entail. For these authors, the problem

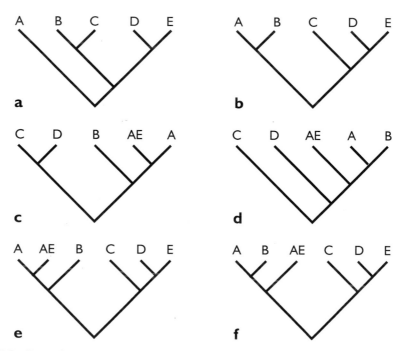

Figure 5.3   Procedure to deal with widespread taxa proposed by van Soest (1996). (a–d) Four taxon–area cladograms, two of them with a widespread taxon in areas A and E; (e–f) general area cladograms obtained: (e) area E is the sister area of A; (f) area E is the sister area of D.

of a widespread taxon is solved by treating it as an unsolved polytomy, consisting of a unit for each of the involved areas. In order to optimize the ancestral distribution of the widespread taxon, three options are applied (fig. 5.4). These options allow different sets of possible ancestral distributions, each with different associated costs. The option "recent" assumes that the wide distribution is recent, due to dispersal. The option "ancient" assumes that the wide distribution is ancestral, due to vicariance and extinction. The option "free" assumes that the wide distribution is unsolved, allowing any combination of biogeographic processes and any resolution of the polytomy, being even more flexible than assumption 2.

Ebach et al. (2005a) suggested that assumptions 1 and 2 may inadvertently use paralogy and widespread taxa or masts and yield spurious results. They proposed that the "transparent method," along with paralogy-free subtree analysis, is appropriate to deal with this problem. This is done by treating all taxon–area cladograms as individual points that may be part of a common pattern (the general area cladogram). Area cladograms with masts are viewed in terms of proximal relationships, and masts are resolved so each area is represented once. For example, if we have the taxon–area cladogram AB (C (D, E)), resolving mast AB implies that two area cladograms are obtained: A (C (D, E)) and B (C (D, F)), which together result in AB (C (D, E)). Ebach et al. (2005a) suggested that the transparent method should be implemented before the paralogy-free subtree analysis.

Redundant Distributions Also known as areas of sympatry (Enghoff 1996), redundant distributions occur when an area appears more than once in a taxon–area

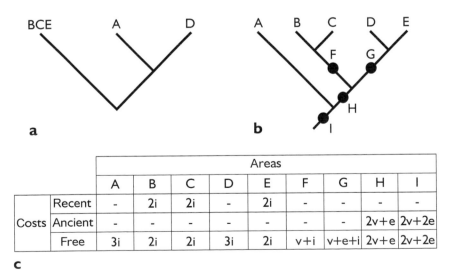

Figure 5.4 table (c):

		Areas								
		A	B	C	D	E	F	G	H	I
Costs	Recent	-	2i	2i	-	2i	-	-	-	-
	Ancient	-	-	-	-	-	-	-	2v+e	2v+2e
	Free	3i	2i	2i	3i	2i	v+i	v+e+i	2v+e	2v+2e

Figure 5.4 Resolution of a widespread taxon under the event-based approach. (a) Taxon–area cladogram with a taxon widespread on areas B, C, and E; (b) general area cladogram; (c) costs of each area under the recent, ancient, and free options. e, extinctions; i, dispersal; v, vicariance.

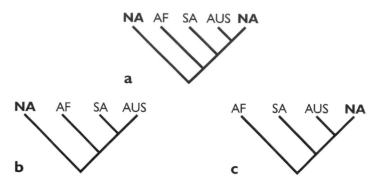

Figure 5.5   Resolutions of a redundant distribution. (a) Taxon with a redundant distribution involving North America; (b–c) two possible solutions deleting one of the distributions each time. AF, Africa; AUS, Australia; NA, North America; SA, South America.

cladogram because in this area, two or more terminal species are distributed. In the taxon (1 (2 (3 (4, 5)))), if species 1 and 5 are distributed in North America when the species are replaced by the areas, this area will appear twice in the taxon–area cladogram (fig. 5.5a). If the species constitute a monophyletic group, obtaining a resolved area cladogram is simple.

There is no special treatment for redundant distributions under assumption 0, although Kluge (1988) proposed a weighting scheme in which a smaller weight is given to the components involving redundant distributions. Under assumption 1, it is interpreted that the redundant distributions are due to duplicated patterns followed by extinction (figs. 5.5b and 5.5c), whereas assumption 2 adds the possibility that sympatry may be due to dispersal (Enghoff 1996; Page 1990a). Most of the authors prefer assumption 2 to treat redundant distributions (Enghoff 1996; Morrone and Carpenter 1994; Nelson and Platnick 1981; Page 1990a).

Missing Areas   When no terminal taxon is distributed in one of the areas analyzed, this area will not be represented in the taxon–area cladogram. In the taxon (1 (2, 3)), if no species inhabits Africa (one of the study areas) when the areas are replaced by the species of the cladogram, this area will not appear in the taxon–area cladogram (fig. 5.6a).

Missing areas, which are caused by extinction or insufficient studies, are treated as noninformative. They are coded with "?" so that they can be placed in all possible positions in the resolved area cladograms (figs. 5.6b–5.6f). Also, it is possible to treat them as primitively absent by coding them with "0" (Kluge 1988).

Combination of Assumptions   Some authors have determined that the assumptions are not mutually exclusive because it would be possible to deal with widespread taxa under one assumption and with redundant distributions under another (Enghoff 1996; Morrone and Crisci 1995; Page 1990a). Van Veller et al. (1999,

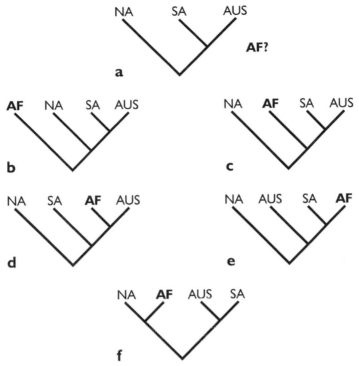

Figure 5.6   Resolutions of a taxon with a missing area. (a) Taxon with a missing area; (b–f) five possible solutions placing it in all possible positions in the cladogram. AF, Africa; AUS, Australia; NA, North America; SA, South America.

2000, 2001) argued that in order to obtain valid general area cladograms, two requirements should be fulfilled:

- Resolved area cladograms from the different taxa analyzed should be obtained under the same assumption because the common pattern for all taxa would have to be explained by the same process.
- The sets of resolved area cladograms obtained under the three assumptions should be successively inclusive, namely, those obtained under assumption 0 would be included within those obtained under assumption 1, and the latter would have to be found within those obtained under assumption 2. The reason is that the processes are additive (each assumption incorporates or includes those of the preceding assumption).

The first requirement is not valid because different taxa can be affected by different processes, although they may show a common pattern. In relation to the second requirement, Ebach and Humphries (2002) suggested that it is violated by the use of assumption 0 when introducing artificial internal nodes for widespread taxa.

General Area Cladograms

On the basis of the information from the different resolved area cladograms, a general area cladogram is derived. It represents a hypothesis on the biogeographic history of the taxa analyzed and the areas where they are distributed.

The general area cladogram that results from the analysis may be falsified with a geological area cladogram or geogram (Swenson et al. 2001). This is an area cladogram based on geological or tectonic data (Morrone and Carpenter 1994; Rosen 1985; Seberg 1991). Cladistic biogeographic analyses that include this step are scarce (Seberg 1991; Swenson et al. 2001; van Welzen et al. 2001). Another way to evaluate general area cladograms is through the calculation of items of error (Morrone and Carpenter 1994; Nelson and Platnick 1981). The procedure consists of determining the terminal number of nodes and areas that are necessary to add to the taxon–area cladogram so that it agrees with the general area cladogram, that is, to map one cladogram onto the other to determine their congruence. The smaller the number of nodes and terminal areas that must be added, the more parsimonious will be the general area cladogram analyzed, and for that reason it will be chosen. Evaluation of items of error can be carried out manually or with the program Component version 1.5 (Page 1989b).

From an epistemological point of view, general area cladograms represent testable hypotheses in the framework of Popper's (1959, 1963) hypothetico-deductive method (Nelson and Platnick 1981; Platnick and Nelson 1978). However, some authors suggest that cladograms are not general hypotheses in Popper's sense (Andersson 1996; Hull 1983). An important aspect of general area cladograms is their retrodictive power (Morrone 1997). When we have obtained a general area cladogram, we may use it to carry out predictions or retrodictions related to taxa still not analyzed (which are expected to agree with the general pattern), with geological or tectonic hypotheses, or the relative ages of biotas (when a molecular clock is available for some of the studied taxa).

Methods

There are fourteen cladistic biogeographic methods (Crisci et al. 2000; Goyenechea et al. 2001; Humphries and Parenti 1999; Morrone 2004a; Morrone and Crisci 1995). I will deal herein with component analysis (Nelson and Platnick 1981), Brooks parsimony analysis (BPA) (Wiley 1987), three area statement analysis (Nelson and Ladiges 1991a, 1991c), tree reconciliation analysis (Page 1990a), paralogy-free subtree analysis (Nelson and Ladiges 1996), dispersal–vicariance analysis (Ronquist 1997), area cladistics (Ebach and Edgecombe 2001), and phylogenetic analysis for comparing trees (Wojcicki and Brooks 2005). Six other methods, namely, reduced consensus cladogram (Rosen 1978), ancestral species maps (Wiley 1980, 1981), quantitative phylogenetic biogeography (Mickevich 1981), component compatibility (Zandee and Roos 1987), quantification of component analysis (Humphries et al. 1988), and vicariance event analysis (Hovenkamp 1997), which have been applied occasionally (Crisci et al. 1991b; de Weerdt 1989; Hovenkamp 2001; Liebherr

1988; Noonan 1988; Patterson 1981; Schuh and Stonedahl 1986; Solervicens 1987; van Soest 1993; van Veller et al. 2000; van Welzen 1992), are not dealt with here. For details on these latter methods, see Humphries and Parenti (1999), Goyenechea et al. (2001), and Morrone (2004a).

A recent biogeographic development that merits special comment here is comparative phylogeography, also known as the regional (Avise 1996) or landscape (Templeton and Georgiadis 1996) approach to phylogeography. It is intended to compare phylogeographic structure exhibited by sympatric species to determine whether they exhibit congruent patterns, geographically structured by vicariance events (Abogast and Kenagy 2001; Bermingham and Moritz 1998; Cunningham and Collins 1998; Riddle and Hafner 2006; Taberlet et al. 1998; Zink 2002). Incongruent patterns may indicate that the species colonized the area more recently, whereas congruent patterns may suggest a longer history of association of the different species (Zink 1996). This approach is similar to cladistic biogeography (Lanteri and Confalonieri 2003; Lieberman 2004; Morrone 2004a; Riddle and Hafner 2006; Santos 2007), so I will not deal with it separately. Some authors have considered BPA to be an appropriate method for comparative phylogeographic studies (Taberlet et al. 1998). Others have not used any formal method for comparison (Costa 2003; Gorog et al. 2004; Mateos 2005; Morales-Barros et al. 2006; Palma et al. 2005; Perdices and Coelho 2006; Riddle et al. 2000a, 2000c; Schäuble and Moritz 2001; Steele and Storfer 2007; Weisrock and Janzen 2000). However, the analysis with any cladistic biogeographic method is possible (Bermingham and Martin 1998; Lapointe and Rissler 2005).

Component Analysis

This method was proposed by Nelson and Platnick (1981). It solves the problems derived from redundant distributions, widespread taxa, and missing areas using assumptions 0, 1, and 2 and then finds the general area cladogram through the intersection of the sets of resolved area cladograms (Biondi 1998; Enghoff 1996; Humphries and Parenti 1986; Humphries et al. 1988; Morrone 1997; Morrone and Carpenter 1994; Nelson 1984; Nelson and Platnick 1981; Page 1988, 1990a; van Veller et al. 2000; Vargas 1992b, 2002). Enghoff (1996) suggested treating widespread taxa under assumption 0 and redundant distributions under assumption 2. The method is illustrated in fig. 5.7, where based on the intersection of the sets of resolved area cladograms of the taxa analyzed, the general area cladogram (NA (AF (SA, AUS))) is found. If it is not possible to find a general area cladogram, it may be possible to find a cladogram shared for some of the sets (Crisci et al. 1991a, 1991b). If more than one general area cladogram is obtained, it is possible to build a consensus cladogram (Morrone and Carpenter 1994).

Zandee and Roos (1987) criticized the use of consensus techniques to obtain the general area cladogram, but Page (1990a) clarified that consensus trees are not the only way to obtain a general area cladogram. Alternatively, component analysis has been criticized for its preference for assumptions 1 and 2 instead of assumption 0,

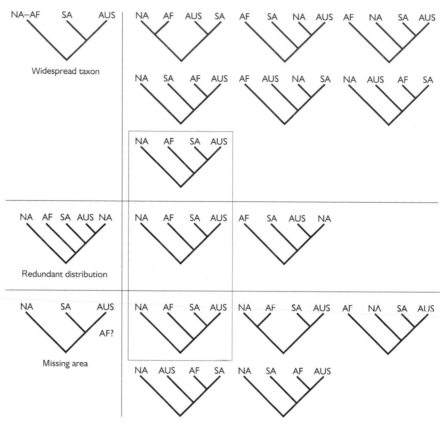

Figure 5.7   Component analysis. *Left column,* taxon–area cladograms with a widespread taxon, a taxon with a redundant distribution, and a taxon with a missing area. *Right column,* resolved area cladograms showing the results of the intersection of the three sets of resolved area cladograms. AF, Africa; AUS, Australia; NA, North America; SA, South America.

which Wiley (1987, 1988a, 1988b) considered more parsimonious. Lieberman (2004) argued that the presence of artificial incongruence due to extinct clades may lead to problems with component analysis because it is not designated to deal with any sort of incongruence. However, I argue that this represents a problem for all cladistic biogeographic methods.

Algorithm   The algorithm consists of the following steps (Biswas and Pawar 2006; Humphries and Parenti 1999; Morrone 1997; Nelson and Platnick 1981; Page 1990a):

1. Obtain the taxonomic cladograms of the taxa distributed in the areas analyzed.
2. Replace the terminal taxa from the taxonomic cladograms with the areas inhabited by them to obtain taxon–area cladograms.

3. Derive resolved area cladograms from each taxon–area cladogram, applying assumptions 0, 1, or 2.

4. Intersect the sets of resolved area cladograms obtained for each taxon–area cladogram in order to find the general area cladogram. If no general area cladogram results from the intersection, check whether an area cladogram is shared by at least some of the sets. If more than one general area cladogram is obtained, build a consensus cladogram.

Software Component version 1.5 (Page 1989b), which constructs sets of resolved area cladograms from taxon–area cladograms under assumptions 0, 1, or 2 (option BUILD) and then determines their intersection (option SHARED TREES). The procedure described by Enghoff (1996) can be applied with Component version 2.0 (Page 1993a).

Empirical Applications Amorim and Tozoni (1994), Andersen (1991), Bremer (1993), Carpenter (1993), Cracraft (1982, 1988), Cracraft and Prum (1988), Crisci et al. (1991b), Enghoff (1995), Flores Villela and Goyenechea (2001), Humphries (1981), Humphries and Parenti (1986), Humphries et al. (1988), Morrone (1993a, 1993c), Morrone and Carpenter (1994), Morrone et al. (1997), Noonan (1988), Page (1988, 1989a), Platnick (1981), Rauchenberger (1988), Roig-Juñent (1994), Roig-Juñent and Flores (2001), Seberg (1991), van Veller et al. (2000), Wallace et al. (1991), and Weston and Crisp (1994).

CASE STUDY 5.1   Cladistic Biogeography of Central Chile

Central Chile traditionally has been considered the area located between 30° and 37° south latitude. The analysis of different plant and animal taxa indicates a high endemism and a close relationship between its austral part and the Subantarctic subregion. Morrone et al. (1997) carried out a cladistic biogeographic analysis and a parsimony analysis of endemicity (PAE) in order to provide a natural regionalization of the area.

Central Chile was divided into smaller areas of endemism on the basis of distributional patterns of species of seven genera of the plant family Asteraceae, a genus of Buprestidae and three genera of Curculionidae (Coleoptera), and two genera of Gnaphosidae (Araneae). Four areas of endemism were recognized: Coquimbo (between 30° and 31°5′), Santiago (between 32° and 34°5′), Curicó (between 35° and 36°), and Ñuble (between 36°2′ and 37°4′). Given its close relationships with the Subantarctic subregion, it was also considered an "external area" (fig. 5.8a). Ten taxon–area cladograms were obtained for *Triptilion*, *Calopappus–Nassauvia sect. Panargyrum*, *Leucheria amoena*, and *L. cerberoana* species groups (Asteraceae); *Mendizabalia* (Buprestidae), *Listroderes nodifer*, and *L. curvipes* species groups; and *Puranius* (Curculionidae), an *Apodrassodes*, and *Echemoides chilensis* species group (Gnaphosidae). Four different methods were applied: component analysis under assumptions 0, 1, and 2, with options BUILD and SHARED of Component version 1.5 (Page 1989b); BPA with Hennig86 1.5 (Farris 1988); three area statement analysis with programs TAS (Nelson and Ladiges 1991b) and Hennig86 1.5

(continued)

CASE STUDY 5.1 Cladistic Biogeography of Central Chile *(continued)*

(Farris 1988); and paralogy-free subtree analysis under assumption 2 with TASS (Nelson and Ladiges 1995) and Hennig86 1.5 (Farris 1988).

Five general area cladograms were obtained. Component analysis under assumptions 0, 1, and 2 produced different sets of resolved area cladograms. The intersection of all sets did not give a result, but the intersection of some sets under assumption 1 produced five general area cladograms (figs. 5.8b–5.8f): 1–3, from most taxa except *Triptilion* and the *Listroderes curvipes* species group; and 4 and 5, from *Triptilion, Calopappus, Nassauvia* sect. *Panargyrum, Leucheria amoena* and *L. cerberoana* species groups, and *Mendizabalia*. BPA led to general area cladograms 4 and 5 (130 steps, consistency index of 0.84, and retention index of 0.62; figs. 5.8e and 5.8f). Three area statement analysis under assumption 1 gave cladogram 5 (707 steps, consistency index of 0.65, and retention index of 0.46; fig. 5.8f), and under assumption 0 it gave general area cladogram 1 (998 steps, consistency index of 0.65, and retention index of 0.47; fig. 5.8b). Paralogy-free subtree analysis under assumption 2 gave general area cladogram 1 (eighteen steps, consistency index

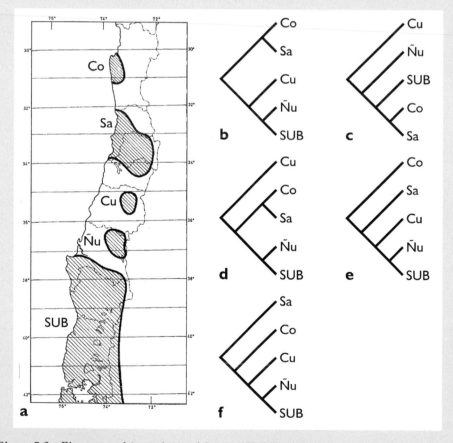

Figure 5.8 Biogeographic analysis of Central Chile by Morrone et al. (1997). (a) Areas of endemism; (b–f) general area cladograms. Co, Coquimbo; Cu, Curicó; Ñu, Ñuble; Sa, Santiago; SUB, Subantarctic.

(continued)

CASE STUDY 5.1   Cladistic Biogeography of Central Chile *(continued)*

of 0.72, and retention index of 0.66; fig. 5.8b). The number of items of error calculated for these general area cladograms was 278 for cladogram 1, 314 for cladogram 2, 326 for cladogram 3, 288 for cladogram 4, and 296 for cladogram 5. General area cladogram 1 was selected because it had the lowest number of items of error. According to clado- gram 1, a first vicariance event isolated the northern areas (Coquimbo and Santiago) from the southern ones (Subantarctic subregion, Curicó, and Ñuble); a second event separated Coquimbo from Santiago; and in the southern areas, Curicó was split first, and the last event separated Ñuble from the Subantarctic subregion.

On the basis of these results, Morrone et al. (1997) concluded that the four areas of Central Chile do not constitute a natural or monophyletic group because Curicó and Ñuble are cladistically more closely related to the Subantarctic subregion than to Coquimbo and Santiago. Consequently, the authors circumscribed the Central Chilean subregion to the northern sector (Coquimbo and Santiago, 30°–34°) and assigned the southern areas (Curicó and Ñuble, south of 34°5´) to the Subantarctic subregion.

References

Farris, J. S. 1988. *Hennig86 reference*. Version 1.5. Port Jefferson, N.Y.: Author.

Morrone, J. J., L. Katinas, and J. V. Crisci. 1997. A cladistic biogeographic analysis of Central Chile. *Journal of Comparative Biology* 2:25–42.

Nelson, G. and P. Y. Ladiges. 1991. *TAS* (MSDos computer program). New York: Authors.

Nelson, G. and P. Y. Ladiges. 1995. *TASS*. New York: Authors.

Page, R. D. M. 1989. *Component user's manual*. Release 1.5. Auckland: Author.

Brooks Parsimony Analysis

BPA was proposed by Wiley (1987, 1988a, 1988b) and posteriorly modified by Brooks (1990). It is based on the ideas developed initially by Brooks (1981, 1985) for historical ecology. It is a parsimony analysis of taxon–area cladograms that are codified as two-state variables and analyzed as characters (Biondi 1998; Brooks 2004; van Veller et al. 2000; Vargas 1992b). In order to apply BPA, a data matrix is constructed on the basis of taxon–area cladograms, and it is analyzed with a parsimony algorithm (fig. 5.9). Brooks (1990) and Brooks and McLennan (1991) proposed another strategy for dealing with parallelisms (dispersal events) that represent falsifications of the null hypothesis. It is named secondary BPA and con- sists of duplicating the involved area and dealing with each of the resulting areas separately. The analysis of the data matrix allows one to determine whether it was really a unique area or whether they were different areas incorrectly treated as a single one (Lomolino et al. 2006).

Four conventions are followed in secondary BPA (Brooks 2004):

- Assumption 0: All species and their distributions in each cladogram must be dealt with without modification, and the result must be logically consis- tent with all input data.

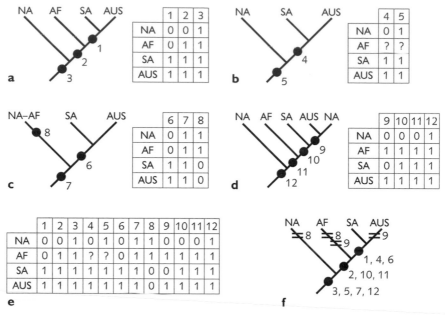

Figure 5.9   Primary Brooks parsimony analysis. (a–d) Taxon–area cladograms and matrices derived from them: (a) trivial case; (b) taxon with a missing area; (c) widespread taxon; (d) taxon with a redundant distribution; (e) data matrix with all the information; (f) general area cladogram obtained. AF, Africa; AUS, Australia; NA, North America; SA, South America.

- Missing data coding: All cases of absence should be coded a priori as missing ("?"), which requires a posteriori interpretations to provide most parsimonious explanations of primitive absences (no member of the clade was ever in the area) or extinctions (secondary loss of a member of a clade).
- Area duplication: Whenever areas have reticulate histories with respect to the species inhabiting them, assumption 0 will be violated unless these areas are represented as separate entities for each separate historical episode. Ambiguous areas should be duplicated until assumption 0 is satisfied.
- The "Threes Rule": In order to distinguish between general and special patterns of historical association based solely on the available data, and, in particular, to determine whether absence from an area is due to secondary extinction or primitive absence, at least three co-occurring clades must be analyzed.

Primary BPA is exemplified in fig. 5.9. Four taxon–area cladograms (figs. 5.9a–5.9d) are analyzed, and a data matrix is derived from the information obtained from them (fig. 5.9e). The parsimony analysis of the data matrix results in a single general area cladogram (fig. 5.9f). In the general area cladogram, there are four parallelisms involving North America and Africa (component 8) and Africa and South America (component 9). For secondary BPA a new data matrix is con-

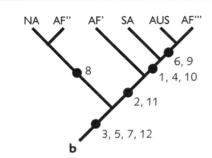

	1	2	3	4	5	6	7	8	9	10	11	12
NA	0	0	1	0	1	0	1	1	0	0	0	1
AF'	0	1	1	?	?	0	1	0	0	0	1	1
AF''	?	?	?	?	?	0	1	1	?	?	?	?
AF'''	?	?	?	?	?	?	?	?	1	1	1	1
SA	1	1	1	1	1	1	1	0	0	1	1	1
AUS	1	1	1	1	1	1	1	0	1	1	1	1

Figure 5.10 Secondary Brooks parsimony analysis. (a) Data matrix with Africa represented three times (AF', AF'', and AF'''); (b) general area cladogram obtained. AF', AF'', and AF''', Africa; AUS, Australia; NA, North America; SA, South America.

structed (fig. 5.10a), where Africa is treated as three different areas: *Africa'* represents the original congruent relationship of Africa with the remaining areas, and *Africa''* and *Africa'''* represent the incongruent relationships due to components 8 and 9, respectively. The analysis of this data matrix results in a single general area cladogram, where *Africa'*, *Africa''*, and *Africa'''* appear in three different positions, depicting the falsifications of the null hypothesis of vicariance.

An alternative implementation of BPA was proposed by Kluge (1988). It differs in three aspects. It considers missing areas to be uninformative, coding them with "0." It considers widespread taxa, caused either by dispersal or by not having responded to vicariance, to be irrelevant and therefore does not take them into account. Because for redundant distributions it is impossible to determine which distribution is irrelevant (by being due to dispersal) and which one is not, Kluge (1988) suggested that they be eliminated one at a time, with the resulting columns weighted in proportion to their number (e.g., if there are two redundant distributions, each of the columns will weigh 0.5, and if there are three, 0.33).

Lieberman (1997, 2000, 2003a, 2004) proposed another modification of BPA, named modified BPA, intended to interpret geodispersal within a cladistic biogeographic framework. The biogeographic analysis is divided into two separate analyses: one to retrieve congruent episodes of vicariance and another to retrieve congruent episodes of geodispersal (Lieberman 2004). The vicariance analysis produces a cladogram that makes predictions about the relative sequence of vicariance events that fragmented biotas. The geodispersal analysis produces a cladogram that provides information about the relative sequence of vicariance events that joined the biotas. Both cladograms provide complementary information about the biotic history and can be placed in a geological framework. The procedure implies optimizing the ancestral states in the area cladograms in order to estimate whether

distributions implied expansions or contractions of the ancestral areas and build-
ing the vicariance and geodispersal data matrices, following the same procedure
as BPA. The best-supported patterns of vicariance and geodispersal emerge from
the parsimony analysis of these matrices. If both cladograms are similar, the same
geological processes may have produced vicariance and geodispersal at different
times (Lieberman 2004) (e.g., cyclical sea level rise and fall). If the cladograms are
very different, they may imply that vicariance and geodispersal have been caused
by noncyclical processes (e.g., continental collisions).

BPA has received some criticism (Carpenter 1992; Cracraft 1988; Ebach and
Edgecombe 2001; Ebach and Humphries 2002; Ebach et al. 2003; Enghoff 2000;
Miranda Esquivel et al. 2003; Nelson and Ladiges 1991c; Page 1993a; Parenti 2007;
Platnick 1988; Ronquist and Nylin 1990; Siddall 2005; Siddall and Perkins 2003;
van Welzen 1992; Warren and Crother 2001). Cracraft (1988) wrote that BPA relies
on a questionable analogy to methods in systematics, so it has the potential to
obscure the history of a biota rather than reveal it. For some authors, it tends to
overestimate dispersal and extinction events (Dowling 2002). Van Welzen (1992)
and Enghoff (2000) stated that BPA sometimes groups areas based on absent taxa.
Ebach and Edgecombe (2001) noted that when taxa are mapped on the general
area cladogram, anomalous reconstructions may appear as descendants dispersing along with their ancestors, thus necessitating a posteriori interpretations. On
one hand, Ebach et al. (2003) found that BPA sometimes gives spurious results. On
the other hand, it has been argued that the parsimony principle should be used for
analyzing the data, not for interpreting the results (Carpenter 1992; Page 1989a).
Miranda Esquivel et al. (2003) concluded that the events and duplication of areas
in secondary BPA are ad hoc, so this method introduces a scheme of verification,
not of falsification. Siddall and Perkins (2003) compared the performance of BPA
and tree reconciliation obtained with TreeMap, finding that sometimes BPA gives
less parsimonious results. Siddall (2005) concluded that BPA lacks an optimality
criterion and the coherence of a research program because published descriptions
of the method are self-contradictory. Furthermore, he suggested that rules for a
posteriori duplication of entities in secondary BPA are not specified clearly and
that both primary and secondary BPA arrive at solutions that may defy logical
or temporally consistent interpretation. Parenti (2007) criticized the codification
strategy used to deal with geodispersal in modified BPA because it specifies a direction that makes it an extension of Hennig's progression rule.

Brooks et al. (2001, 2003) responded to the criticisms suggesting that all the
authors have applied the method incorrectly because they did not take into account the modifications that were performed after the original formulation. These
authors argued that primary BPA finds the most parsimonious general area cladogram, indicating through homoplasy how the null hypothesis of vicariance may
be falsified. Secondary BPA integrates the incongruent elements, choosing the
general area cladogram that postulates the smallest number of duplicated areas,
each one of which represents a falsification of the null hypothesis. Brooks et al.

(2003) responded to Siddall and Perkins (2003) by arguing that TreeMap cannot treat widespread taxa properly and that BPA is a better method.

Algorithm The algorithm of primary BPA consists of the following steps (Dowling 2002; Wiley 1987, 1988a, 1988b):

1. Obtain the taxonomic cladograms of the taxa distributed in the areas analyzed.
2. Replace the terminal species in the taxonomic cladograms with the areas inhabited by them to obtain taxon–area cladograms.
3. Label the components and the widespread terminal species (assumption 0) in the taxon–area cladograms.
4. Construct a data matrix in which areas are the rows and components and widespread terminal species are the columns, coding "1" if the area is present and "0" if it is absent. Use "?" for missing areas. Add a row with all "0" to root the cladogram.
5. Analyze the data matrix with a parsimony algorithm in order to obtain the general area cladogram.
6. Optimize the components in the general area cladogram to identify vicariance events (synapomorphies), dispersal events (parallelisms), and extinctions (reversals).

Software Hennig86 (Farris 1988), PHYLIP (Felsenstein 1993), NONA (Goloboff 1998), PAUP (Swofford 2003), Pee-Wee (Goloboff, retrieved May 25, 2008, from http://www.zmuc.dk/public/phylogeny/Nona-PeeWee/), and TNT (Goloboff et al., retrieved May 25, 2008, from http://www.zmuc.dk/public/phylogeny/TNT/). For reading and editing data files and cladograms: Winclada (Nixon 1999), compatible with NONA, Pee-Wee, and Hennig86; and MacClade (Maddison and Maddison, retrieved May 25, 2008, from http://macclade.org/macclade.html), compatible with PAUP.

Empirical Applications Biondi (1998), Brooks (1990), Brooks and McLennan (2001), Brooks et al. (2003), Contreras-Medina et al. (2007b), Cracraft (1988), Craw (1989b), Crisci et al. (1991b), Crother and Guyer (1996), Dowling (2002), Espinosa Organista et al. (2006), Gesundheit and Macías García (2005), Glasby (2005), Griswold (1991), Halas et al. (2005), Kluge (1988), Ladiges et al. (1992), Lieberman (2000), Liebherr (1991), Lovejoy (1997), Marshall and Liebherr (2000), Mayden (1988), McLennan and Brooks (2002), Morafka et al. (1992), Morrone and Carpenter (1994), Morrone and Urtubey (1997), Morrone et al. (1997), Nihei and de Carvalho (2007), Pinto-da-Rocha and da Silva (2005), Posadas and Morrone (2003), Retana-Salazar (2005), Riddle and Hafner (2006), Rode and Lieberman (2005), Siddall and Perkins (2003), Taberlet et al. (1998), van Soest and Hajdu (1997), van Veller et al. (2000), Van Welzen et al. (2003), and Wiley (1987, 1988a, 1988b).

CASE STUDY 5.2 Cladistic Biogeography of Afromontane Spiders

Several biotic patterns are known for the African forest biota, with disjunct montane and alpine forest "islands" surrounded by lowland forest. The origin of these patterns has been traditionally attributed to climatic changes. Griswold (1991) undertook a cladistic biogeographic analysis based on spider forest taxa.

Griswold (1991) analyzed four taxa for which phylogenetic hypotheses were available: the genus *Microstigmata* (Microstigmatidae), the tribes Vidoleini and Phyxelidini (Amaurobiidae), and the *Moggridea quercina* species group (Migidae). He selected nine areas of endemism: Table Mountains, Knysna Forest, Natal–Zululand Coast, Transkei–Natal midlands, Natal Drakensberg, Transvaal Drakensberg, Eastern Arc Mountains, East African volcanoes, and central Madagascar (fig. 5.11a). The author applied Kluge's (1988) implementation of BPA, constructing a data matrix of nine areas by seventeen components, which he analyzed with Hennig86 (Farris 1988).

The analysis of the data matrix gave eight general area cladograms of twenty-one steps, a consistency index of 0.80, and a retention index of 0.77. The cladograms were identical except for the placement of the Natal–Zululand Coast, which could be placed at different positions (fig. 5.11b). The basal dichotomy leads to the Cape region of South Africa (Table Mountains and Knysna Forest) and the remainder of continental Africa plus Madagascar. A region south of the Limpopo River Basin in northern Africa (Transvaal Drakensberg) shows affinities with tropical Africa (Eastern Arc Mountains), East African volcanoes, and central Madagascar. Eastern South Africa (Natal–Zululand Coast, Transkei–Natal midlands, and Natal Drakensberg) is related to tropical Africa and Madagascar rather than to the nearby Cape region. Two upland areas (Transkei–Natal midlands and Natal Drakensberg) are sister areas. The most striking result is that central Madagascar is related to eastern Africa (Eastern Arc Mountains and East African volcanoes) rather than being the sister area to all Africa.

Griswold (1991) suggested that the general area cladogram was difficult to reconcile with a historical scenario involving primarily Pleistocene vicariance events. On one

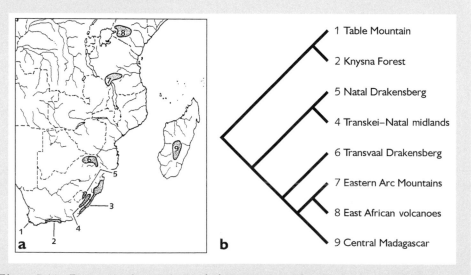

Figure 5.11   Biogeographic analysis of afromontane spiders by Griswold (1991). (a) Areas of endemism analyzed; (b) general area cladogram obtained.

(continued)

CASE STUDY 5.2 Cladistic Biogeography of Afromontane Spiders (*continued*)

hand, a Pleistocene vicariance scenario in which forests expanded and contracted may suggest that geographic proximity should be a good predictor of sister area relationships, which was not the case. On the other hand, the sister area relationship between Madagascar and East Africa led Griswold (1991) to hypothesize a greater age for the taxa involved, probably Mesozoic.

References

Farris, J. S. 1988. *Hennig86 reference.* Version 1.5. Port Jefferson, N.Y.: Author.

Griswold, C. E. 1991. Cladistic biogeography of afromontane spiders. *Australian Systematic Botany* 4:73–89.

Kluge, A. G. 1988. Parsimony in vicariance biogeography: A quantitative method and a greater Antillean example. *Systematic Zoology* 37:315–328.

CASE STUDY 5.3 Biogeographic History of the North American Warm Desert Biota

The deserts of northern Mexico and the southern United States harbor a particular biota that is of great biogeographic interest. Axelrod (1979, 1983) proposed a model to explain the formation of these deserts, where desert floras developed in local dry sites during the Tertiary drying trend, drawing arid-adapted species from boreal shrub steppes, Great Plains grasslands, Mexican highlands, Sinaloan thornscrub, and Californian chaparral. Semideserts attained their maximum extension during the Early Pliocene and were reduced in area during the moist Late Pliocene and during Pleistocene pluvial intervals. Full regional deserts formed during interglacials and reached their maximum extent after the Wisconsinan glacial. As a consequence, sister taxa in alternate desert areas might share common ancestry at one of at least two broad ages: the Late Miocene to Pliocene, along with tectonic events that underlay the early development of regional deserts, or the Pleistocene in association with climatic oscillations. Several authors have postulated models to account for the biotic evolution of these deserts (Grismer 1994; Hubbard 1973; Morafka 1977; Morafka et al. 1992; Murphy and Aguirre-León 2002), discussing specific vicariance events. Riddle and Hafner (2006) developed a five-step approach to analyze the relationships between North American desert areas, combining phylogeographic hypotheses, PAE, and primary and secondary BPA.

Riddle and Hafner (2006) analyzed twenty-two taxon cladograms from mammal, bird, reptile, amphibian, and cactus phylogroups, with species distributed in a subset of the North American warm desert biota that they considered adequate for their study. To identify the areas of endemism, they applied PAE to eleven core warm desert distributional areas. They used primary and secondary BPA to examine biogeographic structure across the areas found with PAE. PAUP version 4.0 (Swofford 1999) was used for both PAE and BPA.

(*continued*)

CASE STUDY 5.3 Biogeographic History of the North American Warm Desert Biota *(continued)*

The PAE of the eleven areas of distribution, which was run removing sequentially areas lacking many endemic species, led to the identification of four areas of endemism (fig. 5.12a): Continental East (Trans-Pecos, Coahuilan, and Zacatecan areas), Continental West (Sonoran and Coloradan areas), Peninsular South (Magdalenan and San Lucan

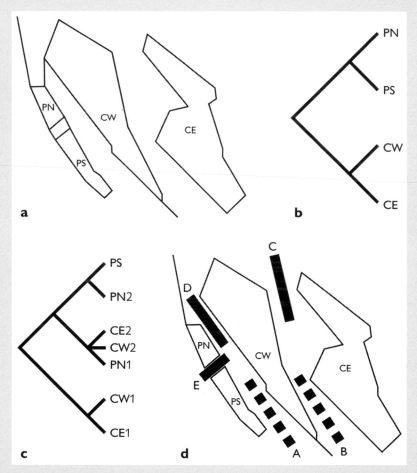

Figure 5.12 Biogeographic analysis of North American deserts by Riddle and Hafner (2006). (a) Areas of endemism analyzed; (b) most parsimonious primary Brooks parsimony analysis cladogram obtained; (c) most parsimonious secondary Brooks parsimony analysis cladogram obtained; (d) model of historical vicariance depicting the vicariance events involved. Areas: CE, Continental East; CW, Continental West; PN, Peninsular North; PS, Peninsular South. Vicariance events: A, Trans-Gulfian opening (earlier); B, Sierra Madre Occidental uplift (earlier); C, Sierra Madre Occidental uplift and ecological barriers (later); D, Trans-Gulfian embayments (later); E, Vizcaíno Seaway.

(continued)

CASE STUDY 5.3 Biogeographic History of the North American Warm Desert Biota (continued)

areas), and Peninsular North (Cirios area). A single most parsimonious primary BPA cladogram (fig. 5.12b) was produced, with 120 steps, a consistency index of 0.70, and a retention index of 0.58. In the cladogram, Continental West and Continental East are sister areas, and Peninsular North and Peninsular South are sister areas. The high number of synapomorphies suggests common vicariance patterns, although there are several homoplasies, especially between Continental West and Peninsular North, and Continental West and Peninsular North and Peninsular South. Secondary BPA resulted in a cladogram (fig. 5.12c) with duplications involving all areas except Peninsular South.

Riddle and Hafner (2006) postulated a vicariance model (fig. 5.12d), correlating the different events that might have been associated with the biotic history of the areas analyzed. It implies an equally deep or perhaps deeper history of vicariance between Continental West and Continental East than between Continental West and one or both peninsular areas. This may imply that the Continental West area has played a central role in diversification between two separate events: one involving the evolution of the Sea of Cortés to the west and another involving the evolution of the Sierra Madre Occidental and the Mexican Plateau to the east.

References

Axelrod, D. I. 1979. Age and origin of Sonoran Desert vegetation. *Occasional Papers of the California Academy of Sciences* 132:1–74.

Axelrod, D. I. 1983. Paleobotanical history of the western deserts. In *Origin and evolution of deserts*, ed. S. G. Wells and D. R. Haragan, 113–129. Albuquerque: University of New Mexico Press.

Grismer, L. L. 1994. The origin and evolution of the peninsular herpetofauna of Baja California, Mexico. *Herpetological Natural History* 2(1):51–106.

Hubbard, J. P. 1973. Avian evolution in the aridlands of North America. *Living Bird* 12:155–196.

Morafka, D. J. 1977. *A biogeographical analysis of the Chihuahuan Desert through its herpetofauna*. The Hague: Junk.

Morafka, D. J., G. A. Adest, and L. M. Reyes. 1992. Differentiation of North American deserts: A phylogenetic evaluation of a vicariance model. *Tulane Studies in Zoology and Botany, Supplementary Publications* 1:195–226.

Murphy, R. W. and G. Aguirre-León. 2002. Nonavian reptiles; origins and evolution. In *A new island biogeography of the Sea of Cortés*, ed. T. J. Case, M. L. Cody, and E. Ezcurra, 181–220. New York: Oxford University Press.

Riddle, B. R. and D. J. Hafner. 2006. A step-wise approach to integrating phylogeographic and phylogenetic biogeographic perspectives on the history of a core North American warm deserts biota. *Journal of Arid Environments* 66:435–461.

Swofford, D. L. 1999. *PAUP*: Phylogenetic analysis using parsimony (*and other methods)*. Version 4.0 Beta. Sunderland, Mass.: Sinauer.

Three Area Statement Analysis

Nelson and Ladiges (1991a, 1991c) developed this method, which is based on the three item statement analysis (Marques 2005; Nelson and Ladiges 1993; Nelson and Platnick 1991; Williams and Humphries 2003). In contrast to component coding, applied in BPA, in three-item coding each node is considered a relationship between branches, where some branches are related more closely than other branches. Each separate relationship is expressed minimally as a three-item statement (Williams and Humphries 2003). All area statements are input in a data matrix and analyzed with a parsimony or compatibility algorithm (fig. 5.13).

The three-item statement analysis has received some criticism (Farris 2000; Farris and Kluge 1998; Harvey 1992; Kluge 1993), and a few authors have defended it (de Pinna 1996; Marques 2005; Scotland 2000; Siebert and Williams 1998). The main reasons for controversy have to do with the relationship of this approach to cladistics, observations and homology, data transformation, parsimony, and synapomorphies (Marques 2005). Discussions are of a highly technical nature, but I concur with de Pinna (1996:11):

> Regardless of the fate of three-item analysis as a method of character analysis, I predict there will be an increase awareness of the problem of ancestor–descendant relationships among character states. Even if three-item analysis turns out not to be an appropriate solution, I think it will have its place in history as the first concrete attempt at terminating ancestor–descendant thinking at the level of character coding.

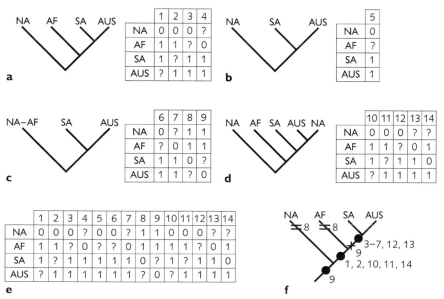

Figure 5.13 Three area statement analysis. (a–d) Taxon–area cladograms and matrices derived from them: (a) trivial case; (b) taxon with a missing area; (c) widespread taxon; (d) taxon with a redundant distribution; (e) data matrix with all the information; (f) general area cladogram obtained. AF, Africa; AUS, Australia; NA, North America; SA, South America.

Algorithm The algorithm consists of the following steps (Nelson and Ladiges 1991a, 1991c, 1993):

1. Obtain the taxonomic cladograms of the taxa distributed in the areas analyzed.

2. Replace the terminal taxa from the taxonomic cladograms with the areas inhabited by them to obtain taxon–area cladograms.

3. Construct a data matrix in which areas are the rows, and components and widespread terminal species are the columns, coding the relationships as three area statements. Add a row with all "0" to root the cladogram.

4. Analyze the data matrix with a parsimony or compatibility algorithm in order to obtain the general area cladogram.

Software TAS (Nelson and Ladiges 1991b), which implements assumptions 0 and 1. The matrix obtained may be analyzed with any parsimony or compatibility algorithm.

Empirical Applications de Meyer (1996), Ladiges et al. (1992), Morrone (1993b), Morrone and Carpenter (1994), Morrone et al. (1994, 1997), van Soest and Hajdu (1997), and van Veller et al. (2000).

CASE STUDY 5.4 Cladistic Biogeography of the "Blue Ash" Eucalypts

The twenty-three species of the plant genus *Eucalyptus* known as "blue ash" constitute a clade distributed in southeastern Australia. Ladiges et al. (1992) undertook their cladistic analysis and based on the cladogram obtained presented a resolved area cladogram using the three area statement approach.

Ladiges et al. (1992) obtained the three area statements under assumptions 0, 1, and 2 and used Hennig86 (Farris 1988) for the parsimony analysis of the data matrices. Additionally, they performed a BPA. The authors represented the distributional areas of the species analyzed on maps, and on the basis of their coincidence they identified twelve areas of endemism (fig. 5.14a). After replacing the terminal species in the consensus cladogram with these areas of endemism, they found some areas repeated across major clades, which were taken as evidence of geographic paralogy. They concluded that three orthologous clades were appropriate units for the analysis: *E. stenostoma–E. haemostoma*, *E. stellulata*, and *E. fraxinoides* groups. The *E. stenostoma–E. haemostoma* group (fig. 5.14b), under assumption 1 generated 160 three area statements, which produced twelve equally parsimonious cladograms (consensus cladogram in fig. 5.14c). Adding 111 three area statements for widespread taxa (assumption 0) resulted in a single most parsimonious cladogram (fig. 5.14d). BPA resulted in seven cladograms (consensus cladogram in fig. 5.14e). Resolution by hand of assumption 2, accepting the position of E_1 as resolved by assumption 1, is presented in fig. 5.14f. The taxon–area cladograms of the *E. stellulata* and *E. fraxinoides* groups (figs. 5.14g and 5.14h) are easier to interpret. The former suggests that E_2 is at the root of that cladogram, whereas the latter suggests that G is related to E_1. Adding them to the cladogram in fig. 5.14d gave a resolved area cladogram (fig. 5.14i) for the "blue ash" clade, where E_2 had two alternative positions, and areas F and A were not considered.   *(continued)*

CASE STUDY 5.4 Cladistic Biogeography of the "Blue Ash" Eucalypts (*continued*)

On the basis of the area cladogram obtained, Ladiges et al. (1992) suggested a historical sequence in which area H, the southern New South Wales (NSW) escarpment, differentiated first from the remaining areas in southeastern Australia, although differentiation of Mount Buffalo, Victoria (E_2) may have been even earlier. Differentiation of the Northern Tablelands (area B) followed. The Victorian part of the Great Dividing Range (area E_1) differentiated from the NSW Tablelands, whereas the North Coast (area J) differentiated from the coastal and tableland regions to the south. Finally, areas in central and southern NSW differentiated, with the Blue Mountains and adjacent coast (areas I and K) differentiating from the central and southern tablelands (areas C and D). This scenario suggests that several clades within the "blue ash" had already evolved before these vicariance processes occurred, that some taxa that are widespread today may have been historically widespread but failed to differentiate, and that some taxa dispersed to other areas.

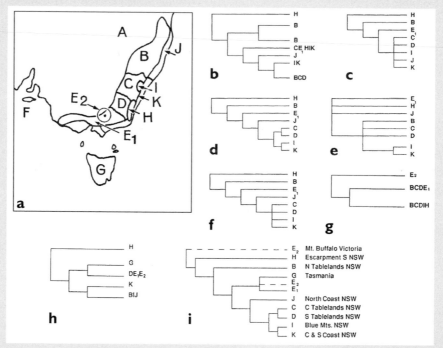

Figure 5.14 Biogeographic analysis of "blue ash" eucalypts by Ladiges et al. (1992). (a) Areas of endemism in southeastern Australia analyzed; (b) *E. stenostoma–E. haemostoma* taxon–area cladogram; (c) resolved area cladogram under assumption 1; (d) resolved area cladogram under assumption 0; (e) resolved area cladogram obtained by Brooks parsimony analysis; (f) resolved area cladogram under assumption 2; (g) *E. stellulata* taxon–area cladogram; (h) *E. fraxinoides* taxon–area cladogram; (i) resolved area cladogram, showing the alternative positions of E_2. A, Queensland Blackdown Tablelands; B, Northern Tablelands; C, Central Tablelands NSW and adjacent western slopes; D, Southern Tablelands NSW; E_1, Great Dividing Range, Victoria; E_2, Mt. Buffalo Victoria; F, Kangaroo Island, South Australia; G, Tasmania; H, escarpment southern NSW; I, Blue Mountains; J, North Coast; K, Central and Southern Coast NSW.

(continued)

CASE STUDY 5.4 Cladistic Biogeography of the "Blue Ash" Eucalypts (*continued*)

References

Farris, J. S. 1988. *Hennig86 reference*. Version 1.5. Port Jefferson, N.Y.: Author.

Ladiges, P. Y., S. M. Prober, and G. Nelson. 1992. Cladistic and biogeographic analysis of the "blue ash" eucalypts. *Cladistics* 8:103–124.

Tree Reconciliation Analysis

Tree reconciliation analysis is also known as maximum cospeciation (Crisci et al. 2000), maximum vicariance (Sanmartín and Ronquist 2002), and parsimony-based tree fitting (Sanmartín and Ronquist 2004). It was developed independently in molecular systematics, parasitology, and biogeography as a way to describe historical associations between genes and organisms (Goodman et al. 1979), parasites and hosts (Mitter and Brooks 1983), and taxa and areas (Page 1990a, 1993b; Page and Charleston 1998), respectively (see table 5.1). It was formalized by Page (1994a, 1994b) as a general method that maximizes the amount of codivergence or shared history between area cladograms from different taxa, minimizing losses (due to extinctions or lack of collection) and duplications (independent vicariance events) when different area cladograms are combined to obtain a general area cladogram (Morrone and Crisci 1995; van Veller et al. 2000; Vargas 2002).

When there is correspondence between the taxa and the components of the area cladograms that are compared, they are reconciled easily (fig. 5.15a). In most cases, there is no complete correspondence between the cladogram topologies, so in order to reconcile them, some components must be duplicated. In fig. 5.15b, component b was duplicated, giving rise to an identical component b' that contains terminal taxa Africa', South America', and Australia'. In biogeography, codivergence between areas and organisms is equivalent to vicariance, duplications are equivalent to speciation independent of the vicariance events, horizontal transference is equivalent to dispersal, and losses are equivalent to extinction events.

Table 5.1

Terms	Parasitology	Biogeography	Molecular systematics
Host	Host	Area	Organism
Associate	Parasite	Taxon	Gene
Codivergence	Cospeciation	Vicariance	Genetic differentiation
Horizontal transmission	Secondary infestation, colonization or host switching	Dispersal	Horizontal gene transfer
Lost	Parasite extinction	Extinction	Gene loss or lineage sorting
Duplication	Parasite speciation without host speciation	Sympatric speciation or geographic paralogy	Paralogous genes or gene duplication without speciation

Figure 5.15   Reconciled tree analysis. (a) Example with total codivergence; (b) example in which there is no total codivergence and we need to duplicate a component to reconcile both cladograms. AF, Africa; AUS, Australia; NA, North America; SA, South America.

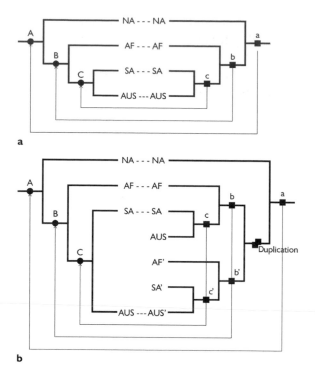

This method has been criticized for not considering host switching or dispersal (Page and Charleston 1998). Charleston (1998) developed a solution using mathematical structures called "jungles" that contain all possible partial orderings in which the associate cladogram may be tracked in the host cladogram, considering codivergence, duplication, sorting, and host-switching events and all the existing known associations. When the costs are calculated for each of these events, it is possible to find the subgraphs that correspond to the least costly reconstructions of the association.

Huelsenbeck et al. (2000) stated that in inferring host-switching events, it is assumed that the host and parasite cladograms are estimated without error. They suggested that a Bayesian estimation can be used in models where host-switching events are assumed to occur at a constant rate over the entire evolutionary history of the association. This method provides information on the probability that an event of host switching is associated with a particular pair of branches and reduces the possibility that a particular phylogenetic hypothesis may be overturned if a reexamination of the group results in a different cladogram.

Crisci et al. (2000) discussed a problem of the tree reconciliation analysis and dispersal–vicariance analysis: that they violate metricity when considering that duplications are less probable than vicariance events. Violation of metricity assumes in the calculations the existence of a nonmetric space whose geometric properties are difficult to explore. Dowling (2002) enumerated some criticisms to the program TreeMap; the most important is that it may overestimate duplications and underestimate dispersal.

Algorithm The algorithm consists of the following steps (Crisci et al. 2000):

1. Obtain the taxonomic cladograms of the taxa distributed in the areas analyzed.
2. Replace the terminal taxa from the taxonomic cladograms with the areas inhabited by them to obtain taxon–area cladograms.
3. Superimpose each component of a cladogram on the components of the other cladograms.
4. Assume maximum cospeciation and no dispersal, attributing the differences between the cladograms to duplication or extinction events.
5. Choose the reconstruction that implies the maximum cospeciation and minimum number of losses and duplications.

Software Component version 2.0 (Page 1993a), TreeMap (Page 1994c), and Tree-Fitter version 1.3 (Ronquist 2002). Component version 2.0 allows users to treat widespread taxa under assumption 1 by not choosing the option "map widespread associates" (equivalent to assumption 0). Enghoff (1998) discovered that implementation of assumption 1 occurs only if the widespread taxon has a terminal position in the cladogram; if it is situated basally, Component version 2.0 implements assumption 2. Assumption 2 may be applied manually, with the areas involved in the widespread taxon deleted one at a time (Humphries and Parenti 1999).

Empirical Applications Biondi (1998), Brooks et al. (2003), Contreras-Medina and Luna-Vega (2002), Crisp et al. (1995), Dowling (2002), Flores Villela and Goyenechea (2001), Liebherr (1997), Liebherr and Zimmermann (1998), Linder and Crisp (1995), Miranda-Esquivel (2001), Morrone and Carpenter (1994), Morrone and Coscarón (1998), Page (1990b, 1994a), Posadas and Morrone (2003), Sanmartín and Ronquist (2004), Sanmartín et al. (2007), Siddall and Perkins (2003), Soares and de Carvalho (2005), Swenson and Bremer (1996), Swenson et al. (2001), Upchurch et al. (2002), van Soest and Hajdu (1997), van Veller et al. (2000), and van Welzen et al. (2001).

CASE STUDY 5.5 Biogeography of South American Assassin Bugs (Hemiptera)

In a preliminary biogeographic analysis of the South American species of the subfamily Peiratinae (Hemiptera: Reduviidae) using a PAE, Morrone and Coscarón (1996) postulated that the gradual development of a diagonal of open vegetation encompassing the Chaco, Cerrado, and Caatinga separated the former Amazonian forest in a northwestern part (Amazonian forest in the strict sense) and another southeastern (Parana–Atlantic forest). Morrone and Coscarón (1998) carried out a cladistic biogeographic analysis to test this hypothesis.

The areas of endemism considered in the analysis (fig. 5.16a) were the Caribbean (CAR) and Amazonian subregions (AMA) and the Coastal Desert (DES), Chaco (CHA), Cerrado (CER), Caatinga (CAA), Parana (PAR), and Atlantic (ATL) provinces. Four taxon–area cladograms were obtained on the basis of the taxonomic cladograms of the genera *Rasahus, Eidmannia, Melanolestes,* and *Thymbreus,* replacing their terminal species with the areas they inhabit. Two methods were used: tree reconciliation analysis, with Component version 2.0 (Page 1993a), and paralogy-free subtree analysis, with TASS (Nelson and Ladiges 1995) and Hennig86 1.5 (Farris 1988).

Component version 2.0 produced one general area cladogram (fig. 5.16b), minimizing the number of terminal taxa added, and two general area cladograms (figs. 5.16c and 5.16d), minimizing the number of losses. The parsimony analysis of the matrix

a

Figure 5.16   Biogeographic analysis of South America by Morrone and Coscarón (1998). (a) Areas of endemism; (b) reconciled trees minimizing the number of terminal taxa added; (c–d) reconciled trees minimizing the number of losses; (e) paralogy-free subtree analysis. AMA, Amazonian subregion; ATL, Atlantic Forest province; CAA, Caatinga province; CAR, Caribbean subregion; CER, Cerrado province; CHA, Chacoan province; DES, Coastal Desert province; PAR, Parana province.

(continued)

CASE STUDY 5.5 Biogeography of South American Assassin Bugs (Hemiptera) *(continued)*

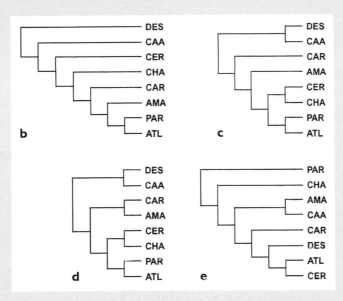

obtained with TASS led to a single general area cladogram (fig. 5.16e), with nineteen steps, a consistency index of 0.73, and a retention index of 0.78. Items of error were calculated for the general area cladograms, and the first one (fig. 5.16b) was chosen for having the smallest number. The sequence of areas in the general area cladogram was as follows: (Coastal Desert (Caatinga (Cerrado (Chaco ((Caribbean, Amazonian) (Parana, Atlantic)))))).

According to this hypothesis, the four areas with open vegetation (Coastal Desert, Caatinga, Cerrado, and Chaco) are situated basally in the general area cladogram, whereas the forest areas, despite today being clearly disjunct, constitute a monophyletic group. This seems to corroborate the preliminary hypothesis (Morrone and Coscarón 1996) on the progressive development of the diagonal of open vegetation, which in turn fragmented the primitive forest into two blocks: Caribbean–Amazonian and Parana–Atlantic.

References

Farris, J. S. 1988. *Hennig86 reference.* Version 1.5. Port Jefferson, N.Y.: Author.

Morrone, J. J. and M. del C. Coscarón. 1996. Distributional patterns of the American Peiratinae (Heteroptera: Reduviidae). *Zoologische Medeligen Leiden* 70:1–15.

Morrone, J. J. and M. del C. Coscarón. 1998. Cladistics and biogeography of the assassin bug genus *Rasahus* Amyot and Serville (Heteroptera: Reduviidae: Peiratinae). *Zoologische Medeligen Leiden* 72:73–87.

Nelson, G. and P. Y. Ladiges. 1995. *TASS.* New York: Authors.

Page, R. D. M. 1993. *Component user's manual.* Release 2.0. London: The Natural History Museum.

CASE STUDY 5.6 Biogeography of Plant and Animal Taxa in the Southern Hemisphere

The biogeographic history of the southern continents represents a classic example of a vicariance scenario, with disjunct distributions explained by the sequential breakup of Gondwana in the last 165 million years. Although most Gondwanan taxa are presumably poor dispersers, unable to cross oceanic barriers, paleogeographic reconstructions indicate that the biogeographic history of the Southern Hemisphere cannot be entirely reduced to a simple sequence of vicariance events. In addition, some molecular studies indicate that dispersal may have played a more important role in the biotic evolution of the area than was previously assumed. Sanmartín and Ronquist (2004) examined a large data set of plant and animal taxa, applying tree reconciliation analysis in order to assess the relative roles of vicariance and dispersal in the biogeographic evolution of the Southern Hemisphere.

Sanmartín and Ronquist (2004) identified six basic biogeographic patterns (figs. 5.17a–5.17f) to test in their analysis. The southern Gondwana pattern, Brundin's (1966) classic pattern based on chironomid midges, which shows Africa diverging early, followed by New Zealand, and finally by southern South America and Australia, is congruent with the breakup of southern temperate Gondwana during the Mesozoic–Tertiary. The tropical Gondwana pattern (Madagascar–Africa–northern South America) is explained by vicariance of the western part of northern Gondwana. The plant southern pattern (New Zealand–Australia) and the inverted southern pattern (New Zealand–southern South America) are both incongruent with the geological scenario and are typically explained by dispersal. The northern Gondwana pattern, which shows the connection between the areas that once formed northeastern tropical Gondwana, is complementary to the tropical Gondwana pattern. The trans-American pattern follows earlier studies that showed the composite nature of South America. The authors analyzed seventy-three taxa for which cladograms were available, which consisted of fifty-four animal taxa and nineteen plant taxa (the fungus genus *Cyttaria* was included in this group because of similar means of dispersal). Areas analyzed corresponded mainly to historically persistent Gondwanan landmasses: Africa (excluding the region north of the Saharan belt), Madagascar (including the Seychelles, Reunion, and Mauritius), India (including Nepal, Tibet, and Sri Lanka), Australia (including Tasmania), New Zealand, New Caledonia, New Guinea, southern South America (the southern temperate region), northern South America (the northern tropical region), southeast Asia (including the Malaysian peninsula, Philippines, Sumatra, and Borneo), southwest Pacific (Melanesian archipelagos), and the Holarctic (Palearctic and Nearctic regions). The authors applied tree reconciliation analysis with TreeFitter version 1.3 (Ronquist 2002).

Sanmartín and Ronquist (2004) obtained best optimal general area cladograms for the whole animal data set (fig. 5.17g), the insects excluding Eucnemidae (fig. 5.17h), the noninsect animals (fig. 5.17i), and the plants (fig. 5.17j). These four general area cladograms were significant in terms of overall cost and had a higher number of vicariance events than expected by chance ($p = .01$ for the plant data set and $p < .01$ for the others). The cladogram based on animals (fig. 5.17g) agrees with the southern Gondwana pattern, except for the position of Africa. The same cladogram was obtained when insects were analyzed separately, but when the Eucnemidae were excluded from the data set, the southern Gondwana pattern became complete in that it also included Africa (fig. 5.17h). The analysis of the animal taxa excluding insects also reflected the

(continued)

CASE STUDY 5.6 Biogeography of Plant and Animal Taxa in the Southern Hemisphere *(continued)*

southern Gondwana pattern (fig. 5.17i). In contrast, the plant area cladogram (fig. 5.17j) supported the plant southern pattern. Both animal and plant cladograms supported a hybrid origin of South America (Crisci et al. 1991b): Northern South America was

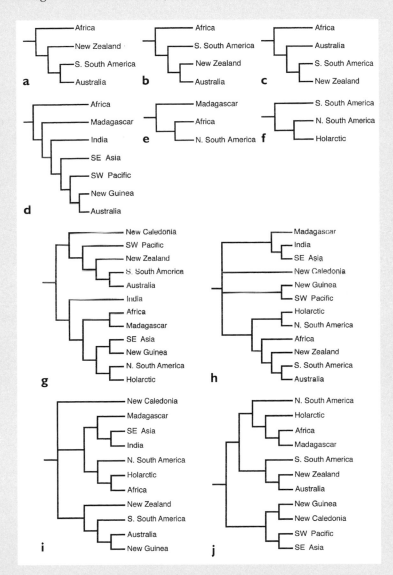

Figure 5.17   Biogeographic analysis of the Gondwanan areas by Sanmartín and Ronquist (2004). (a–f) Area cladograms representing the biogeographic patterns discussed: (a) southern Gondwana pattern; (b) plant southern pattern; (c) inverted southern pattern; (d) northern Gondwana pattern; (e) tropical Gondwana pattern; (f) trans-American pattern; (g–j) general area cladograms obtained: (g) animal data set; (h) insects excluding Eucnemidae; (i) noninsect animals; (j) plants.

(continued)

CASE STUDY 5.6 Biogeography of Plant and Animal Taxa in the Southern Hemisphere *(continued)*

grouped with the Holarctic region, and southern South America appeared more closely related to Australia or New Zealand.

The authors then compared all taxon–area cladograms with each of the six biogeographic patterns initially identified. They found that the animal data set was dominated by the southern Gondwana and northern Gondwana patterns, although the latter was almost exclusively due to the Eucnemidae. The majority of the plant groups supported either the plant southern pattern or the inverted southern pattern. Several groups did not support any clear pattern or supported a pattern different from those analyzed. When the phylogenies of the different taxa were fitted to a geological area cladogram depicting the relationships of the Gondwanan landmasses, a significant distribution pattern ($p < .01$) was displayed. However, the processes involved differed between plants and animals. The animal data set as a whole showed a higher frequency of vicariance, extinction, and duplication events than expected by chance, whereas dispersal events were rare. The plant data set also exhibited a higher frequency of duplications and a lower frequency of dispersals than expected, but the frequencies of vicariance and extinction events did not depart significantly from expected values. This difference was also observed when individual groups were fitted to the geological area cladogram. With the exception of the family Strelitziaceae, none of the plant groups showed a significant frequency of vicariance or extinction, whereas many of the animal taxa did.

Finally, the authors examined the role played by dispersal in shaping the studied patterns by fitting the taxonomic phylogenies to the geological area cladogram. They found that the frequency of terminal dispersals was significantly higher in plants than in animals, but the frequencies of ancestral and total dispersals, although higher in plants, did not differ from expected values. Thus, there was more dispersal in the plant data set, but it appeared to be solely due to dispersal within the widespread species. Several concordant dispersal patterns were inferred, but they were different in animals and plants. In animals, trans-Antarctic dispersal between Australia and southern South America was significantly more frequent than any of the other dispersal events involving the austral landmasses. In contrast, for plants, trans-Tasman dispersal (New Zealand–Australia) was significantly more frequent than trans-Antarctic (Australia–southern South America) and trans-Pacific (New Zealand–southern South America) dispersals. Trans-American dispersal (northern South America–southern South America) in animal taxa was significantly lower than the biotic exchange between northern South America and North America (Holarctic). Dispersal between Madagascar and Africa was significantly more frequent than trans-Atlantic dispersal (northern South America–Africa).

Sanmartín and Ronquist (2004) concluded that the breakup of Gondwana has played an important role in molding the vicariance patterns of the animal Gondwanan groups. The congruence between animal patterns and the Gondwana breakup implies that the taxa may be old, presumably Mesozoic. In contrast, plants did not seem to have been significantly influenced by vicariance induced by the Gondwanan breakup. The plant southern pattern conflicts with continental fragmentation patterns, so it is possible that the plant groups analyzed were too young to have been affected or they once fragmented in response to Gondwanan splits, but their original patterns were

(continued)

CASE STUDY 5.6 Biogeography of Plant and Animal Taxa in the Southern Hemisphere *(continued)*

subsequently lost because of dispersal and extinction events. Concordant dispersal patterns differ between the groups. The animal data set is dominated by trans-Antarctic dispersals, presumably because of the long period of geological contact between Australia and southern South America via Antarctica. In plants, dispersal events seem to have occurred between landmasses that were not connected at the time, as trans-Tasman dispersal between Australia and New Zealand. Both animal and plant data sets point to the hybrid origin of the South American biota, with few dispersal events between southern and northern South America. New Caledonia and New Zealand appear to have retained few old Gondwanan lineages, particularly those of plants.

References

Brundin, L. 1966. Transantarctic relationships and their significance as evidenced by midges. *Kungliga Svenska Vetenskaps Akademien Handlingar (Series 4)* 11:1–472.

Crisci, J. V., M. M. Cigliano, J. J. Morrone, and S. Roig-Juñent. 1991. Historical biogeography of southern South America. *Systematic Zoology* 40:152–171.

Ronquist, F. 2002. *TreeFitter,* version 1.3. Retrieved May 25, 2008, from http://www.ebc.uu.se/systzoo/research/treefitter/treefitter.html.

Sanmartín, I. and F. Ronquist. 2004. Southern Hemisphere biogeography inferred by event-based models: Plant versus animal patterns. *Systematic Biology* 53: 216–243.

Paralogy-Free Subtree Analysis

Paralogous areas are those that conflict with duplications of themselves. Nelson and Ladiges (1996, 2001) suggested that because of geographic paralogy the components may provide biogeographic information but not be directly informative. This means that we may have contradictory relationships due to sympatric speciation, lack of response to vicariance events, incorrect definition of areas, and other explanations, which can lead to erroneous interpretations (Nelson and Ladiges 2001). Paralogy-free subtrees simplify the cladistic biogeographic analysis, so that geographic data need not be associated with paralogous nodes, preventing artifactual results, if not completely then at least to a significant degree (Nelson and Ladiges 2003; Parenti 2007).

Nelson and Ladiges (1996) developed an algorithm that constructs paralogy-free subtrees, starting off at the most terminal groups of the cladogram. The procedure reduces complex cladograms to paralogy-free subtrees, meaning that geographic data are associated only with informative nodes, and areas duplicated or redundant in the descendants of each node do not exist (fig. 5.18). These are the only data relevant for cladistic biogeography. When obtained, paralogy-free subtrees are represented in a component or a three-item matrix and analyzed with a parsimony algorithm. Before paralogy-free subtrees are obtained, the transparent method (Ebach et al. 2005a) may be implemented to resolve widespread taxa.

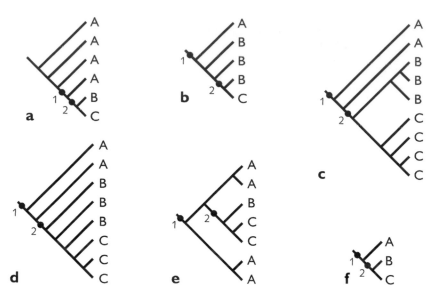

Figure 5.18   Paralogy-free subtree analysis. (a–e) Original taxon–area cladograms, with paralogous nodes; (f) the single paralogy-free subtree that can be derived from them.

Algorithm   The algorithm consists of the following steps (Nelson and Ladiges 1996):

1. Obtain the taxonomic cladograms of the taxa distributed in the areas analyzed.

2. Replace the terminal taxa from the taxonomic cladograms with the areas inhabited by them to obtain taxon–area cladograms.

3. Resolve widespread taxa with the transparent method.

4. Identify the paralogy-free subtrees starting at each terminal node and progressing to the base of each taxon–area cladogram. When a node leads to one or more terminal taxa that are geographically widespread and part of that distribution overlaps with that of another taxon or taxa, reduce the widespread distribution to the nonoverlapping geographic element.

5. Represent the nodes of all the paralogy-free subtrees in a component or three-item matrix.

6. Analyze the data matrix with a parsimony algorithm to obtain the general area cladograms.

Empirical Applications   Brown et al. (2006), Contreras-Medina et al. (2007b), Giribet and Edgecombe (2006), Ladiges et al. (1997), Lattke (2003), Linder (1999), Morrone and Coscarón (1998), Morrone and Urtubey (1997), Morrone et al. (1997), Nelson and Ladiges (1996, 2001), Roig-Juñent et al. (2003, 2006), and Zhang et al. (2002).

Software   TASS (Nelson and Ladiges 1995) and Nelson05 (Cao and Ducasse 2005).

CASE STUDY 5.7 Biogeography of the Northern Andes

The Andes are the most important mountainous system of South America. Classically, they have been divided into the Austral Andes (54°–18° south latitude), Central Andes (18°–3°), and Northern Andes (3°–10°). The Northern Andes run from northern Venezuela to southern Ecuador, and despite having different types of vegetation (super-páramo, páramo, subpáramo, and forest), they were treated as a biogeographic unit by Morrone (1994a, 1996b). From the biogeographic point of view they are very interesting because they have numerous endemic taxa and complex distributional patterns, and their most austral part, the Amotape–Huancabamba zone of southern Ecuador, is transitional between the Northern and Central Andes (Berry 1982). Morrone and Urtubey (1997) carried out a cladistic biogeographic analysis of the Northern Andes to explain their biogeographic history and to determine whether they represent a natural biogeographic unit.

On the basis of the distribution of the species analyzed, seven areas of endemism were identified (fig. 5.19a): A–D, in the Northern Andes, and E–G, "external areas" from the Central Andes. The taxa analyzed were five weevil genera (Coleoptera: Curculionidae) and the plant genus *Barnadesia* (Asteraceae). Paralogy-free subtrees were obtained with program TASS (Nelson and Ladiges 1995), and two data matrices were built, one based on assumption 2 and another for BPA. Data matrices were analyzed with Hennig86 version 1.5 (Farris 1988).

Application of TASS to the six original taxon–area cladograms (figs. 5.19b–5.19g) resulted in eleven paralogy-free subtrees. The parsimony analysis of the matrix of paralogy free subtrees led to three general area cladograms whose strict consensus has the same topology as one of them (fig. 5.19h). The analysis of the BPA matrix resulted in four general area cladograms; their consensus (fig. 5.19i) showed different relationships. The difference between both results may be due to paralogy.

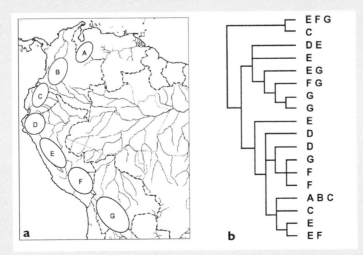

Figure 5.19 Biogeographic analysis of the northern Andes by Morrone and Urtubey (1997). (a) Areas of endemism in the Andes; (b–g) taxon–area cladograms of *Barnadesia* and five genera of Curculionidae; (h) general area cladogram obtained from the three-area statement matrix; (i) general area cladogram obtained from the Brooks parsimony analysis matrix.

(continued)

CASE STUDY 5.7 Biogeography of the Northern Andes *(continued)*

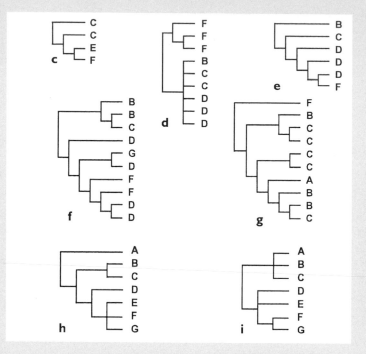

According to both general area cladograms, the Northern Andes do not constitute a natural unit because areas A–D do not constitute a monophyletic group. Furthermore, the Amotape–Huancabamba area (D) is related to the Central Andes. Berry (1982) considered the Amotape–Huancabamba area as a transitional zone; however, in the majority of the taxa analyzed by Morrone and Urtubey (1997), endemic species of the area are related to species from clades in the Central Andes, whereas only species of *Andesianellus* belong to clades of the Northern Andes.

References

Berry, P. E. 1982. The systematics and evolution of *Fuchsia* sect. *Fuchsia* (Onagraceae). *Annals of the Missouri Botanical Garden* 69:1–198.

Farris, J. S. 1988. *Hennig86 reference*. Version 1.5. Port Jefferson, N.Y.: Author.

Morrone, J. J. 1994. Distributional patterns of species of Rhytirrhinini (Coleoptera: Curculionidae) and the historical relationships of the Andean provinces. *Global Ecology and Biogeography Letters* 4:188–194.

Morrone, J. J. 1996. The biogeographical Andean subregion: A proposal exemplified by Arthropod taxa (Arachnida, Crustacea, and Hexapoda). *Neotropica* 42:103–114.

Morrone, J. J. and E. Urtubey. 1997. Historical biogeography of the northern Andes: A cladistic analysis based on five genera of Rhytirrhinini (Coleoptera: Curculionidae) and *Barnadesia* (Asteraceae). *Biogeographica* 73:115–121.

Nelson, G. and P. Y. Ladiges. 1995. *TASS*. New York: Authors.

CASE STUDY 5.8 Biogeography of *Rhododendron* Section *Vireya* in the Malesian Archipelago

The Malesian Archipelago has been of interest to biogeographers since the nineteenth century, when Wallace recognized a biotic discontinuity between the Indo-Malayan and Australian vertebrate faunas. What became known as Wallace's line separates the islands of Bali and Lombock from Borneo and Sulawesi and runs to the southeast of the Philippines. Subsequent to Wallace, other authors proposed other lines to separate the Australian and Oriental biotas. The region where these lines are located is known as Wallacea. Brown et al. (2006) analyzed the evolutionary history of section *Vireya* of the plant genus *Rhododendron* from the Malesian Archipelago in order to elucidate the historical relationships of the areas of endemism inhabited by its species.

The authors analyzed a molecular phylogeny of *Rhododendron* section *Vireya* based on parsimony and Bayesian analyses of cpDNA sequences. Twenty areas of endemism throughout the Malesian Archipelago and neighboring regions (fig. 5.20a) were identified on the basis of geological information, previous biogeographic studies of the region, and distributional maps of the species analyzed. Brown et al. (2006) identified paralogy-free subtrees in the taxon–area cladogram and coded them as characters for parsimony analysis with Hennig86 (Farris 1988), with missing areas coded as question

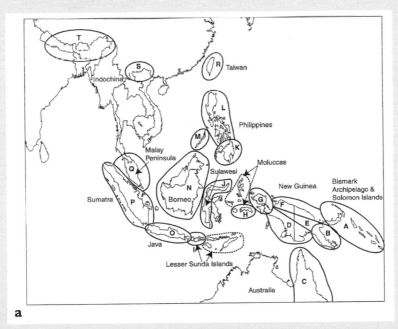

a

Figure 5.20   Biogeographic analysis of Malesia and surrounding areas by Brown et al. (2006). (a) Areas of endemism; (b) general area cladogram. A, Bismarck Archipelago and Solomon islands; B, Papuan Peninsula; C, northeastern Australia; D, New Guinea craton; E, central New Guinea; F, northern New Guinea; G, Vogelkop Peninsula; H, South Moluccas; I, Lesser Sunda Islands; J, north and west Sulawesi; K, southern Philippines; L, northern Philippines; M, Palawan; N, Borneo; O, Java and Bali; P, Sumatra; Q, Malay Peninsula; R, Taiwan; S, north Vietnam and south China; T, Himalayas.   *(continued)*

CASE STUDY 5.8 Biogeography of *Rhododendron* Section *Vireya* in the Malesian Archipelago (*continued*)

marks. Rather than compute the consensus cladogram, the authors presented the minimal tree (the shortest tree with the least resolution).

Paralogy-free subtree analysis of the taxon–area cladogram produced thirteen paralogy-free subtrees. Components from these subtrees were coded as characters in a data matrix, which produced an overflow of trees (7,108+) of twenty-three steps, a consistency index of 1.0, and a retention index of 1.0. The general area cladogram (fig. 5.20b) had a basal polytomy, including the Himalayas (T) and three clades: one related Taiwan (R) to north Vietnam and south China; another related the eastern areas, including the Bismarck Archipelago and the Solomon Islands (A), northeastern Australia (C), and New Guinea areas (B, D, E, F, and G); and the other related the western and middle Malesian areas of southern Moluccas, Lesser Sunda Islands, north and west Sulawesi, southern and northern Philippines, Palawan, Borneo, Java and Bali, Sumatra, and the Malay Peninsula (H to Q).

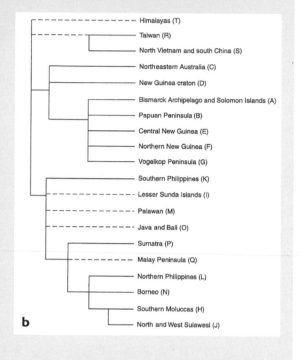

The analysis revealed a distinct biogeographic pattern, with one major clade restricted to the east of Wallace's line and another to the west. In order to interpret it, Brown et al. (2006) offered two alternative hypotheses. Hypothesis 1 considers *Vireya* as an old group, with ancestors present in Gondwana before India rifted north in the Cretaceous. As the islands of Malesia formed and moved to their present positions, species of *Vireya* dispersed farther into Malesia. Similar scenarios have been suggested for the establishment of other angiosperms in tropical Asia–Australasia. Hypothesis 2 considers *Vireya* as a young group, which dispersed eastward from India to Australia and the Solomon Islands because the islands of Malesia were in, or close to, their present-day positions. Two main radiation–dispersal events into Malesia are likely to have occurred, representing the main two lineages resolved in the phylogeny. There was only one eastward radiation into New Guinea, Australia, and the Solomon Islands, followed by mass speciation leading to the present-day diversification, possibly related to orogenic events in New Guinea. A similar hypothesis has been put forward for genera of Dipterocarpaceae. On the basis of the presence of taxa in relictual rain forests that include ancient angiosperms and fossil minimal ages, the authors preferred hypothesis 1. It may be that deep divergences within the taxa analyzed have an old history, but diversification within clade is more recent.

> **CASE STUDY 5.8 Biogeography of *Rhododendron* Section *Vireya* in the Malesian Archipelago (*continued*)**
>
> *References*
>
> Brown, G. K., G. Nelson, and P. Y. Ladiges. 2006. Historical biogeography of *Rhododendron* section *Vireya* and the Malesian Archipelago. *Journal of Biogeography* 33:1929–1944.
> Farris, J. S. 1988. *Hennig86 reference*. Version 1.5. Port Jefferson, N.Y.: Author.

Dispersal–Vicariance Analysis

Proposed by Ronquist (1997a), based on the ideas developed by Ronquist and Nylin (1990) to analyze host–parasite relationships, this method reconstructs ancestral distributions from one given phylogenetic hypothesis, without assuming a particular process a priori, taking into account vicariance, dispersal, and extinction (Miranda Esquivel et al. 2003). It reconstructs the biogeographic history of individual taxa, but it can also be used to find the general relationships of an area, especially when these relationships do not conform to a hierarchical pattern, that is, when there are reticulate relationships due to biogeographic convergence.

The biogeographic reconstruction is based on a cost matrix, which is constructed according to the following premises (Crisci et al. 2000):

- Vicariance events have a null cost of 0. Speciation is assumed to be by vicariance.
- Duplication events receive a null cost of 0. This implies sympatric speciation.
- Dispersal events receive a cost of 1 per area unit added to a distribution.
- Extinction events receive a cost of 1 per area unit deleted from a distribution.

Ronquist (1997b) developed a modification of this method, called constrained dispersal–vicariance analysis (DIVA), which distinguishes between random dispersals (those that imply that the taxon passes through a barrier) and predictable dispersals (those that occur when a barrier disappears). In constrained DIVA, the cost matrix is constructed according to the following rules:

- Vicariance events receive a benefit value of –1.
- Duplication events receive a null cost of 0.
- Extinction events receive a cost of 1.
- Random dispersal events receive a cost of 1.
- Predictable dispersal events receive a benefit value of –1.

Ronquist (1998) maintained that dispersal–vicariance models are oversimplified. However, he predicted that the three-dimensional cost matrix framework is powerful enough to allow optimization of any conceivable biogeographic model regardless of its complexity.

Algorithm The algorithm consists of the following steps (Biswas and Pawar 2006; Ronquist 1997a):

1. Obtain the taxonomic cladograms of the taxa distributed in the areas analyzed.
2. Replace the terminal taxa from the taxonomic cladograms with the areas inhabited by them, to obtain taxon–area cladograms.
3. Construct a cost matrix assigning values to vicariance, duplication, dispersal, and extinction events.
4. Obtain the general area cladogram that optimizes these values.

Software DIVA version 1.2 (Ronquist 1996), which generates ancestral reconstructions for each node and gives a series of statistics indicating the frequencies of the different events.

Empirical Applications Bessega et al. (2006), Biondi (1998), Biswas and Pawar (2006), Braby and Pierce (2007), Brooks and McLennan (2001), Dávalos (2006), Donato (2006), Donato et al. (2003), Donoghue et al. (2001), Jansa et al. (2006), Matthee et al. (2004), Miranda-Esquivel (2001), Moore et al. (2006), Perret et al. (2006), Posadas and Morrone (2003), Roig-Juñent (2002, 2004), Ronquist (1997a), Sanmartín (2003), Sanmartín et al. (2001), Sereno et al. (1998), Voelker (1999), Xiang and Soltis (2001), Yuan et al. (2005), and Zink et al. (2000).

CASE STUDY 5.9 Historical Biogeography of the Subantarctic Subregion

The Subantarctic subregion includes the Austral Andes, from 37° south latitude to Cape Horn, also including the archipelago of southern Chile and Argentina and the Falklands, South Georgia, and Juan Fernandez Islands (Morrone 2001a). Several authors have indicated the relationships of its biota with the austral continents; within the Andean region, the Subantarctic subregion is more closely related to the Central Chilean subregion (Marino et al. 2001; Morrone 2001a; Morrone et al. 1997). Morrone (1994a) undertook a panbiogeographic analysis based on species of Rhytirrhinini (Coleoptera: Curculionidae), finding two generalized tracks that share their first portion in the Subantarctic and Central Chilean subregions. He postulated that the austral biota originally was restricted to the southern portion of the Andean region and later it dispersed, occupying the rest of the Andean region. Posadas and Morrone (2003) carried out a cladistic biogeographic analysis of the Subantarctic subregion based on taxa of the family Curculionidae (Coleoptera).

Areas of endemism analyzed (fig. 5.21a) were the Maule, Valdivian Forest, Magellanic Forest, Magellanic Moorland, and Falkland Islands provinces, as well as the Central Chilean subregion. The taxa analyzed were six weevil genera (*Alastoropolus*, *Aegorhinus*, *Puranius*, *Germainiellus*, *Antarctobius*, and *Rhyephenes*) and the *Falklandius* generic group, with a total of sixty-nine species. Methods applied were tree reconciliation analysis, with Component version 2.0 (Page 1993a); BPA, with Hennig86 version 1.5 (Farris 1988); and dispersal–vicariance analysis, with DIVA version 1.0 (Ronquist 1996).

(continued)

CASE STUDY 5.9 Historical Biogeography of the Subantarctic Subregion (*continued*)

The tree reconciliation analysis, minimizing duplications, losses, and both simultaneously, gave a single general area cladogram (fig. 5.21b), which shows the Central Chilean subregion as the sister area to the two northern Subantarctic provinces (Maule and Valdivian Forest). This group is the sister group to the three remaining provinces (Falkland Islands, Magellanic Forest, and Magellanic Moorland). The separation sequence of these three latter shows the Falkland Islands as the sister area to the Magellanic Forest–Magellanic Moorland pair. BPA produced a single cladogram (135 steps, consistency index of 0.74, and retention index of 0.55), with the same topology. The dispersal–vicariance analysis showed that the most frequent dispersal events involved the following areas: Maule–Valdivian Forest (21.40%), Valdivian Forest–Maule (12.06%), Maule–Central Chilean (18.63%), and Central Chilean–Maule (9.00%). The most frequent vicariance events involved the separation of the Falkland Islands from the Magellanic Forest–Magellanic Moorland pair.

Posadas and Morrone (2003) concluded that the complex relationships between the north of the Subantarctic subregion and the Central Chilean subregion could be due to dispersal because this process includes about 61% of the dispersal events. Furthermore, the most frequent dispersal events always involved the Maule province in relation to the Central Chilean subregion or the Valdivian Forest, implying the capacity of its biota

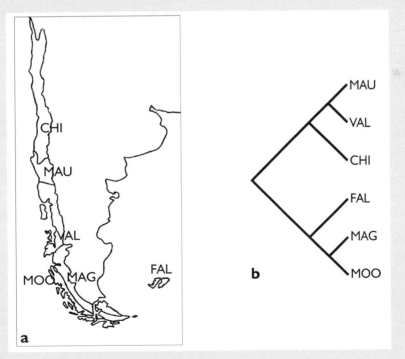

Figure 5.21   Biogeographic analysis of the Subantarctic and Central Chilean subregions by Posadas and Morrone (2003). (a) Areas of endemism; (b) general area cladogram. CHI, Central Chilean subregion; FAL, Falklands; MAG, Magellanic Forest; MAU, Maule; MOO, Magellanic Moorland; VAL, Valdivian Forest.   (*continued*)

CASE STUDY 5.9 Historical Biogeography of the Subantarctic Subregion *(continued)*

to colonize other areas. On one hand, the apparent relationship between the Valdivian Forest and the Maule with the Central Chilean subregion in the general area clado-grams obtained may be a distortion because these methods are unable to discriminate the dispersal phenomena that could generate reticulate area relationships. On the other hand, the relationship between the three southern provinces of the Subantarctic sub-region may be caused by vicariance events, which are supported by the application of DIVA, which showed low dispersal frequencies between these areas.

A PAE based on flies of the family Ceratopogonidae (Marino et al. 2001) postulated that the provinces of the Subantarctic subregion constitute a monophyletic unit, which is the sister area to the Central Chilean subregion. This difference may be due to the influence of ecological factors because the taxa analyzed were aquatic. Two cladistic biogeographic analyses (Amorim and Pires 1996; Soares and de Carvalho 2005) found that the Central Chilean subregion was paraphyletic in relation to the Subantarctic sub-region. The different patterns found can be linked to the complex geological history of the area, with events of marine introgressions between 15 and 11 mya that changed the current conformation of southern and central Chile (Soares and de Carvalho 2005). Additionally, southern Chile was covered with ice during the last glacial age, impos-ing limitations on biotic expansions, and some distributional patterns may be due to vicariance and others to geodispersal after the retreat of the ice sheets and the reestab-lishment of forests.

References

Amorim, D. de S. and M. R. S. Pires. 1996. Neotropical biogeography and a method for maximum biodiversity estimation. In *Biodiversity in Brazil: A first approach*, ed. C. E. M. Bicudo and N. A. Menezes, 183–219. São Paulo: CNPq.

Farris, J. S. 1988. *Hennig86 reference*. Version 1.5. Port Jefferson, N.Y.: Author.

Marino, P. I., G. R. Spinelli, and P. Posadas. 2001. Distributional patterns of species of Ceratopogonidae (Diptera) in southern South America. *Biogeographica* 77:113–122.

Morrone, J. J. 1994. Distributional patterns of species of Rhytirrhinini (Coleoptera: Curculionidae) and the historical relationships of the Andean provinces. *Global Ecology and Biogeography Letters* 4:188–194.

Morrone, J. J. 2001. *Biogeografía de América Latina y el Caribe*. Saragossa, Spain: Manu-ales y Tesis SEA, no. 3.

Morrone, J. J., L. Katinas, and J. V. Crisci. 1997. A cladistic biogeographic analysis of Central Chile. *Journal of Comparative Biology* 2:25–42.

Page, R. D. M. 1993. *Component user's manual*. Release 2.0. London: The Natural His-tory Museum.

Posadas, P. and J. J. Morrone. 2003. Biogeografía histórica de la familia Curculioni-dae (Coleoptera) en las subregiones Subantártica y Chilena Central. *Revista de la Sociedad Entomológica Argentina* 62(1–2):75–84.

Ronquist, F. 1996. *DIVA, version 1.0: Computer program for MacOS and Win32*. Retrieved May 25, 2008, from http://www.ebc.uu.se/systzoo/research/diva/manual/dmanual.html.

(continued)

> **CASE STUDY 5.9 Historical Biogeography of the Subantarctic Subregion (*continued*)**
>
> Soares, E. D. G. and C. J. B. de Carvalho. 2005. Biogeography of *Palpibrachus* (Diptera: Muscidae): An integrative study using panbiogeography, parsimony analysis of endemicity, and component analysis. In *Regionalización biogeográfica en Iberoamérica y tópicos Afines: Primeras Jornadas Biogeográficas de la Red Iberoamericana de Biogeografía y Entomología Sistemática (RIBES XII.I–CYTED)*, ed. J. Llorente Bousquets and J. J. Morrone, 485–494. Mexico, D.F.: Las Prensas de Ciencias, UNAM.

Area Cladistics

Area cladistics (Ebach 2003; Ebach and Edgecombe 2001; Ebach and Humphries 2002) is derived from component and three area statement analyses, although it uses a different approach for resolving problems of the taxon–area cladograms. Area cladistics begins by replacing the names of the terminal taxa of two or more taxon–area cladograms and deriving areagrams and then resolves paralogy using the transparent method (Ebach et al. 2005a). Areagrams of different taxa inhabiting the same areas are combined, and patterns are searched (fig. 5.22). When geographic congruence between the different areagrams is found, we may interpret the biotic history. Biotic congruence is evidence of vicariance, and we can infer that biotic components inhabiting sister areas were once part of the same biotic component, which posteriorly diverged.

Ebach and Humphries (2002) summarized the underlying rationale of area cladistics with six axioms:

- An endemic area is the known distribution of a taxon.
- Congruence is a result of allopatric speciation, the direct or indirect consequence of continental drift.
- The relationship of areas equates to the relative geographic proximity of areas.

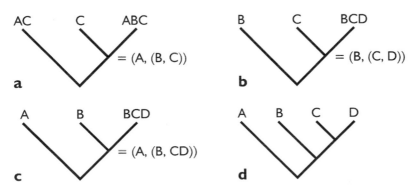

Figure 5.22   Area cladistics. (a–c) Three taxon–area cladograms with paralogous nodes, each showing the application of the transparent method; (d) general area cladogram obtained.

- Area cladistic data may not necessarily be consistent with other forms of data.
- Sampling of different data sets will yield consistent results.
- Paleomagnetic and area cladistic data provide independent evidence of continental drift.

Williams and Ebach (2004) categorized cladistic biogeographic methods as either transformational or taxic. The latter, which includes area cladistics, is no more than proximal relationships of two areas in relation to a third (area homology). A combination of area homologies forms a general areagram or general area cladogram, which represents geographic homology consisting of area clades that share a common history. Williams and Ebach (2004) argued that the alternatives (e.g., identification of dispersal routes, centers of origin, or even vicariance events) are themselves artifacts of the transformational perspective.

Algorithm The algorithm consists of the following steps (Ebach 2003; Ebach and Humphries 2002):

1. Obtain the taxonomic cladograms of the taxa distributed in the areas analyzed.
2. Replace the terminal taxa from the taxonomic cladograms with the areas inhabited by them to obtain taxon–area cladograms.
3. Resolve widespread taxa with the transparent method.
4. Represent the nodes of the taxon–area cladograms in either a component or a three-item matrix.
5. Analyze the data matrix with a parsimony or compatibility algorithm to obtain the general area cladogram.

Software 3item (Ebach et al. 2005b) and Nelson05 (Cao and Ducasse 2005).

Empirical Applications Ebach and Edgecombe (2001), Escalante et al. (2007a), and Humphries and Ebach (2004).

CASE STUDY 5.10 Cladistic Biogeography of the Hawaiian Islands

The Hawaiian Archipelago, comprising the islands of Kaua'i, O'ahu, Moloka'i, Lana'i, East and West Maui, Kaho'olawe, and Hawai'i (fig. 5.23a), is situated in the North Pacific Ocean, between latitudes of 19° and 29° N, about 4,000 km from the nearest continent. Its biota contains a high percentage of endemic species. Humphries and Ebach (2004) undertook an area cladistics analysis of the Hawaiian Islands.

Humphries and Ebach (2004) analyzed fourteen plant and animal taxon–area cladograms, taken from Funk and Wagner (1995), with software 3item (Ebach et al. 2005). Additionally, they used TASS (Nelson and Ladiges 1996) to obtain a data matrix of paralogy-free subtrees under assumption 2 and NONA (Goloboff 1998) to analyze the matrix.

(continued)

CASE STUDY 5.10 Cladistic Biogeography of the Hawaiian Islands
(continued)

Area cladistics with 3item and TASS/NONA showed a basally resolved cladogram but with a terminal polytomy for Moloka'i, Lana'i, East and West Maui, and Hawai'i (fig. 5.23b). When Maui was treated as a single area, TASS/NONA found a single area cladogram (fig. 5.23c), which is completely dichotomous, showing Lana'i and Maui as sister areas. This relationship is consistent with Carlquist's (1995) reconstruction, where the Maui Nui island complex gave rise to Lana'i and Moloka'i during the time Hawaii was formed. The divergence of Lana'i and Maui suggested that the Hawai'i biota was already established on Maui Nui 1 mya and that the current Maui and Lana'i biotas developed more recently (0.5 mya to present). The analysis with 3item treating Maui as a single area produced a single areagram (fig. 5.23d), where Kaua'i is the sister area to Moloka'i and Hawai'i is the sister area to Lana'i and Maui. This second analysis suggested a more recent history (0.5 mya to present) because both main clades

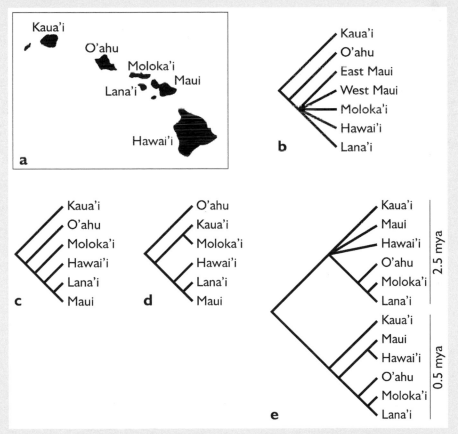

Figure 5.23 Biogeographic analysis of the Hawaiian islands by Humphries and Ebach (2004). (a) Map of the islands; (b) general area cladogram obtained with 3item and TASS/NONA; (c) general area cladogram obtained with TASS/NONA treating Maui as a single area; (d) general area cladogram obtained with 3item treating Maui as a single area; (e) general area cladogram including a time slice.

(continued)

CASE STUDY 5.10 Cladistic Biogeography of the Hawaiian Islands (*continued*)

match the present current proximity of the Kaua'i, O'ahu, and Moloka'i biota and of the Hawai'i, Maui, and Lana'i biota. Humphries and Ebach (2004) concluded that a time slice should be implemented for the areas. They presented a general area-gram (fig. 5.23e) incorporating a time slice. This general areagram suggested that two different biotic divergences took place during the formation of the Hawaiian island chain.

Liebherr (1997) wrote that although many Hawaiian biogeographic patterns are due to overwater dispersal, the subsidence of the Maui Nui Island affords the search for vicariance patterns among the resulting islands of Moloka'i, Lana'i, and Maui. Further-more, there is geological evidence of former islands that existed in the vicinity of the Hawaiian Islands since the Late Paleocene (Craw et al. 1999). Thus, some of the Hawai-ian cenocrons may be derived from even earlier islands (Funk and Wagner 1995). Addi-tional evidence, from molecular clock estimates that have indicated 40 to 23 mya for the divergence of Hawaiian Drosophilidae from their North American relatives (Powell and DeSalle 1995), seems to support the idea that the Hawaiian biota is much older than the oldest present-day islands. According to Rotondo et al. (1981), the high degree of endemism recorded on some of the Hawaiian Islands cannot be explained by jump dispersal. On the basis of the reconstructions that postulate the integration into the Hawaiian Island chain of some islands that originated outside the present Hawaiian hotspot, these authors proposed the mixing of two different cenocrons, Hawaiian and non-Hawaiian, to explain the high endemism.

References

Carlquist, S. 1995. Introduction. In *Hawaiian biogeography: Evolution on a hotspot archi-pelago,* ed. W. L. Wagner and V. A. Funk, 1–13. Washington, D.C.: Smithsonian Institution Press.

Ebach, M. C., R. A. Newman, C. J. Humphries, and D. M. Williams. 2005. *3item version 2.0: Three-item analysis for cladistics and area cladistics.* Oxford: Author.

Funk, V. A. and W. L. Wagner. 1995. Biogeographic patterns in the Hawaiian islands. In *Hawaiian biogeography: Evolution on a hotspot archipelago,* ed. W. L. Wagner and V. A. Funk, 379–419. Washington, D.C.: Smithsonian Institution Press.

Goloboff, P. 1998. *NONA.* Version 2.0. Retrieved May 25, 2008, from http://www.cladistics.com/about_nona.htm.

Humphries, C. J. and M. C. Ebach. 2004. Biogeography on a dynamic Earth. In *Fron-tiers of biogeography: New directions in the geography of nature,* ed. M. V. Lomolino and L. R. Heaney, 67–86. Sunderland, Mass.: Sinauer.

Liebherr, J. K. 1997. Dispersal and vicariance in Hawaiian Platynine carabid beetles (Coleoptera). *Pacific Science* 51:424–439.

Nelson, G. and P. Y. Ladiges. 1996. Paralogy in cladistic biogeography and analysis of paralogy-free subtrees. *American Museum Novitates* 3167:1–58.

Rotondo, G. M., V. G. Springer, G. A. J. Scott, and S. O. Schlanger. 1981. Plate move-ment and island integration: A possible mechanism in the formation of endemic biotas, with special reference to the Hawaiian islands. *Systematic Zoology* 30: 12–21.

Phylogenetic Analysis for Comparing Trees

Wojcicki and Brooks (2005) described phylogenetic analysis for comparing trees (PACT), which is aimed at generating general area cladograms that provide accurate representations of the information contained in taxon–area cladograms. PACT operates with Venn diagrams obtained from the taxon–area cladograms, which are compared by looking for common elements. According to Wojcicki and Brooks (2005), this algorithm is most useful as an a posteriori method, but it is also superior to all previous a priori methods because it does not specify costs, weights, or probabilities of any biogeographic process. Brooks (2005) wrote that PACT works implicitly under the taxon pulse model (Erwin 1979, 1981).

Algorithm The algorithm consists of the following steps (Brooks 2005; Wojcicki and Brooks 2004, 2005):

1. Obtain the taxonomic cladograms of the taxa distributed in the areas analyzed.
2. Replace the terminal taxa from the taxonomic cladograms with the areas inhabited by them to obtain taxon–area cladograms.
3. Convert the taxon area cladograms into Venn diagrams and represent them in parenthetic format.
4. Choose any taxon–area cladogram and determine its elements (components and terminal taxa). It will be the template area cladogram.
5. Select a second taxon–area cladogram and determine its elements as in the template area cladogram, and document which elements occur in the latter (denoted by Y) and which do not (denoted by N). Each Y represents a match with a previous pattern and is combined, whereas each N represents a new element and is attached to the template cladogram at the component where it is linked with a Y. This requires two rules: (1) $Y + Y = Y$ (combine common elements), as long as they are connected at the same component; and (2) $Y + N = YN$ (add new elements to the template area cladogram at the component where they first appear).
6. When the new elements from the second taxon–area cladogram have been added to the template area cladogram, see whether any of them can be further combined. This requires three rules: (1) $Y(Y-) = Y(Y-)$ (do not combine Ys if they are attached at different components on the template area cladogram); (2) $Y + YN = YN$ (Y is part of YN); and (3) $YN + YN = YNN$ (Y is the same for each, but each N is different).
7. Repeat steps 5 and 6 to incorporate all available taxon–area cladograms in order to obtain the general area cladogram(s).

Software Software for PACT (Wojcicki and Brooks 2004, 2005) is under development.

Empirical Applications Folinsbee and Brooks (2007).

CASE STUDY 5.11 Dispersal of Hominines in the Old World

The family Hominidae (Primates), comprising apes and humans, exhibits a complex history of dispersal and speciation events over large areas of the Old World in the last 25 million years. Folinsbee and Brooks (2007) integrated phylogenetic data of hominines, hyaenids (hyenas), and proboscideans (elephants) with paleoclimatological, paleoenvironmental, and geological evidence in order to provide a general hypothesis explaining their biogeographic history.

Folinsbee and Brooks (2007) analyzed the cladograms of the superfamily Hominoidea, the family Hyaenidae, and the order Proboscidea. These taxa share broadly similar habitat preferences and are frequently found at the same fossil localities, indicating that they might have shared biogeographic distribution in the past. The authors used PACT software (Wojcicki and Brooks 2004, 2005) to obtain a general area cladogram.

Analysis of the three area cladograms (figs. 5.24a–5.24c) with PACT produced a single general area cladogram (fig. 5.24d), which is rather complex, with many area reticulations and widespread taxa, especially among the proboscideans. When first appearance dates in the fossil record are mapped onto the general area cladogram to provide minimum ages for each node, usually the oldest taxa appear at the base and the youngest at the apical end; however, some of the dates do not correlate at all. Of the nodes in the general area cladogram, seventeen are associated with concurrent events in all three clades, nine are associated with events in two clades, and eight are associated with an event involving a single clade. Some of the twenty-six nodes involving at least two of the three clades, designated by the authors as general nodes, are vicariance nodes; others are biotic expansion nodes and widespread taxa. Hominoids, proboscideans, and hyaenids may have been associated in Africa at least as early as the early Miocene, followed by an "out of Africa" expansion into Europe, Asia, and North America; a second general episode of species formation in Asia in the mid-Miocene;

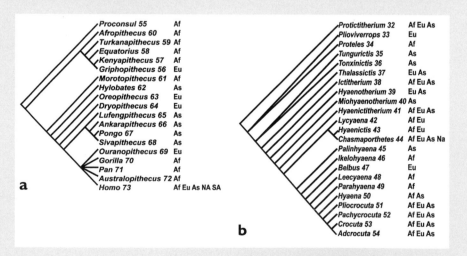

Figure 5.24   Biogeographic analysis of the Hominoidea, Hyaenidae, and Proboscidea by Folinsbee and Brooks (2007). (a) Taxon–area cladogram of the superfamily Hominoidea; (b) taxon–area cladogram of the family Hyaenidae; (c) taxon–area cladogram of the order Proboscidea; (d) general area cladogram obtained with PACT. Af, Africa; As, Asia; Eu, Europe; NA, North America; SA, South America.

(continued)

CASE STUDY 5.11 Dispersal of Hominines in the Old World *(continued)*

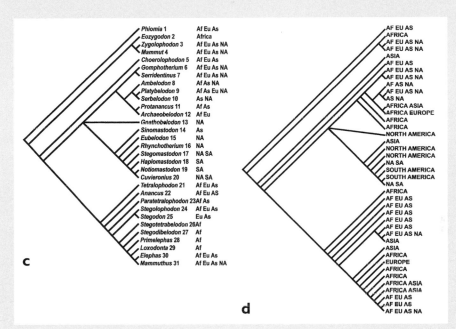

and another "out of Asia" expansion into Africa, Europe, and North America. Finally, there were two additional "out of Africa" events in the Late Miocene and Pliocene, the last one setting the stage for the emergence and spread of the genus *Homo*. In addition to the common episodes of vicariance and dispersal, each group exhibits clade-specific within-area and peripatric speciation events.

Folinsbee and Brooks's (2007) analysis situated hominoid diversification within the complex history of biotic diversification and extinction of large land vertebrates in the Miocene. The dispersal of *Homo* out of Africa in the Pliocene is not a unique event in human evolution but another episode in the taxon pulse diversification of Old World biotas.

Andrews (2007) analyzed paleoecological evidence, concluding that Folinsbee and Brooks's (2007) hypothesis was unsupported. He stated that the early expansion of *Homo erectus* into Asia by the beginning of the Pleistocene could have been possible only for a large, carnivorous species. Additionally, in the Late Miocene, when Folinsbee and Brooks (2007) suggested ancestral hominines reentered Africa along with hyaenids and Proboscidea, fossil apes were small, partly arboreal frugivores or herbivores that lacked the dispersal means enjoyed by the other two groups that enabled them to achieve intercontinental geographic ranges.

References

Andrews, P. 2007. The biogeography of hominid evolution. *Journal of Biogeography* 34:381–382.

Folinsbee, K. E. and D. R. Brooks. 2007. Miocene hominoid biogeography: Pulses of dispersal and differentiation. *Journal of Biogeography* 34:383–397.

(continued)

> **CASE STUDY 5.11 Dispersal of Hominines in the Old World *(continued)***
>
> Wojcicki, M. and D. R. Brooks. 2004. Escaping the matrix: A new algorithm for phylo-
>   genetic comparative studies of coevolution. *Cladistics* 20:341–361.
> Wojcicki, M. and D. R. Brooks. 2005. PACT: An efficient and powerful algorithm for
>   generating area cladograms. *Journal of Biogeography* 32:755–774.

Evaluation and Classification of the Methods

Morrone and Carpenter (1994) compared empirically four cladistic biogeographic
methods: component analysis, BPA, three area statement analysis, and tree reconciliation analysis. They analyzed ten different data sets from the literature, calculating the items of error necessary to reconcile the original taxon–area cladograms
with the different general area cladograms obtained. They concluded that none of
the methods was consistently better than the others, producing results that might
be considered the least ambiguous; the main sources of ambiguity were dispersal
and the existence of duplicated lineages combined with extinction. Apparently,
the different methods are affected differentially by them; for example, BPA is more
affected by dispersal, whereas component analysis is more affected by redundant
distributions (Morrone and Crisci 1995). Another comparative analysis (Biondi
1998) basically agreed with these conclusions. After these contributions, several
new methods have been described, so we lack a complete comparative study.

Van Veller et al. (2000) and van Veller and Brooks (2001) compared component
compatibility, BPA, component analysis, tree reconciliation analysis, and three area
statement analysis to evaluate the implementation of assumptions 0, 1, and 2 in
agreement with both requirements previously indicated. In relation to the requirement that the solved sets of cladograms of areas are successively inclusive, van Veller et al. (2000) detected violations by component analysis, tree reconciliation analysis, and three area statement analysis when there are redundant distributions or a
combination of widespread taxa and redundant distributions. Van Veller and Brooks
(2001) concluded that when vicariance and extinction are the most probable explanations, BPA, component analysis, and tree reconciliation analysis would give the same
general area cladogram, but when dispersal is the most reasonable explanation, secondary BPA represents each dispersal event as a falsification of the null hypothesis,
whereas component analysis and tree reconciliation analysis remove a priori data, or
duplicate lineages, and postulate extinctions a posteriori to avoid falsification.

Some authors (Crisci 2001; Crisci et al. 2000; Ronquist and Nylin 1990;
Sanmartín and Ronquist 2002) have considered the existence of two types of cladistic biogeographic methods:

- Pattern-based methods: those that search for general patterns of relationships
  between areas, without initial assumptions about particular biogeographic
  processes. These methods would belong to cladistic biogeography in the strict
  sense (e.g., Crisci 2001; Crisci et al. 2000). They include component analysis,
  BPA, component compatibility, and paralogy-free subtree analysis.

- Event-based methods: those that derive explicit models of particular bio-geographic processes. These methods are excluded from cladistic biogeography by Crisci et al. (2000) and Crisci (2001). They include tree reconciliation analysis, vicariance events analysis, and dispersal–vicariance analysis.

Van Veller et al. (2000; see also Biswas and Pawar 2006; Brooks and van Veller 2003; Ebach 2001; Morrone 2005a; van Veller and Brooks 2001; van Veller et al. 2003; Wojcicki and Brooks 2005; Zandee and Roos 1987) consider the existence of two groups of methods, whose purpose is to implement different research programs:

- A posteriori methods: those that deal with dispersal, extinction, and du-plicated lineages after the parsimony analysis of a data matrix based on the unmodified taxon–area cladograms. They are intended to implement "vicariance biogeography" sensu Zandee and Roos (1987) or "phylogenetic biogeography" sensu van Veller et al. (2003). (The name *phylogenetic biogeography* is inappropriate because for decades it has been the name of Hennig's [1950] and Brundin's [1966] approach.) A posteriori methods include BPA and component compatibility.
- A priori methods: those that allow modification of the area relationships in the taxon–area cladograms to deal with dispersal, extinctions, or dupli-cated lineages in order to obtain resolved area cladograms and provide the maximum fit to a general area cladogram. They are intended to implement "cladistic biogeography" sensu Zandee and Roos (1987) and van Veller et al. (2003). A priori methods include component analysis, tree reconciliation analysis, and three area statement analysis.

Ebach and Humphries (2002) stated that a posteriori methods correspond to a generation paradigm, whereas a priori methods correspond to a discovery para-digm. Generation paradigm methods are based on previous beliefs rather than facts. Additionally, these authors criticized the generation methods for immuniz-ing their results to avoid falsification, whereas discovery methods are superior because they allow the free exploration of the data. Ebach et al. (2003) also noticed both uses of the term *phylogenetic biogeography*, considering that Van Veller et al.'s (2003) use corresponded to Hennig's (1950) *parasitological method*, which is inap-propriate for implementing cladistic biogeography.

Brooks (2004) suggested that, in addition to using different methods, cladis-tic and phylogenetic biogeographies are research programs justified by different ontologies. Cladistic biogeographic methods modify the original data in order to provide maximum fit to the null hypothesis of vicariance, amounting to an a priori parsimony criterion. This reflects the ontology of simplicity. Phylogenetic bioge-ography falsifies the null hypothesis of a vicariance explanation when the data do not support it, amounting to an a posteriori parsimony criterion. Data that appear to conflict with the null hypothesis indicate that it is flawed, determining a pos-teriori the minimum number of falsifications that are necessary to explain all the data. This reflects the ontology of complexity.

Humphries and Ebach (2004) postulated that the dichotomy between cladistic and phylogenetic biogeographies has to do with the interpretation of cladograms as homology-based hierarchies or as phylogenetic trees, respectively. This interpretation can be extended to other approaches that also treat cladograms as phylogenetic trees: ancestral area analysis and intraspecific phylogeography. In this respect, it would be pertinent to refer to O'Hara's (1988) distinction between chronicle and history. A chronicle is a description of a series of events, arranged in chronological order, and not accompanied by any causal statements or explanations, whereas a history contains statements about causal connections. Cladistic biogeography estimates a chronicle, and phylogenetic biogeography estimates a history. Another way to distinguish both groups of methods is to apply the distinction between the philosophical positions of reciprocal illumination and total evidence (Rieppel 2004). A posteriori methods work under the principle of reciprocal illumination (Hennig 1950), searching for consilience or congruence between data from different taxa. A priori methods work under the principle of total evidence (Kluge 1989), relying on the largest set of data analyzed simultaneously.

I am unconvinced of the existence of two different research programs (Ebach and Morrone 2005; Morrone 2005a). Although the distinction between the two groups of methods is valid, I suggest that they still can be considered to implement cladistic biogeography. The question of which method to apply is not easy to answer; all have their supporters and their critics. A practical approach is to apply more than one and then compare the differences between the results obtained (Contreras-Medina et al. 2007b).

For Further Reading

Brooks, D. R. 1990. Parsimony analysis in historical biogeography and coevolution: Methodological and theoretical update. *Systematic Zoology* 39:14–30.

Brooks, D. R. and D. A. McLennan. 2001. A comparison of a discovery-based and an event-based method of historical biogeography. *Journal of Biogeography* 28: 757–767.

Ebach, M. C. and C. J. Humphries. 2002. Cladistic biogeography and the art of discovery. *Journal of Biogeography* 29:427–444.

Humphries, C. J. and L. R. Parenti. 1999. *Cladistic biogeography: Interpreting patterns of plant and animal distributions*. 2nd ed. Oxford: Oxford University Press.

Nelson, G. and P. Y. Ladiges. 1996. Paralogy in cladistic biogeography and analysis of paralogy-free subtrees. *American Museum Novitates* 3167:1–58.

Page, R. D. M. 1990. Component analysis: A valiant failure? *Cladistics* 6:119–136.

Wiley, E. O. 1988. Vicariance biogeography. *Annual Review of Ecology and Systematics* 19:513–542.

Problems

Problem 5.1

On the basis of the distributions on the map (fig. 5.25a) and the four hypothetical taxonomic cladograms (figs. 5.25b–5.25e),

    a. Obtain the four taxon–area cladograms by replacing each terminal species with the areas inhabited by it.

    b. Determine which problem (widespread taxa, redundant distributions, and missing areas) applies to each taxon–area cladogram.

    c. If possible, obtain a general area cladogram reflecting their common history.

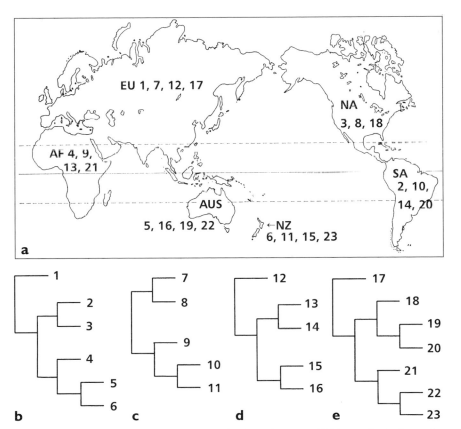

Figure 5.25   (a) Map of the world showing the distribution of 23 hypothetical species; (b–e) taxonomic cladograms of four taxa. AF, Africa; AUS, Australia; EU, Eurasia; NA, North America; NZ, New Zealand; SA, South America.

Problem 5.2

On the basis of the map of areas of endemism in the Neotropical region (fig. 5.26a) and the taxon–area cladograms of eight bird genera and species groups (figs. 5.26b–5.26i; Cracraft 1988),

a. Obtain the general area cladogram by applying two or more of the cladistic bio-geographic methods available.

b. Compare the results of the different analyses.

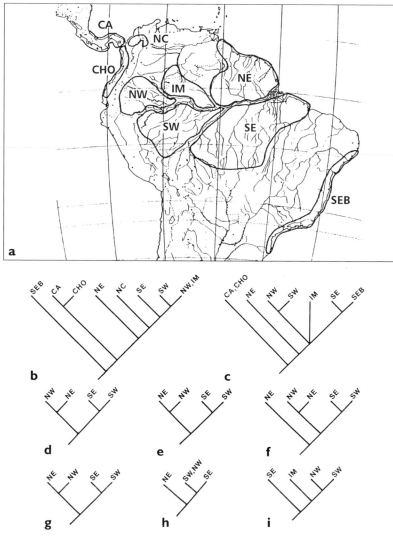

Figure 5.26 (a) Map of some areas of endemism in the Neotropical region; (b–i) taxon–area cladograms: (b) *Pionopsitta;* (c) *Selenidera;* (d) *Psophia;* (e) *Lanio;* (f) *Pipra;* (g) *Pionites;* (h) *Pteroglossus viridis* species group; (i) *P. bitorquatus* species group. CA, Central America; CHO, Chocó; IM, Imerí; NC, Nechí; NE, northeast; NW, northwest; SE, southeast; SEB, southeastern Brazil; SW, southwest.

Problem 5.3

Southern South America, basically the area south of 30° south parallel, harbors an interesting biota, and several authors have discussed its relationships with other austral areas (e.g., Australia, New Zealand, and New Caledonia). On the basis of the taxon–area cladograms of seventeen plant and animal taxa distributed in southern South America and related areas of the world (figs. 5.27a–5.27q; Crisci et al. 1991a),

    a. Obtain the general area cladogram by applying two or more of the cladistic biogeographic methods available.

    b. Compare the results of the different analyses.

For Discussion

1. Imagine you are studying the biotas inhabiting an archipelago consisting of three large islands near a continent. You find that several taxa from the two islands situated nearest the continent are similar to continental taxa, whereas other taxa inhabiting the more distant island are rather different, apparently not related to the remaining taxa. You suspect that this latter island is not geologically linked to the other two islands. How could you test this hypothesis? What alternative hypotheses can you think of? What are the relevant data that may falsify any of these hypotheses?

2. One way of evaluating a biogeographic article is to postulate alternative explanations that the authors might have overlooked. From a cladistic biogeographic viewpoint, try to think of other explanations for the data presented in the following:

    Escalante, T., G. Rodríguez, and J. J. Morrone. 2004. The diversification of the Nearctic mammals in the Mexican transition zone. *Biological Journal of the Linnean Society* 83:327–339.

    Morrone, J. J. 1998. On Udvardy's Insulantarctica province: A test from the weevils (Coleoptera: Curculionoidea). *Journal of Biogeography* 25:947–955.

    da Silva, J. M. C. and D. C. Oren. 1996. Application of parsimony analysis of endemicity in Amazonian biogeography: An example with primates. *Biological Journal of the Linnean Society* 39:427–437.

3. Carefully read the following article:

    Ebach, M. C. and C. J. Humphries. 2002. Cladistic biogeography and the art of discovery. *Journal of Biogeography* 29:427–444.

    a. Recognize and extract the main ideas.

    b. Transform each main idea in a question.

4. Carefully read the following articles:

    Brooks, D. R. and D. A. McLennan. 2001. A comparison of a discovery-based and an event-based method of historical biogeography. *Journal of Biogeography* 28:757–767.

    Brooks, D. R., M. G. P. van Veller, and D. A. McLennan. 2001. How to do BPA, really. *Journal of Biogeography* 28:345–358.

Siddall, M. E. 2005. Bracing for another decade of deception: The promise of secondary Brooks parsimony analysis. *Cladistics* 21:90–99.

Siddall, M. E. and S. L. Perkins. 2003. Brooks parsimony analysis: A valiant failure. *Cladistics* 19:554–564.

a. Establish the arguments favoring BPA and those criticizing it.

b. What are your conclusions?

CHAPTER 6

Regionalization

One of the most striking facts about the geographic distributions of taxa is that they have limits. Because these limits are repeated for different taxa, they allow the recognition of biotic components. Biotic components are nested within other larger components, so they can be ordered hierarchically in a system of realms, regions, dominions, provinces, and districts. In this chapter I discuss biogeographic regionalization and present a case study from Latin America and the Caribbean.

Biogeographic Classification

Once biotic components have been identified, they may be ordered hierarchically and used to provide a biogeographic classification. Given the historical and logical primacy of classification over process explanations (Rieppel 1991, 2004), this stage of the analysis takes place before cenocrons are elucidated and a geobiotic scenario is proposed (see chapters 7 and 8).

Viloria (2005) found a lack of uniformity and consensus in the criteria that have been used for regionalization. Analyzing Venezuela as a case study, he detected a mixture of classificatory systems based on distinct criteria, and a lack of rigor in the equivalences or synonymies between these regionalizations. He concluded that two different classificatory systems should be recognized: one based on areas of endemism and other on biotic elements. In order to have a single, standardized system of regionalization, another important thing is a unified nomenclature, paralleling the different codes used in systematics (Viloria 2005). An international group of specialists on different fossil taxa discussed the principles of classification and nomenclature of marine paleobiogeographic classification in order to reach a consensus on the nomenclature of biogeographic areas (Westermann 2000). A similar effort is taking place for neobiogeographic classification (see http://www.sebasite.org).

Realms, Regions, and Transition Zones

Biogeographic regionalization implies the recognition of successively nested areas. These areas should be natural; that is, they should correspond to biotic components. Classically, the following five categories have been used: *realm, region, dominion, province,* and *district.* Whenever more categories are necessary, the prefix *sub-* may be added to them (e.g., *subregion, subprovince*).

Sometimes it is more difficult to determine the exact boundaries of two realms or regions. For example, Müller (1979) found eighteen proposed boundaries between the Palearctic and Ethiopian regions (fig. 6.1). As a result, authors have described transition zones (Darlington 1957; Halffter 1987; Ruggiero and Ezcurra 2003), which represent events of biotic hybridization, promoted by historical and ecological changes that allowed the mixture of different biotic components. Transition zones may have depauperate biotas, but in some cases they harbor a particularly high biodiversity. From the evolutionary viewpoint, transition zones deserve special attention because they represent areas of intense biotic interaction. The analysis of transition zones by ecological biogeographers is mostly quantitative, whereas for evolutionary biogeographers it is qualitative. In panbiogeographic analyses, transition zones are indicated by the presence of nodes (Escalante et al. 2004), whereas in cladistic biogeographic analyses, putative transition zones give conflicting results because they appear to be sister areas to different areas.

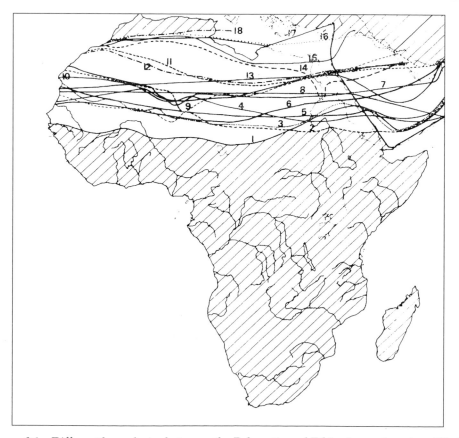

Figure 6.1   Different boundaries between the Palearctic and Ethiopian regions (modified from Müller 1979:61).

Ruggiero and Ezcurra (2003) analyzed the compatibility between historical and ecological analyses in South America. They found consistent similarities, which point to the possibility of reciprocal illumination (in the sense of Hennig 1950) between evolutionary and ecological biogeography, in order to provide a better understanding of the factors that explain the origin and maintenance of continental biotas.

Regionalization of the World

Modern biogeographic classification began with de Candolle (1820, 1838) for plants (phytogeography) and Wallace (1876) for animals (zoogeography). De Candolle (1820) recognized twenty botanical regions: northern Asia, Europe, and America; southern Europe and north of the Mediterranean; Siberia; Mediterranean area; eastern Europe to the Black and Caspian seas; India; China, Indochina, and Japan; Australia; south Africa; east Africa; tropical west Africa; Canary Islands; northern United States; northwest coast of North America; the Antilles; Mexico; tropical America; Chile; southern Brazil and Argentina; and Tierra del Fuego. De Candolle (1838) added another twenty regions, making a total of forty. They have remained the basis for some more modern treatments (e.g., Good 1974; Takhtajan 1969). Wallace's (1876) system of six zoogeographic regions, built on a previous work by Sclater (1858), is probably the most generally known biogeographic global terrestrial regionalization (Bartholomew et al. 1911; Lankester 1905). It consists of six regions: Nearctic, Neotropical, Palearctic, Ethiopian, Oriental, and Australian. These large zoogeographic regions were divided into subregions, which basically correspond to de Candolle's (1820) regions.

Alongside the development of these systems were efforts to construct ecogeographic systems (Allen 1871; Udvardy 1969). They are based on the assumption that adaptations to natural surroundings generally confine species in definite areas. Thus, ecogeographic areas or biomes can be delineated by correlating plant and animal distributions with climatic, edaphic, and other environmental factors (Cox and Moore 1998).

Craw and Page (1988) considered Wallace's system defective because it was based on a belief in the permanence of continents and oceans since the time of origin of modern life. They proposed instead a biogeographic system in which areas now widely separated by oceans and sea basins are related to one another by generalized tracks. De Candollean and Wallacean regions are not parts of this system but boundaries where different biotic components interrelate in space and time. In fact, Craw and Page (1988) concluded that the natural biogeographic regions for terrestrial and freshwater taxa are not present-day landmasses but the world's major ocean basins. Parenti (1991) reviewed the relevance of ocean basin evolution to the distribution of freshwater fishes and terrestrial organisms, modifying slightly Craw and Page's (1988) regions to include the Tethys Sea, which she considered important in understanding the composite Indo-Australian area (fig. 6.2a).

Cox (2001) examined the floral kingdoms or realms from de Candolle, Engler, and Takhtajan and the zoogeographic regions from Sclater and Wallace and analyzed their differences. He reviewed levels of endemism of Takhtajan's system, concluding that the Cape realm should be treated as a region of the African realm and that the Antarctic realm should be divided and transferred to the Neotropical and Australian realms. He also considered the names *Neotropical, Nearctic,* and *Palearctic,* used for both floral realms and faunal regions, to be cumbersome and unnecessary, replacing them with the names *South American, North American,* and *Eurasian,* respectively (fig. 6.2b). Morrone (2002a) stated that a single biogeographic scheme for all organisms, to serve as a general reference system, would be a desirable goal. Based on several panbiogeographic and cladistic biogeographic papers that have shown that some of the units recognized in traditional phytogeographic and zoogeographic systems do not represent natural units, he presented a general system of biogeographic realms and regions, intending to incorporate their conclusions. This biogeographic regionalization (fig. 6.2c) is as follows:

1. Holarctic realm: Europe, Asia north of the Himalayan mountains, northern Africa, North America (excluding southern Florida), and Greenland. From a paleogeographic viewpoint, it corresponds to the paleocontinent of Laurasia.

   1.1. Nearctic region: New World, namely Canada, most of the United States, and northern Mexico.

   1.2. Palearctic region: Old World, namely Eurasia and Africa north of the Sahara.

2. Holotropical realm: basically the tropical areas of the world, between 30° south latitude and 30° north latitude, that correspond to the eastern portion of the Gondwana paleocontinent (Crisci et al. 1993).

   2.1. Neotropical region: tropical South America, Central America, south central Mexico, the West Indies, and southern Florida.

   2.2. Afrotropical or Ethiopian region: central Africa, the Arabian peninsula, Madagascar, and the West Indian Ocean islands.

   2.3. Oriental region: India, Himalaya, Burma, Malaysia, Indonesia, the Philippines, and the Pacific islands. Despite the obvious tropical biota of this region, it has been placed in earlier paleogeographic reconstructions as part of Laurasia.

   2.4. Australian Tropical region: northwestern Australia.

3. Austral realm: southern temperate areas in South America, South Africa, Australasia, and Antarctica that correspond to the western portion of the paleocontinent of Gondwana (Crisci et al. 1993; Moreira-Muñoz 2007).

   3.1. Andean region: southern South America below 30° south latitude, extending through the Andean highlands north of this latitude, to the Puna and North Andean Paramo.

   3.2. Antarctic region: Antarctica.

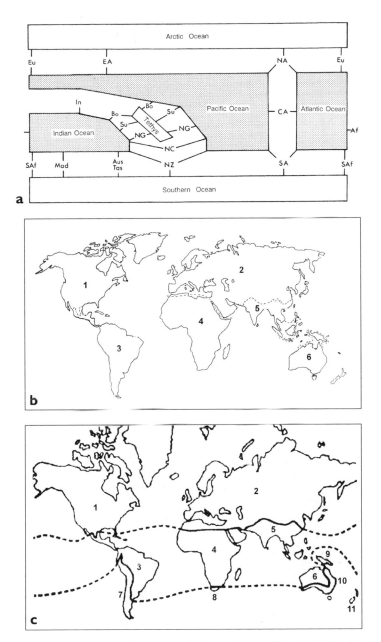

Figure 6.2 Biogeographic regionalizations of the world. (a) Parenti (1991); (b) Cox (2001); (c) Morrone (2002a). (a) Af, Africa; Aus, Australia; Bo, Borneo; CA, Central America; EA, Eurasia; Eu, Europe; In, India; Mad, Madagascar; NA, North America; NC, New Caledonia; NG, New Guinea; NZ, New Zealand; SA, South America; Saf, South Africa; Su, Sumatra; Tas, Tasmania. (b) 1, North American region; 2, Eurasian region; 3, South American region; 4, African region; 5, Oriental region; 6, Australian region. (c) 1, 2, Holarctic realm; 1, Nearctic region; 2, Palearctic region; 3–6, Holotropical realm, 3, Neotropical region; 4, Afrotropical region; 5, Oriental region; 6, Australian Tropical region; 7–11, Austral realm; 7, Andean region; 8, Cape region; 9, Neoguinean region; 10, Australian Temperate region; 11, Neozealandic region.

3.3. Cape or Afrotemperate region: South Africa.

3.4. Neoguinean region: New Guinea plus New Caledonia.

3.5. Australian Temperate region: southeastern Australia.

3.6. Neozealandic region: New Zealand.

Marine regionalization began with Ortmann (1896). In the twentieth century, important contributions included Ekman (1935), Briggs (1974, 1995), and Pierrot-Bults et al. (1986). Briggs (1995) recognized twenty-three marine biogeographic regions: Indo-West Pacific, Eastern Pacific, Western Atlantic, Eastern Atlantic, Southern Australian, Northern New Zealand, Western South America, Eastern South America, Southern Africa, Mediterranean–Atlantic, Carolina, California, Japan, Tasmanian, Southern New Zealand, Antipodean, Subantarctic, Magellan, Eastern Pacific Boreal, Western Atlantic Boreal, Eastern Atlantic Boreal, Antarctic, and Arctic.

CASE STUDY 6.1   Regionalization of Latin America

In recent years panbiogeographic and cladistic biogeographic analyses have dealt with areas in Latin America and the Caribbean. On the basis of these analyses, seventy biotic components were recognized and treated as biogeographic provinces (Morrone 2001a, 2006). In addition, they were grouped hierarchically in regions, subregions, and dominions. Also, two transition zones between the biogeographic regions were recognized (Morrone 2004c, 2005b, 2006). I describe briefly these provinces (fig. 6.3), along with the regions, subregions, dominions, and zones of transition in which they are classified.

The Nearctic region basically corresponds to the cold temperate areas of North America, in Canada, the United States (excluding southern Florida), and northern Mexico. The Latin American Nearctic provinces are found in Mexico, and with the exception of the Baja California province they all extend north into the United States. They have been assigned to the North American Pacific subregion. On the basis of parsimony analyses of endemicity (PAEs) (Katinas et al. 2004; Morrone et al. 1999; Rojas Soto et al. 2003) and panbiogeographic analyses (Contreras-Medina and Eliosa León 2001; Escalante et al. 2004), it is possible to group them into two dominions (Morrone and Márquez 2003): the California and Baja California provinces in the Californian Nearctic dominion, and the Sonora, Mexican Plateau, and Tamaulipas provinces in the Continental Nearctic dominion.

The Mexican Transition Zone basically corresponds to central Mexico. Darlington (1957) defined broadly the Central American–Mexican Transition Zone, and Halffter (1962, 1964, 1965, 1972, 1974, 1976, 1978, 1987) worked extensively in this area, defining the Mexican Transition Zone as "a complex and varied area in which the Neotropical and the Nearctic faunas overlap" (Halffter 1987:95). Morrone and Márquez (2001) undertook a panbiogeographic analysis based on beetle (Coleoptera) taxa, finding a northern generalized track, basically comprising mountain areas (Sierra Madre Occidental, Sierra Madre Oriental, Transmexican Volcanic Belt, Balsas Basin, and Sierra Madre del Sur), and a southern generalized track (Sierra Madre de Chiapas and lowland areas in Chiapas, the Mexican Gulf, and the Mexican Pacific Coast, reaching south to the Isthmus of Panama). The northern generalized track is the place with the highest latitudinal and altitudinal mixture of Nearctic and Neotropical cenocrons, with a major

(continued)

CASE STUDY 6.1 Regionalization of Latin America *(continued)*

Figure 6.3   Biogeographic provinces of Latin America and the Caribbean. 1, California; 2, Baja California; 3, Sonora; 4, Mexican Plateau; 5, Tamaulipas; 6, Sierra Madre Occidental; 7, Sierra Madre Oriental; 8, Transmexican Volcanic Belt; 9, Balsas Basin; 10, Sierra Madre del Sur; 11, Mexican Pacific Coast; 12, Mexican Gulf; 13, Yucatan Peninsula; 14, Chiapas; 15, Eastern Central America; 16, Western Panamanian Isthmus; 17, Bahama; 18, Cuba; 19, Cayman Islands; 20, Jamaica; 21, Hispaniola; 22, Puerto Rico; 23, Lesser Antilles; 24, Chocó; 25, Maracaibo; 26, Venezuelan Coast; 27, Trinidad and Tobago; 28, Magdalena; 29, Venezuelan Llanos; 30, Cauca; 31, Galápagos Islands; 32, Western Ecuador; 33, Arid Ecuador; 34, Tumbes-Piura; 35, Napo; 36, Imeri; 37, Guyana; 38, Humid Guyana; 39, Roraima; 40, Amapa; 41, Varzea; 42, Ucayali; 43, Madeira; 44, Tapajos-Xingu; 45, Para; 46, Pantanal; 47, Yungas; 48, Caatinga; 49, Cerrado; 50, Chaco; 51, Pampa; 52, Monte; 53, Brazilian Atlantic Forest; 54, Parana Forest; 55; *Araucaria angustifolia* Forest; 56, North Andean Paramo; 57, Coastal Peruvian Desert; 58, Puna; 59, Atacama; 60, Prepuna; 61, Coquimbo; 62, Santiago; 63, Juan Fernandez Islands; 64, Maule; 65, Valdivian Forest; 66, Magellanic Forest; 67, Magellanic Paramo; 68, Falkland Islands; 69, Central Patagonia; 70, Subandean Patagonia.

(continued)

CASE STUDY 6.1 Regionalization of Latin America *(continued)*

Nearctic influence at higher altitudes and a higher Neotropical influence at lower altitudes. Because of its mixed biota and its placement between the other regions, it represents the Mexican Transition Zone in the strict sense (Escalante et al. 2004; Morrone 2004c, 2005b, 2006). A study based on mammals (Ortega and Arita 1998) arrived at a similar transition zone. Submontane areas of this transition zone were connected by a generalized track based on species of Pieridae and Nymphalidae (Lepidoptera) (Ochoa et al. 2003). The Mexican Transition Zone comprises five provinces: Sierra Madre Occidental, Sierra Madre Oriental, Transmexican Volcanic Belt, Balsas Basin, and Sierra Madre del Sur.

The Neotropical region corresponds to the tropics of the New World, in most of South America, Central America, southern Mexico, the West Indies, and southern Florida. It does not include the Andean portion of South America, which is assigned to the Andean region and the South American Transition Zone. In pre-Quaternary times, the South American Neotropical biota expanded north, to Central America and Mexico, and south, to the Andean region.

The Caribbean subregion extends through southern Mexico, Central America, the Antilles, and northwestern South America. It has an extremely complex geobiotic history, which is reflected in its multiple relationships with other areas of the Neotropics, the Nearctic, and the tropics of the Old World. A South American–Caribbean generalized track (Greater Antilles, Lesser Antilles, and South America) and a North American–Caribbean generalized track (Greater Antilles and Central America) have been postulated (Rosen 1976). According to cladistic biogeographic analyses based on animal taxa (Amorim 2001; Amorim and Pires 1996), a PAE based on anurans (Ron 2000), and the phylogenetic analyses of some insect taxa (Morrone 2002c; Nihei and Carvalho 2004, 2007), the Caribbean subregion is hypothesized to be the sister area to the remaining Neotropical subregions. It comprises twenty-four provinces, which are arranged in three dominions: Mesoamerican, Antillean, and Northwestern South American. The Mesoamerican dominion, in addition to Central America and the Sierra Madre de Chiapas, extends to the lowlands of the Mexican Gulf and the coast of the Pacific Ocean; it includes five provinces: Mexican Pacific Coast, Mexican Gulf, Chiapas, Eastern Central America, and Western Panamanian Isthmus. The Antillean dominion extends to areas of the Caribbean Basin and comprises eight provinces: Yucatan Peninsula, Bahama, Cuba, Cayman Islands, Jamaica, Hispaniola, Puerto Rico, and Lesser Antilles. A cladistic biogeographic analysis based on vertebrate, crustacean, insect (Coleoptera and Trichoptera), and arachnid taxa (Crother and Guyer 1996) showed the close relationship between the Cuba, Hispaniola, and Puerto Rico provinces. The Northwestern South American dominion comprises eleven provinces: Chocó, Maracaibo, Venezuelan Coast, Trinidad and Tobago, Magdalena, Venezuelan Llanos, Cauca, Western Ecuador, Arid Ecuador, Tumbes-Piura, and Galápagos Islands.

The Amazonian is the largest subregion of the Neotropical region, extending to Brazil, the Guyanas, Venezuela, Colombia, Ecuador, Peru, Bolivia, Paraguay, and Argentina. Thirteen provinces are assigned to the Amazonian subregion: Napo, Imeri, Guyana, Humid Guyana, Roraima, Amapa, Varzea, Ucayali, Madeira, Tapajos-Xingu, Para, Pantanal, and Yungas. Several PAEs and cladistic biogeographic analyses have reconstructed the relationships between Amazonian provinces (Amorim and Pires 1996; Cracraft and Prum 1988; Hall and Harvey 2002; Patton et al. 2000; Racheli and Racheli

(continued)

CASE STUDY 6.1 Regionalization of Latin America *(continued)*

2004; Ron 2000; da Silva and Oren 1996), but there is little agreement between the alternative hypotheses. A recent cladistic biogeographic analysis (Nihei and Carvalho 2007) suggested that the Amazonian subregion may be a composite area.

The Chacoan subregion occupies northern and central Argentina, southern Bolivia, western and central Paraguay, Uruguay, and central and northeastern Brazil. Evidence of a "savanna corridor" or "diagonal of open formations" (Prado and Gibbs 1993) led to the hypothesis that the Cerrado province, formerly assigned to the Amazonian domain, connected the Caatinga with the other Chacoan provinces. The Chacoan subregion is closely related to the Amazonian and Parana subregions. The development of the Chacoan subregion during the Tertiary split the former continuous Amazonian–Parana forest, representing an example of dynamic vicariance (Zunino 2003). Some insect taxa show the Amazonian–Parana disjunction, whereas others, which probably belong to a more recent cenocron, are found in both the Chacoan and Parana subregions. A phylogeographic analysis of small mammal species (Costa 2003) showed that the central Brazilian gallery and dry forests play an important role as present and past habitats for forest species from the Amazonian and Parana subregions. Populations of Chacoan mammals have their closest relatives in the Amazonian subregion or in the Parana subregion or are basal to both subregions. Four provinces are assigned to the Chacoan subregion: Caatinga, Cerrado, Chaco, and Pampa.

The Parana subregion is situated in northeastern Argentina, eastern Paraguay, and southern and eastern Brazil. In addition to its relationships with the other Neotropical subregions, several authors have discussed its relationships with the Subantarctic subregion. Paleontological, paleoclimatological, and geological evidence indicates that a temperate climate prevailed in southern South America during the Tertiary, allowing the existence of a continuous cloud forest that extended further south than today. Cooling and aridification began in the Oligocene and Miocene, and later the forest fragmented simultaneously with the climatic changes induced by the uplift of the Andes and the expansion of the Chacoan biota (Kuschel 1969). A PAE based on species of anurans (Ron 2000) showed the close relationship between the Parana and Amazonian subregions. The Parana subregion comprises three provinces: Brazilian Atlantic Forest, Parana Forest, and *Araucaria angustifolia*.

The South American Transition Zone extends along the highlands of the Andes between western Venezuela and northern Chile and west central Argentina. The Prepuna, Coastal Peruvian Desert, and Monte provinces, previously assigned to the Neotropical region, are assigned to this transition zone on the basis of their close biotic links with the Puna and North Andean Paramo. Because of the preeminence given to some tropical cenocrons, Cabrera and Willink (1973) assigned the North Andean Paramo to the Neotropical region; its close relationships with other Paramo Punan provinces led other authors to place it in the Andean region. Six provinces are assigned to the South American Transition Zone: North Andean Paramo, Puna, Coastal Peruvian Desert, Atacama, Prepuna, and Monte.

The Andean region extends to central Chile and Patagonia. Many insect species are common to this region and the South American Transition Zone. Most of the Andean biota originally evolved in Patagonia and then gradually spread north into the South American Transition Zone during the Tertiary and Pleistocene, when tropical forests evolved to temperate and arid communities. Brundin (1966) documented

(continued)

CASE STUDY 6.1 Regionalization of Latin America *(continued)*

the relationships between the Austral continents in his phylogenetic analysis of some Chironomidae (Diptera) from New Zealand, Australia, Patagonia, and South Africa. Edmunds (1972) corroborated these connections on the basis of phylogenetic evidence from mayflies (Ephemeroptera). More recent biogeographic studies searched for congruence between distributional patterns of insects and other animal and plant taxa. Two cladistic biogeographic analyses based on plant, fungal, and animal taxa (Crisci et al. 1991b; Sanmartín and Ronquist 2004) showed that South America is a composite area because southern South America is more closely related to the southern temperate areas that correspond to the Austral realm, whereas tropical South America is more closely related to Africa and North America. Other cladistic and panbiogeographic studies (Katinas et al. 1999; Lopretto and Morrone 1998; Morrone 1996; Patterson 1981) also support the hypothesis that South America is a composite area, with the Andean region closely related to the southern temperate areas and the Neotropical region closely related to the Old World tropics. The Andean region comprises the Central Chilean, Subantarctic, and Patagonian subregions.

The Central Chilean subregion extends to central Chile, between latitudes 30° and 34° south. PAE and cladistic biogeographic analyses based on extant arthropod and plant taxa (Carvalho and Couri 2002; Marino et al. 2001; Morrone 1994a; Morrone et al. 1997; Posadas and Morrone 2003) showed that this subregion is closely related to the Subantarctic subregion. The close relationship between the Central Chilean subregion and the northern part of the Subantarctic subregion (Maule and Valdivian Forest provinces) may result from dispersal rather than vicariance, according to a cladistic biogeographic analysis based on weevil taxa (Posadas and Morrone 2003). Two provinces are assigned to the Central Chilean subregion: Coquimbo and Santiago.

The Subantarctic subregion comprises the austral Andes, from latitude 36° south to Cape Horn, including the archipelago of southern Chile and Argentina. Several authors have emphasized the distinctive character of the Subantarctic biota and its links with the biota of the Austral continents, especially Australia and New Zealand. Within the Andean region, the Subantarctic subregion is more closely related to the Central Chilean subregion, although it has some Patagonian cenocrons. In addition, the Subantarctic subregion shows relationships with the Parana subregion that indicate a former connection. Six provinces are assigned to the Subantarctic subregion: Juan Fernandez Islands, Maule, Valdivian Forest, Magellanic Forest, Magellanic Moorland, and Falkland Islands.

The Patagonian subregion extends to southern Argentina, from central Mendoza, widening through Neuquén, Río Negro, Chubut, and Santa Cruz, to northern Tierra del Fuego, and reaches Chile in Aisén and Magallanes. The Patagonian subregion comprises two provinces: Subandean Patagonia and Central Patagonia.

References

Amorim, D. de S. 2001. Dos Amazonias. In *Introducción a la biogeografía en Latino-américa: Teorías, conceptos, métodos y aplicaciones*, ed. J. Llorente Bousquets and J. J. Morrone, 245–255. Mexico, D.F.: Las Prensas de Ciencias, UNAM.

Amorim, D. de S. and M. R. S. Pires. 1996. Neotropical biogeography and a method for maximum biodiversity estimation. In *Biodiversity in Brazil: A first approach*, ed. C. E. M. Bicudo and N. A. Menezes, 183–219. São Paulo: CNPq.

Brundin, L. 1966. Transantarctic relationships and their significance as evidenced by midges. *Kungliga Svenska Vetenskaps Akademien Handlingar (Series 4)* 11:1–472.

(continued)

CASE STUDY 6.1 Regionalization of Latin America *(continued)*

Cabrera, A. L. and A. Willink. 1973. *Biogeografía de América Latina*. Washington, D.C.: Organización de Estados Americanos.

Carvalho, D. J. B. de and M. S. Couri. 2002. A cladistic and biogeographic analysis of *Apsil* Malloch and *Reynoldsia* Malloch (Diptera, Muscidae) of southern South America. *Proceedings of the Entomological Society of Washington* 104:309–317.

Contreras-Medina, R. and H. Eliosa León. 2001. Una visión panbiogeográfica preliminar de México. In *Introducción a la biogeografía en Latinoamérica: Conceptos, teorías, métodos y aplicaciones*, ed. J. Llorente Bousquets and J. J. Morrone, 197–211. Mexico, D.F.: Las Prensas de Ciencias, UNAM.

Costa, L. P. 2003. The historical bridge between the Amazon and the Atlantic forest of Brazil: A study of molecular phylogeography with small mammals. *Journal of Biogeography* 30:71–86.

Cracraft, J. and R. O. Prum. 1988. Patterns and processes of diversification: Speciation and historical congruence in some Neotropical birds. *Journal of Biogeography* 12:603–620.

Crisci, J. V., M. M. Cigliano, J. J. Morrone, and S. Roig-Juñent. 1991. Historical biogeography of southern South America. *Systematic Zoology* 40:152–171.

Crother, B. I. and C. Guyer. 1996. Caribbean historical biogeography: Was the dispersal-vicariance debate eliminated by an extraterrestrial bolide? *Herpetologica* 52:440–465.

Darlington, P. J. Jr. 1957. *Zoogeography: The geographical distribution of animals*. New York. Wiley.

Da Silva, J. M. C. and D. C. Oren. 1996. Application of parsimony analysis of endemicity in Amazonian biogeography: An example with primates. *Biological Journal of the Linnean Society* 39:427–437.

Edmunds, G. F. Jr. 1972. Biogeography and evolution of Ephemeroptera. *Annual Review of Entomology* 17:21–42.

Escalante, T., G. Rodríguez, and J. J. Morrone. 2004. The diversification of the Nearctic mammals in the Mexican transition zone. *Biological Journal of the Linnean Society* 83:327–339.

Halffter, G. 1962. Explicación preliminar de la distribución geográfica de los Scarabaeidae mexicanos. *Acta Zoológica Mexicana* 5:1–17.

Halffter, G. 1964. La entomofauna americana, ideas acerca de su origen y distribución. *Folia Entomológica Mexicana* 6:1–108.

Halffter, G. 1965. Algunas ideas acerca de la zoogeografía de América. *Revista de la Sociedad Mexicana de Historia Natural* 26:1–16.

Halffter, G. 1972. Eléments anciens de l'entomofaune Neotropicale: Ses implications biogéographiques. *Biogeographie et Liaisons Intercontinentales au cours du Mésozoique, 17me Congr. Int. Zool., Monte Carlo* 1:1–40.

Halffter, G. 1974. Eléments anciens de l'entomofaune Neotropicale: Ses implications biogéographiques. *Quaestiones Entomologicae* 10:223–262.

Halffter, G. 1976. Distribución de los insectos en la zona de transición mexicana: Relaciones con la entomofauna de Norteamérica. *Folia Entomológica Mexicana* 35:1–64.

Halffter, G. 1978. Un nuevo patrón de dispersión en la zona de transición mexicana: El mesoamericano de montaña. *Folia Entomológica Mexicana* 39–40:219–222.

Halffter, G. 1987. Biogeography of the montane entomofauna of Mexico and Central America. *Annual Review of Entomology* 32:95–114.

(continued)

CASE STUDY 6.1 Regionalization of Latin America *(continued)*

Hall, J. P. W. and D. J. Harvey. 2002. The phylogeography of Amazonia revisited: New evidence from Riodinid butterflies. *Evolution* 56:1489–1497.

Katinas, L., J. V. Crisci, W. L. Wagner, and P. C. Hoch. 2004. Geographical diversification of tribes Epilobieae, Gongylocarpeae, and Onagreae (Onagraceae) in North America, based on parsimony analysis of endemicity and track compatibility analysis. *Annals of the Missouri Botanical Garden* 91:159–185.

Katinas, L., J. J. Morrone, and J. V. Crisci. 1999. Track analysis reveals the composite nature of the Andean biota. *Australian Journal of Botany* 47:111–130.

Kuschel, G. 1969. Biogeography and ecology of South American Coleoptera. *Biogeography and ecology in South America*, Vol. 2, ed. E. Fittkau, J. J. Illies, H. Klinge, G. H. Schwabe, and H. Sioli, 709–722. The Hague: Junk.

Lopretto, E. C. and J. J. Morrone. 1998. Anaspidacea, Bathynellacea (Syncarida), generalised tracks, and the biogeographical relationships of South America. *Zoologica Scripta* 27:311–318.

Marino, P. I., G. R. Spinelli, and P. Posadas. 2001. Distributional patterns of species of Ceratopogonidae (Diptera) in southern South America. *Biogeographica* 77: 113–122.

Morrone, J. J. 1994. Distributional patterns of species of Rhytirrhinini (Coleoptera: Curculionidae) and the historical relationships of the Andean provinces. *Global Ecology and Biogeography Letters* 4:188–194.

Morrone, J. J. 1996. Austral biogeography and relict weevil taxa (Coleoptera: Nemonychidae, Belidae, Brentidae, and Caridae). *Journal of Comparative Biology* 1:123–127.

Morrone, J. J. 2001. *Biogeografía de América Latina y el Caribe*. Saragossa, Spain: Manuales y Tesis SEA, no. 3.

Morrone, J. J. 2002. The Neotropical weevil genus *Entimus* (Coleoptera: Curculionidae: Entiminae): Cladistics, biogeography, and modes of speciation. *Coleopterists Bulletin* 56(4):501–513.

Morrone, J. J. 2004. La Zona de Transición Sudamericana: Caracterización y relevancia evolutiva. *Acta Entomológica Chilena* 28:41–50.

Morrone, J. J. 2005. Hacia una síntesis biogeográfica de México. *Revista Mexicana de Biodiversidad* 76(2):207–252.

Morrone, J. J. 2006. Biogeographic areas and transition zones of Latin America and the Caribbean Islands, based on panbiogeographic and cladistic analyses of the entomofauna. *Annual Review of Entomology* 51:467–494.

Morrone, J. J., D. Espinosa Organista, C. Aguilar-Zúñiga, and J. Llorente Bousquets. 1999. Preliminary classification of the Mexican biogeographic provinces: A parsimony analysis of endemicity based on plant, insect, and bird taxa. *Southwestern Naturalist* 44:508–515.

Morrone, J. J., L. Katinas, and J. V. Crisci. 1997. A cladistic biogeographic analysis of Central Chile. *Journal of Comparative Biology* 2:25–42.

Morrone, J. J. and J. Márquez. 2001. Halffter's Mexican Transition Zone, beetle generalised tracks, and geographical homology. *Journal of Biogeography* 28:635–650.

Morrone, J. J. and J. Márquez. 2003. Aproximación a un atlas biogeográfico Mexicano: Componentes bióticos principales y provincias bigeográficas. In *Una perspectiva latinoamericana de la biogeografía*, ed. J. J. Morrone and J. Llorente Bousquets, 217–220. Mexico, D.F.: Las Prensas de Ciencias, UNAM.

(continued)

CASE STUDY 6.1 Regionalization of Latin America *(continued)*

Nihei, S. S. and C. J. B. de Carvalho. 2004. Taxonomy, cladistics and biogeography of *Coenosopia* Malloch (Diptera, Anthomyiidae) and its significance to the evolution of anthomyiids in the Neotropics. *Systematic Entomology* 29:260–275.

Nihei, S. S. and C. J. B. de Carvalho. 2007. Systematics and biogeography of *Polietina* Schnabl and Dziedzicki (Diptera, Muscidae): Neotropical area relationships and Amazonia as a composite area. *Systematic Entomology* 32:477–501.

Ochoa, L., B. Cruz, G. García, and A. Luis Martínez. 2003. Contribución al atlas panbiogeográfico de México: Los géneros *Adelpha* y *Hamadryas* (Nymphalidae), y *Dismorphia, Enantia, Leinix* y *Pseudopieris* (Pieridae) (Papilionoidea; Lepidoptera). *Folia Entomológica Mexicana* 42(1):65-77.

Ortega, J. and H. T. Arita. 1998. Neotropical–Nearctic limits in Middle America as determined by distributions of bats. *Journal of Mammalogy* 79(3):772–781.

Patterson, C. 1981. Methods of paleobiogeography. In *Vicariance biogeography: A critique*, ed. G. Nelson and D. E. Rosen, 446–489. New York: Columbia University Press.

Patton, J. L., M. N. F. da Silva, and J. R. Malcolm. 2000. Mammals of the Rio Juruá and the evolutionary and ecological diversification of Amazonia. *Bulletin of the American Museum of Natural History* 244:1–306.

Posadas, P. and J. J. Morrone. 2003. Biogeografía histórica de la familia Curculionidae (Coleoptera) en las subregiones Subantártica y Chilena Central. *Revista de la Sociedad Entomológica Argentina* 62(1–2):75–84.

Prado, D. E. and P. E. Gibbs. 1993. Patterns of species distributions in the dry seasonal forests of South America. *Annals of the Missouri Botanical Garden* 80:902–927.

Racheli, L. and T. Racheli. 2004. Patterns of Amazonian area relationships based on raw distributions of papilionid butterflies (Lepidoptera: Papilioninae). *Biological Journal of the Linnean Society* 82:345–357.

Rojas Soto, O. R., O. Alcántara Ayala, and A. G. Navarro. 2003. Regionalization of the avifauna of the Baja California peninsula, Mexico: A parsimony analysis of endemicity and distributional modelling approach. *Journal of Biogeography* 30:449–461.

Ron, S. R. 2000. Biogeographic area relationships of lowland Neotropical rainforest based on raw distributions of vertebrate groups. *Biological Journal of the Linnean Society* 71:379–402.

Rosen, D. E. 1976. A vicariance model of Caribbean biogeography. *Systematic Zoology* 24:431–464.

Sanmartín, I. and F. Ronquist. 2004. Southern Hemisphere biogeography inferred by event-based models: Plant versus animal patterns. *Systematic Biology* 53(2):216–243.

Zunino, M. 2003. Nuevos conceptos en la biogeografía histórica: Implicancias teóricas y metodológicas. In *Una perspectiva latinoamericana de la biogeografía*, ed. J. J. Morrone and J. Llorente, 159–162. Mexico, D.F.: Las Prensas de Ciencias, UNAM.

For Further Reading

Cox, C. B. 2001. The biogeographic regions reconsidered. *Journal of Biogeography* 28:511–523.

Müller, P. 1973. *The dispersal centres of terrestrial vertebrates in the Neotropical realm: A study in the evolution of the Neotropical biota and its native landscapes.* The Hague: Junk.

Problems

Problem 6.1

Espinosa Pérez and Huidobro Campos (2005) analyzed the distribution of freshwater fishes in Mexico. Based on a PAE of 450 species distributed in eight hydrological basins, they found a cladogram (fig. 6.4) that has a basal dichotomy between a northern and a southern biotic component, with their separation placed at the eastern part of the Transmexican Volcanic Belt. The authors noted that most of the fish species supporting the northern clade have Nearctic affinities, whereas those from the southern clade had mostly Neotropical affinities.

   The Mexican Gulf biogeographic province, which basically corresponds with the area analyzed by Espinosa Pérez and Huidobro Campos (2005), is currently assigned to the Neotropical region, Caribbean subregion, and Mesoamerican dominion. How would you regionalize this province into districts on the basis of the results of their study?

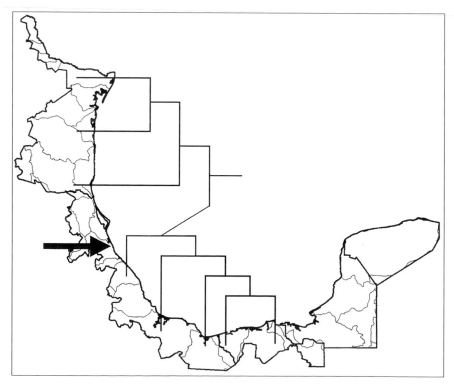

Figure 6.4   Cladogram obtained in a parsimony analysis of endemicity of freshwater fishes from hydrological basins of the Mexican Gulf by Espinosa Pérez and Huidobro Campos (2005).

Problem 6.2

Soares and Carvalho (2005) undertook a cladistic biogeographic analysis of the Andean region. Based on the taxon–area cladograms of four insect genera, they found a general area cladogram (fig. 6.5), which does not agree with the current classification of the provinces into subregions (Morrone 2006).

How would you regionalize the area on the basis of these results?

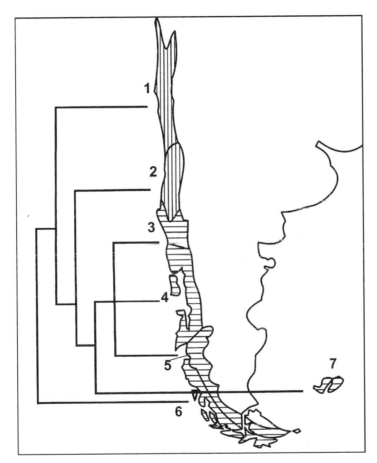

Figure 6.5   Cladogram obtained in a cladistic biogeographic analysis of insect taxa from the Andean region by Soares and Carvalho (2005). 1–2, Central Chilean subregion; 3–7, Subantarctic subregion; 1, Coquimbo province; 2, Santiago province; 3, Maule province; 4, Valdivian Forest province; 5, Magellanic Forest province; 6, Magellanic Moorland province; 7, Falkland Islands province.

For Discussion

1. Do you think biogeographic regionalizations are useful? If so, why?
2. What are the implications of the recognition of transition zones for biogeographic regionalizations? Discuss the relevance of transition zones for panbiogeographic and cladistic biogeographic analyses.
3. Carefully read the following article:

    Craw, R. C. and R. Page. 1988. Panbiogeography: Method and metaphor in the new biogeography. In *Evolutionary processes and metaphors,* ed. M.-W. Ho and S. W. Fox, 163–189. Chichester: Wiley.

    a. Identify the main arguments related to biogeographic regionalizations.
    b. Contrast them with your personal opinion.

CHAPTER 7

Identification of Cenocrons

Dispersal explanations traditionally have rested on narrative frameworks, lacking a general theory to explain distributional patterns, so they have been rejected by panbiogeographers and cladistic biogeographers as ad hoc explanations. After establishing biogeographic homology patterns, however, dispersal explanations can help establish when the cenocrons assembled in the identified components, incorporating a time perspective in the study of biotic evolution. In this chapter I explore how time slicing, intraspecific phylogeography, and molecular clocks may be used to incorporate temporal information in the biotic components recognized previously, helping identify the cenocrons that they comprise and how they dispersed and integrated.

Time Slicing

Hunn and Upchurch (2001) emphasized the relevance of time in evolutionary biogeography because data on temporal distribution may provide important constraints in biogeographic analyses, helping reinforce or overturn specific hypotheses. The incorporation of temporal information entails assigning time values to cladogenetic events (Crisci et al. 2000). Lieberman (2004) highlighted the relevance of the deep time perspective in biogeography, based on three arguments: The fossil record is the only primary and direct chronicle of the history of life, the phenomenon of extinction can influence our ability to retrieve biogeographic patterns, and the fossil record may provide examples of dispersal events.

Upchurch and Hunn (2002) stated that the extent to which temporal data are incorporated into biogeography depends on the researcher's choice of sources of spatial data and analytical methods. It has been suggested that the temporal ranges of organisms are important because distribution patterns seem to decay through time as new ones are superimposed (Grande 1985; Hunn and Upchurch 2001; Upchurch and Hunn 2002; Upchurch et al. 2002). Donoghue and Moore (2003) postulated that cladistic biogeographic methods are susceptible to the confounding effects of pseudo-incongruence and pseudo-congruence if they do not incorporate information on the absolute timing of the diversification of the lineages. *Pseudo-incongruence* means that different area cladograms may show conflict when the taxa evolved at the same time but diversified in response to different events.

Pseudo-congruence means that different area cladograms may show the same area relationships although the taxa diversified at different times, presumably under different underlying causes. However, Riddle and Hafner (2006) argued that time alone might not necessarily cause us to resort to an explanation of pseudo-congruence.

Cladistic biogeographers avoid using temporal data because of the risk of incorporating ideas of unobserved processes in the elucidation of biogeographic patterns. This would imply unverifiable assumptions, with the risk of falling back on narrative scenarios. However, the need to consider time in biogeography becomes clearer in cases of biogeographic convergence. The terms *convergence* and *divergence* were proposed by Hallam (1974) to distinguish two extreme biogeographic patterns. Widespread taxa and redundancy identify biogeographic convergence, whereas vicariance is the most common interpretation of divergence patterns. Convergence can be the result of area coalescence (due to the elimination of geographic barriers). Analyses of biogeographic convergence are unlikely to show congruence.

Upchurch et al. (2002) noted that biogeographic analyses over an extensive stratigraphic range may fail to find the correct area relationships. This point is illustrated by the hypothetical succession of area separations followed by area coalescence described by Upchurch and Hunn (2002), where branching relationships are evident only when extensive dispersal has not yet overwhelmed the original vicariance pattern. In fact, if biotic components A and C, formerly separated by vicariance following the pattern (A (B, C)), merge while the third biotic component B remains isolated, the final relation will be uninformative because we will have two biotas and two areas: A1 (= A + C) and B. Area coalescence causes vicariance patterns to fade out through time (Grande 1985). Biogeographic convergence leads to a sort of biogeographic overprinting, and therefore analytical techniques based on parsimony algorithms may be inappropriate (Young 1995). When previously merged biotic components subsequently undergo vicariance, patterns may need to be treated within a multiple time plane approach (Rosen and Smith 1988) or time slicing (Upchurch et al. 2002). The latter term corresponds to the analysis of biotic distributional data according to a sequence of individual stratigraphic intervals (time slices).

Let us consider an example. Originally (time t_0), we have a biotic component A. Then a vicariance event at time t_1 produces two biotic components endemic to areas B and C. At time t_2 these biotic components converge in a single biotic component in area D because of geographic coalescence, but a trace of the previous endemism may be still recognizable. Finally, at time t_3 two new endemic biotic components occur in areas E and F after another vicariance event that subdivided the former component inhabiting area D. How does one recognize the vicariance event at t_1 if the effects of biogeographic convergence that formed the biotic component of area D overwhelmed the biotic fingerprint on the two original biotic components B and C? How does one recognize what happened at times t_1 and t_2 if one studies only the latest time plane (t_3) and the lineages occurring in areas E and F?

Events of biogeographic convergence such as those described for Mesozoic ammonites (Cecca 2002) or for plant and animal taxa from the Mexican Transition Zone (Halffter 1987) produce a sort of overprinting of past biogeographic histories by more recent patterns (e.g., reticulated area histories). This overprinting lowers the chances of establishing area relationships through congruence, which is the final goal of cladistic biogeography. The solution to problems posed by instances of biogeographic convergence is time slicing (Grande 1985; Upchurch and Hunn 2002). Although assessments of faunal similarity usually are undertaken with faunas of successive geological ages, traditional cladistic biogeography has used data on organism relationships and spatial distributions only on a single time plane (usually the present). Time slicing may reconcile the use of time and a synchronic approach. Ideally, paleobiogeographers should be able to use a synchronic approach for each time slice they identify. This is difficult because of the limits imposed by geological constraints (e.g., insufficient precision or resolution of chronological correlations, incompleteness of the fossil record).

Methods: Temporally Partitioned Component Analysis

In order to apply time slicing, three methods are available. Two of them, parsimony analysis of endemicity and area cladistics, have been already dealt with. The third one, temporally partitioned component analysis (TPCA), is dealt with herein.

Upchurch and Hunn (2002) proposed this method, also known as chronobiogeography, to incorporate explicitly temporal data into a cladistic biogeographic analysis (see also Hunn and Upchurch 2001 and Upchurch et al. 2002). After geodispersal, due to area coalescence events (biogeographic convergence), "the histories of areas and biotas will have a reticulated rather than a branching structure, raising the question as to how well cladistic biogeographic techniques will be able to accurately analyze and depict a reticulate system" (Upchurch and Hunn 2002:280). In short, the starting point of TPCA is the existence of taxon–area cladograms for the taxonomic groups on which the analysis is based.

Although the introduction of time slicing may tend to uncover reticulate histories, it is interesting to note that, ideally, synchronic relations would be found for each time slice on the basis of phylogenetic relations. TPCA is a pragmatic analytical method that allows reticulate histories to be explained, where assumptions are minimized, and the area and lineage duplications that may be produced by Brooks parsimony analysis are avoided.

Algorithm It consists of the following steps (Upchurch and Hunn 2002):
1. Prune or temporally partition the taxon cladograms by deleting all taxa that did not exist at a particular designated time slice.
2. Find optimal area cladograms for each particular time slice by determining which area relationships provide the best (under some designated optimality criterion) explanation for the spatial distributions observed in the taxon cladogram.

3. Use a randomization test to determine whether the degree of fit between area and taxon cladogram for each time slice is greater than would be expected by chance.

Empirical Applications Hunn and Upchurch (2001) and Upchurch and Hunn (2002).

CASE STUDY 7.1 Dinosaurian Biogeography

Dinosaurs were diverse, geographically widespread and stratigraphically long lived terrestrial taxa, which provide an ideal case study for Mesozoic biogeography. Some authors have proposed vicariance, driven by continental fragmentation, as the key factor to explain their biotic evolution, especially during the Cretaceous (Russell 1993; Sampson et al. 1998); however, Sereno (1997, 1999) concluded that continent-level vicariance was rare and unimportant to explain dinosaurian biogeographic patterns. Upchurch et al. (2002) undertook a TPCA to analyze the biogeography of dinosaurs in order to detect the presence of repeated area relationships that may indicate vicariance.

Upchurch et al. (2002) used TreeMap (Page 1994c) to analyze the taxon–area cladograms of Ornitischia and Saurischia. They analyzed nine time slices of various durations: Mesozoic, Late Triassic, Jurassic, Early Jurassic, Middle Jurassic, Late Jurassic, Cretaceous, Early Cretaceous, and Late Cretaceous. For each time slice, the taxon–area cladograms were pruned so that only taxa present at the relevant time were retained.

Only three analyses passed a randomization test, indicating that the patterns for the Middle Jurassic, Late Jurassic, and Early Cretaceous were highly unlikely to have arisen by chance (fig. 7.1). General area cladograms for Middle and Late Jurassic (figs. 7.1a–d) contained only four areas, reflecting the poor sampling of taxa for several continents. They display the same relationships for the areas in common (Asia, Europe, and Africa). The Early Cretaceous general area cladogram (figs. 7.1e and 7.1f) contains six areas and differs from that of the Late Jurassic regarding the relationships of Europe. Incongruities between the three general area cladograms result from the conflicting relationships of Europe and South America relative to the other continents. According to the authors, the failure of the Late Triassic and Early Jurassic analyses may reflect the current selection of areas that are suitable for the detection of continent-level vicariance, given that Pangaea remained largely intact until the Middle Jurassic. Additionally, poor sampling may be largely responsible for obscuring area relationships.

When Upchurch et al. (2002) compared the results of their analysis with paleogeographic reconstructions, they found that the greatest incongruence concerns the position of South America in the Middle Jurassic. According to the general area cladogram, Europe and Africa were more recently in contact with each other than either was with South America, conflicting with paleogeographic reconstructions that suggest that Africa and South America constituted a continuous area throughout the Jurassic and into the Cretaceous. It is conceivable that some nonmarine barrier separated South America and Africa during the Middle Jurassic, but there is no geological evidence. It seems more probable that the problem lies with the apparent biogeographic signal. Another incongruity is the lack of a clear group of Laurasian areas in the general area cladograms, but this may be more apparent than real because Laurasia may have constituted a single continuous continent divided by epicontinental seas into a number of

(continued)

CASE STUDY 7.1 Dinosaurian Biogeography *(continued)*

separate landmasses. The different positions of Europe in the general area cladograms of the Late Jurassic and Early Cretaceous may indicate the convergence of two previously depauperate areas. This is consistent with previous studies that showed that the regression of the Turgai Sea during the Early Cretaceous may have allowed the formation of a bridge between Europe and Asia. Finally, although the existence of Gondwana is consistent with the general area cladogram from the Early Cretaceous, Upchurch et al. (2002) noted a discrepancy in the timing of the separation between Africa and South America that may reflect errors in geological dating or may indicate that the initial

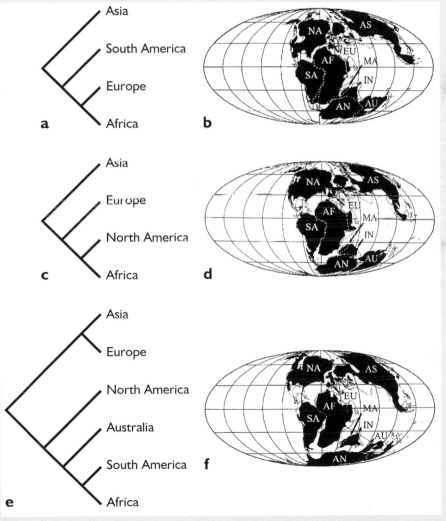

Figure 7.1 Biogeographic analysis of dinosaurs by Upchurch et al. (2002). (a) Middle Jurassic time slice; (b) paleocoastline reconstruction for the Middle Jurassic; (c) Late Jurassic time slice; (d) paleocoastline reconstruction for the Late Jurassic; (d) Early Cretaceous time slice; (e) paleocoastline reconstruction for the Early Cretaceous.

(continued)

CASE STUDY 7.1 Dinosaurian Biogeography *(continued)*

stages of vicariance between Africa and South America occurred soon after the South Atlantic Ocean started to open up.

Upchurch et al. (2002) concluded that the presence of statistically robust biogeographic patterns suggests that vicariance was not overwhelmed by dispersal or extinction, at least during the Middle Jurassic to Middle Cretaceous. The discovery of four vicariance events (Asia–Pangaea in the Early or Middle Jurassic, North America–Gondwana before the "Callovian," eastern–western Gondwana during the "Valanginian," and Africa–South America in the Early or Middle Cretaceous) also suggests that vicariance had the major role in determining dinosaurian biotic evolution.

References

Russell, D. A. 1993. The role of central Asia in dinosaurian biogeography. *Canadian Journal of Earth Sciences* 30:2001–2012.

Sampson, S. D., L. M. Witmer, C. A. Foster, D. M. Krause, P. M. O'Connor, P. Dodson, and F. Ravoavy. 1998. Predatory dinosaur remains from Madagascar: Implications for the Cretaceous biogeography of Gondwana. *Science* 280:1048–1051.

Sereno, P. C. 1997. The origin and evolution of dinosaurs. *Annual Review of Earth and Planetary Sciences* 25:435–489.

Sereno, P. C. 1999. The evolution of dinosaurs. *Science* 284:2137–2147.

Upchurch, P., C. A. Hunn, and D. B. Norman. 2002. An analysis of dinosaurian biogeography: Evidence for the existence of vicariance and dispersal patterns caused by geological events. *Proceedings of the Royal Society of London, Series B* 269:613–621.

Intraspecific Phylogeography

Intraspecific phylogeography studies the principles and processes governing the geographic distribution of genealogic lineages, especially those within and between closely related species, based on molecular data (Avise 2000; Avise et al. 1987; Lanteri and Confalonieri 2003; Lomolino et al. 2006). The maternal, nonrecombinant mode of inheritance of mitochondrial DNA (mtDNA) and the rapid evolution of mtDNA sequences make it possible to obtain haplotypes (combinations of alleles at multiple linked loci), which can be used to obtain intraspecific phylogenetic hypotheses (Crisci et al. 2000). Once the population genetic structure has been assessed on the basis of mtDNA, it is possible to obtain a network or cladogram of haplotypes, which allows one to analyze historical patterns and the processes that have shaped them, such as dispersal, vicariance, range expansion, and colonization, sometimes under a statistical framework (Knowles and Maddison 2002; Templeton 2004). This knowledge can offer insight into when recent cenocrons incorporated to a biotic component.

The results of a phylogeographic analysis are usually represented as a parsimony network connecting the studied haplotypes (fig. 7.2a), where the number of restriction site differences indicates the relative distance between haplotypes

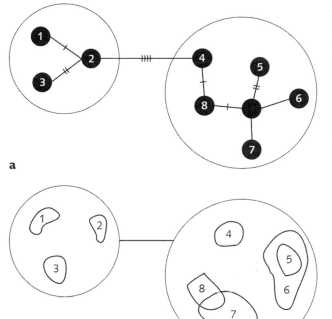

Figure 7.2 Representation of a phylogeographic hypothesis. (a) Parsimony network connecting the haplotypes, where the dashes indicate restriction site differences between haplotypes and groups of haplotypes; (b) geographic distribution of the haplotypes.

and groups of haplotypes. When mapped (fig. 7.2b), disjunctions and sympatry between the haplotypes may give us clues on their evolutionary histories. Highly divergent groups inhabiting disjunct areas can indicate independent evolutionary histories for a long period of time, usually due to vicariance, whereas lack of geographic structure may indicate recent dispersal.

Intraspecific phylogeography is conceptually positioned between macroevolution and microevolution (Lomolino et al. 2006; Riddle and Hafner 2004). With its focus on historical processes, it contextualizes and balances the ecogeographic perspective that tend to emphasize natural selection's role in microevolution (Avise 2000). In addition, it provides the promising basis for a bridge between ecological and evolutionary biogeography.

One promising area of interaction between phylogeography and ecological biogeography concerns the study of range dynamics. The range of a species changes through time, with a set of trajectories that it can take between speciation and ultimate extinction (Gaston 2003). Intraspecific phylogeography may be useful for helping determine which of the broad classes of dynamics that have been identified (Gaston 2003) correspond to the species analyzed (fig. 7.3):

- Stasis: The geographic range remains largely unchanged (fig. 7.3a).
- Stasis postexpansion: The geographic range increases rapidly, and the resultant area is maintained for a while (fig. 7.3b).

Figure 7.3   Types of species range dynamics. (a) Stasis; (b) stasis postexpansion; (c) age and area; (d) cyclic; (e) idiosyncratic.

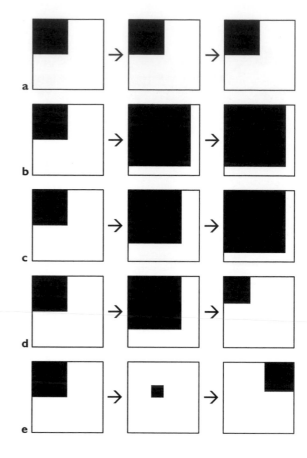

- Age and area: The geographic range increases in size progressively (fig. 7.3c).
- Cyclic: There is a cycle in which the species pass through a sequence of range expansion, evolutionary differentiation between populations, and extinction (fig. 7.3d).
- Idiosyncratic: The species range exhibits an entirely idiosyncratic trajectory, with no clear change (fig. 7.3e).

The time lapse of phylogeographic analyses is skewed to Pleistocene and Holocene time frames, with some studies extending back into Pliocene and Miocene. A recent development, paleophylogeography (Betancourt 2004), has revealed phylogeographic structure in fossil populations. An analysis (Kuch et al. 2002) of ancient cytochrome b fragments from rodent fecal pellets from the Atacama Desert (Chile) has provided interesting insights on paleophylogeography.

In animal species, mtDNA has proven useful for phylogeographic studies, but in plants it evolves quickly with respect to gene order but slowly in nucleotide sequence, so it is of limited utility (Crisci et al. 2000). Chloroplast DNA (cpDNA) has shown to be structured geographically in several plant species; it is also transmitted maternally and has exhibited intraspecific variation (Avise 2000, 2004; Soltis et al. 1992).

An explicit genealogical approach to intraspecific phylogeography implies extending the lineage sorting theory to sister populations (Avise 2000). In terms of maternal genealogy, there are three possible situations: polyphyly, in which some but not all extant matrilines in one population join with some but not all extant matrilines in the other to form a clade; paraphyly, in which all matrilines in one population from a clade are nested within the broader matrilineal history of the other population; and reciprocal monophyly, in which all extant matrilines in each sister population are closer genealogically to each other than to any matriline in the other. These three situations often characterize the same pair of sister populations at different time depths (fig. 7.4).

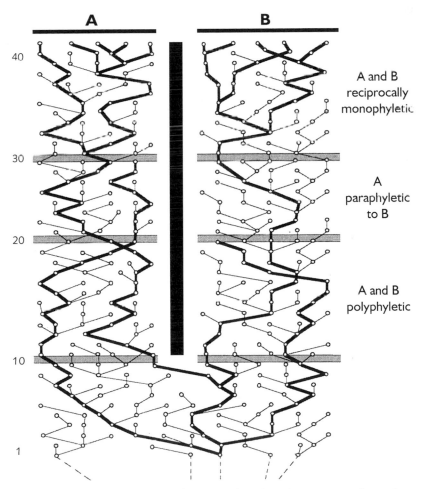

Figure 7.4   Schematic representation of the matrilineal sorting process of two sister populations, along 40 generations, separated by a barrier to gene flow.

Hypotheses and Corollaries Three phylogeographic hypotheses and four corollaries have been formulated by Avise et al. (1987):

1. Most species are composed of geographic populations whose members occupy recognizable matrilineal branches of an extended intraspecific pedigree. Populations of most species display significant phylogeographic structure supported by mtDNA data.

2. Species with limited or shallow phylogeographic population structure have life histories conducive to dispersal and have occupied ranges free of firm, long-standing impediments to gene flow. Nonsubdivided, high-dispersal species may have limited phylogeographic structure.

3. Intraspecific monophyletic groups distinguished by large genealogical gaps usually arise from long-term extrinsic biogeographic barriers to gene flow. Major phylogeographic units within a species reflect long-term historical barriers to gene flow. This hypothesis has four corollaries representing four aspects of genealogical agreement (fig. 7.5):

   a. Agreement across sequence characters within a gene (fig. 7.5a): Every deep phylogenetic split in the intraspecific gene tree is supported concor-

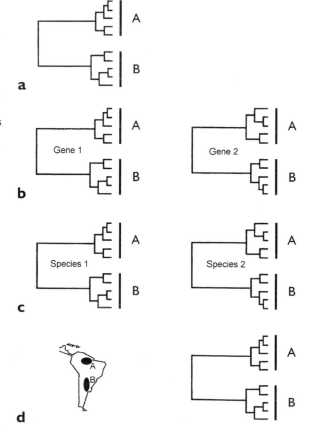

Figure 7.5 Schematic representation of four different aspects of genealogical agreement. (a) Agreement across sequence characters within a gene; (b) agreement in significant genealogical partitions across two different genes within a species; (c) agreement in the geography of gene tree partitions across two codistributed species; (d) agreement of gene tree partitions with spatial boundaries between two biotic components.

dantly by multiple diagnostic characters (e.g., nucleotides or restriction sites) within the mitochondrial genome. If this is not the case, such matrilineal splits would not be evident in the analysis, nor would they receive significant support.

b. Agreement in significant genealogical partitions across multiple genes within a species (fig. 7.5b): Empirical examples show general agreement between deep phylogeographic topologies in multiple gene trees (e.g., mitochondrial and nuclear) within the species analyzed. These deep branch separations characterize the same sets of geographic populations.

c. Agreement in the geography of gene tree partitions across multiple co-distributed species (fig. 7.5c): Several sympatric species with comparable natural histories or habitat needs proved to be phylogeographically structured in similar fashion. In particular, divergent branches in the intraspecific gene trees might map consistently to the same geographic regions.

d. Agreement of gene tree partitions with spatial boundaries between traditionally recognized biogeographic units (fig. 7.5d): An emerging generality from phylogeographic analyses is that deeply separated phylogroups at the intraspecific level are confined to biogeographic provinces or districts as identified by systematic biogeography.

Phylogeographic Patterns Five different phylogeographic patterns can be characterized for mtDNA gene cladograms (Avise 2000) (fig. 7.6):

Category I (deep gene tree, major lineages allopatric) (fig. 7.6a): It is characterized by the presence of spatially circumscribed haplotypes, separated by large mutational distances. This pattern appears commonly in phylogeographic patterns of mtDNA. A long-term extrinsic barrier to genetic exchange is the most commonly invoked explanation.

Category II (deep gene tree, major lineages broadly sympatric) (fig. 7.6b): It is characterized by pronounced phylogenetic gaps between some branches in a gene tree, with main lineages codistributed over a wide area. It could arise in a species of which some anciently separated lineages might have been retained by chance, whereas many intermediate lineages were lost over time by gradual lineage sorting.

Category III (shallow gene tree, lineages allopatric) (fig. 7.6c): Most or all haplotypes are related closely yet are localized geographically. Contemporary gene flow has been low enough in relation to population size to have permitted lineage sorting and random drift to promote genetic divergence of populations that were in historical contact recently.

Category IV (shallow gene tree, lineages sympatric) (fig. 7.6d): It is expected for high–gene-flow species of small effective size whose populations have not been separated by long-term barriers.

Category V (shallow gene tree, major distributions varied) (fig. 7.6e): This category is intermediate between Categories III and IV and involves common lineages that are widespread plus closely related lineages that are confined to one or a few

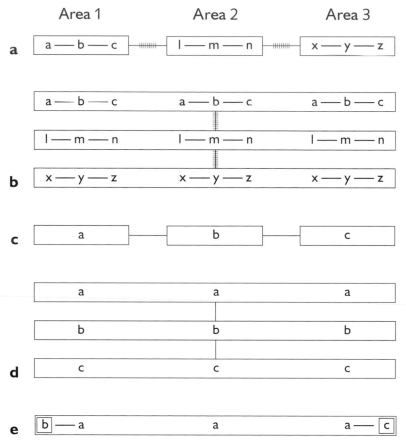

Figure 7.6   Different phylogeographic patterns for mtDNA gene cladograms. (a) Category I (deep gene tree, major lineages allopatric); (b) Category II (deep gene tree, major lineages broadly sympatric); (c) Category III (shallow gene tree, lineages allopatric); (d) Category IV (shallow gene tree, lineages sympatric); (e) Category V (shallow gene tree, major distributions varied).

nearby localities. It implies low contemporary gene flow between populations that are connected tightly in history. Common haplotypes are often plesiomorphic, and rare haplotypes are the presumed apomorphic conditions.

Phylogeographic patterns exhibiting large genetic gaps between phylogroups (groups of closely related haplotypes separated from other phylogroups by large genetic gaps), such as those represented in Categories I and II, are amenable to phylogenetic analyses using cladistic methods (Lomolino et al. 2006). This can be done simply by treating each phylogroup as a terminal unit in the cladistic analysis. When variable haplotypes are closely related and groups of populations are not clustered within the clearly reciprocally monophyletic phylogroups, such as those in Categories III, IV, and V, one can still present an unrooted cladogram to summarize haplotype relationships.

Algorithm It consists of the following steps (Holsinger 2006; Lanteri and Confalonieri 2003; Templeton et al. 1987, 1992):

1. Construct a data matrix in which the different haplotypes of a species or closely related species represent the terminal units and gene sequences from mtDNA (animals) or cpDNA (plants) represent the characters, and use cladistic software to obtain a haplotype phylogram or network:

   a. Estimate the probability P_1 that haplotype pairs differing by a single change are the result of a single substitution. If $p_1 > .95$, connect all pairs of haplotypes that differ by a single change. There may be ambiguities, including loops. Keep them in the network.

   b. Identify the products of recombination by inspecting the 1-step network to determine whether postulating recombination between a pair of sequences can remove ambiguity previously identified.

   c. Increase j by 1 and estimate pj. If $pj > .95$, join $j - 1$-step networks into a j-step network by connecting the two haplotypes that differ by j steps. Repeat until either all haplotypes are included in a single network or you are left with two or more nonoverlapping networks.

   d. If you have two or more networks left to connect, estimate the smallest number of nonparsimonious changes that will occur with greater than 95% probability and connect the networks.

2. Translate the phylogram into a network of nested clades in which each successive hierarchical level is considered to be more ancient than the subordinate clades or phylogroups:

   a. Each haplotype in the sample comprises a 0-step clade (it is separated by 0 evolutionary steps from other copies of the same haplotype). Tip haplotypes are those that are connected to only one other haplotype, whereas interior haplotypes are those that are connected to two or more haplotypes. Set $j = 0$.

   b. Pick a tip haplotype that is not part of any $j + 1$–step network.

   c. Identify the interior haplotype with which it is connected by $j + 1$ nucleotide differences.

   d. Identify all tip haplotypes connected to that interior haplotype by $j + 1$ nucleotide differences.

   e. The set of such tip and interior haplotypes constitutes a $j + 1$–step clade.

   f. If there are tip haplotypes remaining that are not part of a $j + 1$–step clade, return to Step b.

   g. Identify any internal j-step clades that are not part of a $j + 1$–step clade and are separated by $j + 1$ steps.

   h. Designate these clades as terminal and return to Step b.

   i. Increase j by one and return to Step b.

3. Represent the phylogroups on a map and calculate the following:

a. The clade distance (D_c): average distance of each haplotype in the particular clade from the center of its geographic distribution, measured as the great circle distance, to represent the distribution of the phylogroups.

b. The nested clade distance (D_n): average distance of the center of distribution for this haplotype from the center of distribution for the haplotype within which it is nested.

c. The average distance between D_c and D_n for the terminal phylogroups in order to identify whether they are structured geographically.

4. Use a statistical test to evaluate whether the haplotypes of the nested clades are distributed randomly. If this hypothesis is rejected, it means that there is a significant geographic association between them.

Software TCS 1.18 (Clement et al. 2000) and GeoDis (Posada et al. 2000).

Empirical Applications Many phylogeographic papers have been published in recent decades; this is a nonexhaustive list: Arnaiz-Villena et al. (2001), Ávila et al. (2006), Avise (1989, 1992), Avise et al. (1992), Ball and Avise (1992), Bermingham and Martin (1998), Bernatchez (2001), Bernatchez and Dodson (1990), Bohme et al. (2007), Buhay et al. (2007), Burridge et al. (2007), Chen et al. (2007), Comes and Kadereit (1998), Confalonieri et al. (1998), Dawson et al. (2001), Demesure et al. (1996), Demasts et al. (2002), DeSalle and Hunt (1987), Domínguez-Domínguez et al. (2007), Dumolin-Lapegue et al. (1997), Dutton et al. (1999), Emerson et al. (2006), Fedorov et al. (2003), Fordyce and Nice (2003), González-Rodríguez et al. (2004), Gorog et al. (2004), Gows et al. (2006), Hafner et al. (2005), Hall and Harvey (2002), Hewitt (1996, 1999, 2000), Irwin (2002), Juan et al. (2000), Kozac et al. (2006), Lapointe and Rissler (2005), Linz et al. (2007), Liu and Hershler (2007), Ludt et al. (2004), MacHugh et al. (1997), Mateos (2005), Mejía-Madrid et al. (2007), Mock et al. (2007), Morales-Barros et al. (2006), Moritz (1994, 1995), Mortimer and van Vuuren (2007), Neiman and Lively (2004), Nersting and Arctander (2001), O'Donnell et al. (1998), Orange et al. (1999), Osentoski and Lamb (1995), Palma et al. (2005), Pauly et al. (2007), Perdices and Coelho (2006), Picard et al. (2007), Presa et al. (2002), Recuero et al. (2006), Richards et al. (1998), Riddle et al. (2000a, 2000b, 2000c), Roderick and Gillespie (1998), Russell et al. (2005), Schaal et al. (2003), Schäuble and Moritz (2001), Steele and Storfer (2007), Sullivan et al. (1997), Taberlet et al. (1998), Taylor and Hellberg (2003), Tchaicka et al. (2007), Templeton et al. (1995), Vila et al. (1999), and Weisrock and Janzen (2000).

CASE STUDY 7.2 Phylogeography of Red Deers in Eurasia

Red deers have been assigned to two different species, *Cervus elaphus* in Eurasia and *C. canadensis* in North America. The recognition of subspecies within them is questioned, as is the assignment of some Central Asian subspecies to both species. The present classification is based on morphological characters, such as body and antler size, antler shape, and cranial measurements. Ludt et al. (2004) undertook a phylogeographic analysis in order to investigate whether red deers represent one or two different species and whether the named subspecies are consistent with knowledge based on molecular data.

Ludt et al. (2004) analyzed samples of tissue from fifty localities, which represent most living species and subspecies of *Cervus elaphus* and *C. canadensis* across most of their range, as well as Sika deer (*C. nippon*), Thorold's white lipped deer (*C. albirostris*), sambars (*C. unicolor* and *C. timorensis*), and hog deer (*Axis porcinus*). Sequences of *Dama dama*, *Bos taurus*, and *Moschus moschiferus* were used as outgroups. Sequences were aligned with ClustalX version 1.83 (Thompson et al. 1997) and checked visually. Initial sequence comparisons and measures of variability were performed with Mega version 2.1 (Kumar et al. 2001). MODELTEST version 3.06 (Posada and Crandall 1998) was used to determine the model of sequence evolution and PAUP version 4.10b (Swofford 2003) to compute the maximum likelihood cladograms. Dates of evolutionary divergence between lineages were estimated using molecular clock calibrations with the fossil record of *Bos* and *Dama*. Nested clade analysis was performed with GeoDis version 2.0 (Posada et al. 2000).

The maximum likelihood search found a cladogram (fig. 7.7a), which clearly shows eleven distinct groups: Western Europe, Balkan, Middle East, Africa, Tarim, North Asia and America, South Asia, East Asia, and the three outgroups. The nested clade analysis (fig. 7.7b) indicated that haplotypes from western red deer were subdivided into Western Europe (H), Africa (I), and South Europe (FG), whereas haplotypes from eastern red deer formed a single network (BCDE). The results of the nested clade analysis indicate vicariance for the eastern red deer (Clade 5.1) and for the North America and Asia branch (Clade 3.5), restricted gene flow with isolation by distance for the East Asia branch (Clade 2.4), and contiguous range expansion for the western red deer (FGHI).

Ludt et al. (2004) concluded that western red deer (Eurasia) and eastern red deer (North America and Asia) clearly represent two species, *Cervus elaphus* and *C. canadensis*, respectively. Their subdivision into subspecies was not supported by the analysis. Western red deer can be subdivided into four subgroups (Western Europe, Balkan, Middle East, and Africa) and eastern red deer can be subdivided into three subgroups (North Asia and America, South Asia, and East Asia).

(continued)

CASE STUDY 7.2 Phylogeography of Red Deers in Eurasia *(continued)*

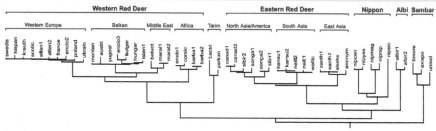

a

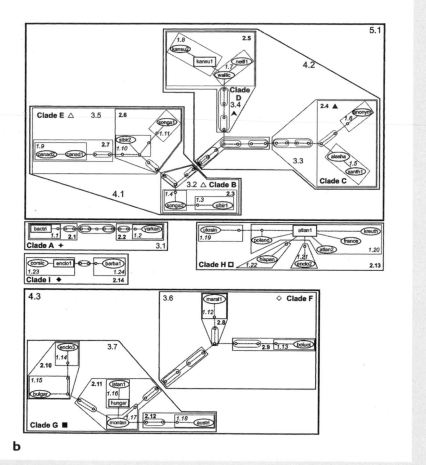

b

(continued)

CASE STUDY 7.2 Phylogeography of Red Deers in Eurasia *(continued)*

Figure 7.7   Biogeographic analysis of *Cervus elaphus* by Ludt et al. (2004; reproduced with permission of *Molecular Phylogenetics and Evolution*). (a) Maximum likelihood cladogram; (b) nested haplotype network. Dotted lines connecting haplotypes represent single substitutions, boxes and numbered clades represent hierarchical nesting levels, and circles represent unsampled haplotypes. alasha, *C. elaphus alashanicus* (China); albir1 and albir2, *C. albirostris* (China); anonym, *C. elaphus xanthopygus* (China); atlan1 and atlan2, *C. elaphus atlanticus* (Norway); austri, *C. elaphus hippelaphus* (Austria); axispo, *Axis porcinus* (India); bactri, *C. elaphus bactrianus* (Tadzikistan); barba1 and barba2, *C. elaphus barbarus* (Tunisia); boluot, *C. elaphus maray* (Turkey); bulgar, *C. elaphus hippelaphus* (Bulgaria); canad1 and canad2, *C. elaphus canadensis* (North America); corsic, *C. elaphus corsicanus* (Corsica); enclo1, *C. elaphus* (Germany); enclo2, *C. elaphus hispanicus* (Spain); enclo3, *C. elaphus hippelaphus* (Germany); france, *C. elaphus hippelaphus* (France); hispan, *C. elaphus hispanicus* (Spain); hungar, *C. elaphus hippelaphus* (Hungary); istan1, *C. elaphus hippelaphus* (Turkey); kansu1 and kansu2, *C. elaphus kansuensis* (China); kreuth, *C. elaphus hippelaphus* (Germany); maral1 and maral2, *C. elaphus maral* (Iran); montan, *C. elaphus montanus* (Romania); neill1 and neill2, *C. elaphus macneilli* (China); nipcen, *C. nippon centralis*; nipmag, *C. nippon mageshima*; nipnip, *C. nippon nippon*; nipsic, *C. nippon sichuanicus* (China); nipyes, *C. nippon yesoensis*; poland, *C. elaphus hippelaphus* (Poland); scotic, *C. elaphus scoticus* (Scotland); sibir1 and sibir2, *C. elaphus sibericus* (China and Mongolia); songa1 and songa2, *C. elaphus songaricus* (China); ukrain, *C. elaphus brauneri* (Ukraine); swedis, *C. elaphus elaphus* (Sweden); timore, *C. timorensis macassanicus*; unicol, *C. unicolor cambojensis* (China); wallic, *C. elaphus wallichi* (China and Tibet); yarkan, *C. elaphus yarkandensis* (China); yugosl, *C. elaphus hippelaphus* (Yugoslavia); xanth1 and xanth2, *C. elaphus xanthopygus* (Russia).

References

Kumar, S., K. Tamura, I. B. Jakobsen, and M. Nei. 2001. MEGA2: Molecular evolutionary genetics analysis software. *Bioinformatics* 17:1244–1245.

Ludt, C. J., W. Schröder, O. Rottmann, and R. Kühn. 2004. Mitochondrial DNA phylogeography of red deer (*Cervus elaphus*). *Molecular Phylogenetics and Evolution* 31:1064–1083.

Posada, D. and K. A. Crandall. 1998. MODELTEST: Testing the model of DNA substitution. *Bioinformatics* 14(9):817–818.

Posada, D., K. A. Crandall, and A. R. Templeton. 2000. GeoDis: A program for the cladistic nested analysis of the geographical distribution of genetic haplotypes. *Molecular Ecology* 9:487–488.

Swofford, D. L. 2003. *PAUP*: Phylogenetic analysis using parsimony (*and other methods)*. Version 4. Sunderland, Mass.: Sinauer. Available at http://paup.csit.fsu.edu/.

Thompson, J. D., T. J. Gibson, F. Plewniak, F. Jeanmougin, and D. G. Higgins. 1997. The ClustalX windows interface: Flexible strategies for multiple sequence alignment aided by quality analysis tools. *Nucleic Acids Research* 25:4876–4882.

CASE STUDY 7.3 Phylogeographic Predictions of a Weevil Species of the Canary Islands

Oceanic islands are attractive for evolutionary biogeography for several reasons: They present discrete geographic entities within defined oceanic boundaries, gene flow between individual islands is reduced by oceanic barriers, their often small geographic size has made the cataloguing of their biotas easier than that of continental systems, despite their small geographic size they can contain a diversity of habitats, and they are often geologically dynamic, with historical and contemporary volcanic and erosional activity (Emerson et al. 2006). In addition to the molecular phylogenetic analyses that have focused on the relationships between species occurring on two or more oceanic islands, these features make them attractive for studies of intraspecific phylogeography within individual islands. Emerson et al. (2006) analyzed the intraspecific phylogeography of the weevil species *Brachyderes rugatus* (Coleoptera: Curculionidae) in La Palma (Canary Islands) in order to test some predictions about the biotic evolution of this taxon.

The geological history of La Palma is fairly well understood (Carracedo and Day 2002). The island consists of the northern shield, composed of older volcanic terrains, and the southern ridge, constituted by terrains of more recent volcanic origin. Subaerial development of the northern shield began about 1.7–2.0 mya and continued until about 0.55 mya, being dominated by two volcanos, Garafia (active from 1.7 to 1.2 mya) and Taburiente (active from 1.2 to 0.4 mya). From 0.8 to 0.7 mya, in the final stages of the Taburiente volcano, the southward migration of volcanism began through the southern, or Cumbre Nueva, rift of the volcano. At about 0.56 mya this rift became unstable, and its western flank collapsed into the sea. After this collapse, the Bejenado volcano dominated activity from 0.56 to 0.40 mya, with possible minor activity continuing until about 0.2 mya. After the end of the growth of the Bejenado volcano, the entire northern shield became quiescent, and the island may have entered a period of volcanic calm, but volcanism in the south of La Palma may have built its submarine and core parts, now concealed by younger lavas of the eruptive period in the northern shield. It seems reasonable to assume that a widespread species with limited dispersal abilities, which was present in the island during its volcanic history, would have been influenced by these events. Genetic footprints of past range expansions and fragmentations of a species such as *Brachyderes rugatus* should be consistent with predictions from volcanic and erosional events. Emerson et al. (2006) hypothesized that ancestral haplotypes should be located predominantly in the northern shield, with the newer Bejenado and Cumbre Vieja volcanic terrains having more derived haplotypes, located peripherally in the haplotype network. They also predicted signatures of population expansion in the areas of the geologically young Cumbre Vieja volcanic terrains: haplotypes occurring in multiple sampling localities and haplotypes with multiple derivatives differing by only one or few mutations.

The authors sampled 138 specimens from eighteen localities across the distribution of the species (eleven in the northern shield, two in the Bejenado volcanic terrain, and five in the Cumbre Vieja volcanic terrain) to reconstruct a haplotype network and apply nested clade phylogeographic analysis for 570 base pairs of sequence data for the mtDNA COII gene. DNA sequences were aligned by eye and a haplotype network was constructed using the parsimony criterion, as implemented in TCS 1.18 (Clement et al. 2000). A nested clade phylogeographic analysis was undertaken with GeoDis (Posada

(continued)

CASE STUDY 7.3 Phylogeographic Predictions of a Weevil Species of the Canary Islands *(continued)*

et al. 2000), quantifying the geographic distribution of the haplotypes through two distance measures: clade distance (D_c), which measures the spatial spread of a clade, and nested clade distance (D_n), which measures how far a clade is from the clades with which it is nested into a higher-level clade.

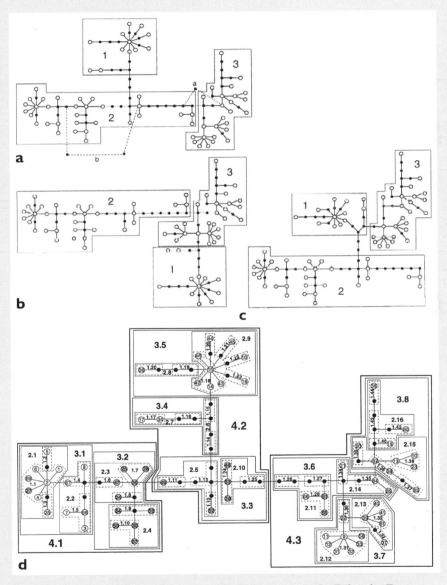

Figure 7.8   Biogeographic analysis of *Brachyderes rugatus* in La Palma by Emerson et al. (2006; reproduced with permission of *Molecular Ecology*). (a–c) Three equally probable arrangements of the haplotype network; (d) nesting design of the haplotype network, where numbered circles refer to specific haplotypes.

(continued)

CASE STUDY 7.3 Phylogeographic Predictions of a Weevil Species of the Canary Islands *(continued)*

Sequences analyzed yielded sixty-nine different haplotypes, with eight haplotypes reported for more than one locality. Network estimation with TCS resulted in a single network, which has a loop with three possible arrangements of three phylogroups (figs. 7.8a–c). These three possible arrangements are equally probable when one considers cladistic information alone but differ when the geography is also considered. Network A (fig. 7.8a) juxtaposes Phylogroups 1 and 3 with Phylogroup 2, a biologically realistic arrangement that does not imply long-distance dispersal of the species through, but not including, an already inhabited area. Networks B and C (figs. 7.8b and 7.8c) juxtapose Phylogroups 1 and 3 only by two mutational differences, implying a biotic connection between the geographically disparate areas of 1 and 3 over a short time period, at the exclusion of the geographically intermediate area 2. Figure 7.8d shows the haplotype network A and its nesting design. Based on the statistics provided by the contingency analysis with GeoDis, Emerson et al. (2006) suggested that a history of contiguous range expansion was the best explanation for the distribution of *B. rugatus* as a whole, but other processes were also present in the three major clades. Clade 4.1 is best explained by past fragmentation followed by range expansion, and its internal Clades 1.1 and 3.2 conform to a contiguous range expansion. Clade 4.2 is also explained by contiguous range expansion, with none of its subclades having any significant geographic structure. Clade 4.3 conforms to a history of past fragmentation followed by range expansion, but internally Subclade 3.6 is consistent with allopatric speciation, and Subclade 3.7 is best explained by restricted gene flow with isolation by distance. The phylogeographic history for *B. rugatus* in La Palma appears complex. The prediction that ancestral haplotypes should occur predominantly in the northern shield, with derived haplotypes featuring more on El Bejenado and Cumbre Vieja terrains, was falsified. The prediction that the newer Bejenado and Cumbre Vieja terrains would have more derived haplotypes, located terminally in a haplotype network, than the northern shield terrain was also falsified. The presence of three distinct phylogroups is clearly at odds with a scenario of a long-term residency in the northern shield, followed by a more recent expansion into the southern terrains. The authors concluded that rather than a single origin in the northern shield followed by a more recent expansion into the Cumbre Vieja region, a series of range expansions has occurred, one into the northern shield and two into the Cumbre Vieja. The authors speculated that these three range expansions may have tracked the past range expansions of the host species *Pinus canariensis*.

References

Carracedo, J. C. and S. Day. 2002. *Canary Islands*. Harpenden, Hertfordshire: Terra.

Clement, M., D. Posada, and K. A. Crandall. 2000. TCS: A computer program to estimate gene genealogies. *Molecular Ecology* 9:1557–1659.

Emerson, B. C., S. Forgie, S. Goodacre, and P. Oromí. 2006. Testing phylogeographic predictions on an active volcanic island: *Brachyderes rugatus* (Coleoptera: Curculionidae) on La Palma (Canary Islands). *Molecular Ecology* 15:449–458.

Posada, D., K. A. Crandall, and A. R. Templeton. 2000. GeoDis: A program for the cladistic nested analysis of the geographical distribution of genetic haplotypes. *Molecular Ecology* 9:487–488.

Molecular Clocks

Cladograms based on molecular data may be used as raw data in cladistic bioge-ography and intraspecific phylogeography. In addition, the assumption that the rate of molecular evolution is approximately constant over time for proteins in all lineages allows one to infer a clock-like accumulation of molecular changes (Brom-ham and Woolfit 2004; Zuckerland and Pauling 1962). The "ticks" of the molecular clock, which correspond to mutations, do not occur at regular intervals but rather at random points in time (Gillespie 1991). This time is measured in arbitrary units and then calibrated in millions of years by reference to the fossil record or geo-logical data (Benton and Donoghue 2007; Magallón 2004; Sanderson 1998), giving minimum estimates of the age of a clade, which in turn may help elucidate the rel-ative minimum ages of the cenocron to which it belongs. Additionally, knowledge of relative minimum ages of divergence may indicate whether a specific disper-sal or vicariance event hypothesis better explains the patterns observed. If clock calibrations provide estimates smaller than those proposed by vicariance events, dispersal may be a better explanation.

The calibration of a molecular clock requires that we find two extant species for which the date of speciation can be determined from the fossil record, to estab-lish the time since the speciation event. Then we compare the DNA sequences of the same gene of both species and count the number of nucleotide substitutions. If all the substitutions are assumed to have arisen after the speciation event, the rate of DNA evolution for the gene under study is obtained by dividing the number of DNA differences between both species by the time since speciation. Assuming a constant mutation rate, we can extrapolate the approximate dates of speciation for other species, for which no fossil dates are available (Crisci et al. 2000). In order to test the molecular clock hypotheses, three tests are available: the likelihood ratio test, the dispersion index, and the relative rate test (Page and Holmes 1998).

As analyses from several taxa began to accumulate in the 1970s, it became apparent that the molecular clock is not always a good model for the process of molecular evolution (Rutschmann 2006). If the null hypothesis of a constant rate is rejected or if we have evidence that rates vary across the tree, we may have to use methods that correct for rate heterogeneity (Rambaut and Bromham 1998) or that estimate divergence times by incorporating rate heterogeneity (Kishino et al. 2001; Sanderson 2002).

Some problems associated with molecular methods include the stochastic na-ture of molecular substitution, the assumption of rate constancy among lineages when such constancy is absent, and the link between substitution rate and elapsed time on the branches of a cladogram (Magallón 2004). Calibration made by refer-ence to geological events runs the risk of circular reasoning because the clock is used to test biogeographic hypotheses that involve an event potentially caused by a geological process (Crisci et al. 2000). Bromham and Woolfit (2004) noted that sometimes molecular date estimates are notoriously at odds with other lines of evidence (e.g., the "Cambrian explosion" of Metazoan phyla and the ordinal-level radiations of mammals and birds are almost twice as old as the available fossils suggest). This discrepancy may have occurred because systematic biases in the fossil record left particular taxa, regions, or periods unrecorded or because explosive radiations could speed the molecular clock, causing dates for these radiations to be overestimated consistently. Magallón (2004) and Benton and Donoghue (2007) clarified the relationship of the fossil record and molecular dating methods, the former documenting first appearances of morphological features and the latter dating splits of molecular lineages (fig. 7.9).

Heads (2005a) offered some criticisms of molecular calibrations:

- Calibration based on fossils: Taking the fossil record at face value and equating the age of the oldest known fossil with the age of the taxon is simplistic and misleading. Although it is recognized that the age of fossilization is less or much less than the age of the taxon, in the practice some authors take the age of the fossil as absolute, not minimum.
- Calibration using age of strata where endemic taxa occur: A common mistake is to assume that a geologist's suggestion that there is no evidence of land in a region and particular period means that there is evidence of no land (e.g., when calculating the age of an island and setting a maximum age limit for a taxon inhabiting it, ignoring the possibility of other former islands in the area or older strata buried beneath the present topography).
- Calibration using paleogeographic events: Most of the standard correlations between distributions and paleogeographic events are highly simplistic. For example, the rise of the Isthmus of Panama, calculated to be 3 mya, which caused vicariance between Pacific and Atlantic marine taxa and also marks the beginning of the Great American Biotic Interchange, is often used as the basis of clock calibrations. However, many authors have pointed out that several Pacific–Atlantic pairs of taxa may have diverged well before the final rise of the isthmus.

Methods   There are several molecular dating methods, grouped into three main classes: methods that use a molecular clock and one global rate of substitution,

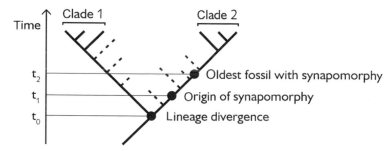

Figure 7.9   Relationship of lineage divergence, origin of a synapomorphy, and occurrence of the oldest fossil with the synapomorphy, according to Magallón (2004). There is a temporal gap between the divergence of a taxon and its sister taxon (t_0), the origin of the synapomorphy (t_1), and the occurrence of the oldest fossil bearing such synapomorphy (t_2).

methods that correct for rate heterogeneity, and methods that try to incorporate rate heterogeneity. Each method has it own algorithm. For a revision, see Rutschmann (2006).

Software DNAMLK of package PHYLIP (Felsenstein 1993), BASEML of package PAML (Yang 1997), QDATE (Rambaut and Bromham 1998), PHYBAYES (Aris-Brosou and Yang 2001), RHINO (Rambaut 2001), maximum likelihood clock optimization method of PAUP (Swofford 2003), PATH (Britton et al. 2002), BEAST (Drummond and Rambaut 2003), MULTIDIVTIME (Thorne and Kishino 2002), TREEEDIT (Rambaut and Charleston 2002), and R8s (Sanderson 2003).

Empirical Applications Bossuyt et al. (2006), Caccone et al. (1997), Chen et al. (2007), Doadrio and Carmona (2003), Fritsch (2001), Hewitt (2004), Hibbett (2001), Jansa et al. (2006), Krzywinski et al. (2001), Lalueza-Fox et al. (2005), León-Paniagua et al. (2007), Magallón and Sanderson (2005), Morell et al. (2000), Olmstead and Palmer (1997), Poux et al. (2005, 2006), Renner et al. (2001), Ribas et al. (2005), Robalo et al. (2006), Tavares et al. (2006), Voelker (1999), Waters et al. (2000), Won and Renner (2006), Xiang et al. (1996, 1998), Yuan et al. (2005), Zakharov et al. (2004), and Zink et al. (2002).

CASE STUDY 7.4 The Mediterranean Lago Mare Theory and the Speciation of European Freshwater Fishes

Primary freshwater fishes, those that are intolerant to marine conditions, have had an important role in the development of evolutionary biogeographic hypotheses. Most authors have used them as a model for studying the paleobiogeography of European rivers, usually from a dispersalist perspective. Banarescu (1992) wrote that the current distribution of European freshwater fishes resulted from dispersal through eastern Asia into Siberia toward Central Europe and the Mediterranean, after the Turgai Strait was closed 35 mya. In contrast with these ideas, Bianco (1990) suggested that the Mediterranean peninsulas were colonized from the south, by crossing of the Mediterranean Sea during the Messinian salinity crisis (5.3 mya), when the Mediterranean basin was almost completely dried up and then refilled with freshwater from the Sarmatic Sea. This is known as the Lago Mare hypothesis. Doadrio and Carmona (2003) analyzed the phylogenetic relationships of the Cyprinid genera *Chondrostoma* and *Squalius*, both widely distributed in Eurasia, on the basis of the mitochondrial cytochrome b gene in order to test the effect of the Mediterranean Lago Mare hypothesis on their speciation and geographic distribution. For each genus, complete nucleotide sequences of cytochrome b were recovered from populations of different species. The Bayesian analysis was performed with MrBayes version 3.0 (Huelsenbeck and Ronquist 2001). In order to examine whether lineages evolved according to a molecular clock, the Likelihood Radio test was conducted with and without molecular clock constraints using Puzzle 4.0.1 (Strimmer and von Haeseler 1996).

The Bayesian analysis on forty-seven specimens of *Squalius* and sixty-one specimens of *Chondrostoma* identified different mtDNA lineages. For *Squalius*, previously recognized Mediterranean Eurasiatic and Paratethys lineages were recovered, and

(continued)

CASE STUDY 7.4 The Mediterranean Lago Mare Theory and the
Speciation of European Freshwater Fishes *(continued)*

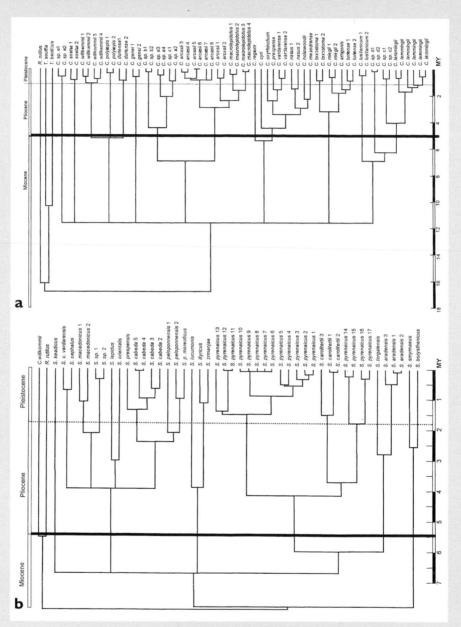

Figure 7.10 Biogeographic analysis of European freshwater fishes by Doadrio and
Carmona (2003; reproduced with permission of *Graellsia*). (a) Maximum likelihood
cladogram of *Chondrostoma*; (b) maximum likelihood cladogram of *Squalius*. Scale
bars on the right show the timescale that resulted from the application of a molecular
clock. Horizontal black bars represent the time of occurrence of the Lago Mare phase.

(continued)

CASE STUDY 7.4 The Mediterranean Lago Mare Theory and the Speciation of European Freshwater Fishes *(continued)*

within the Mediterranean lineage, relationships between the Iberian, Greek, and Italo-Balkanic species were unresolved, indicating a divergence in a short period of time (fig. 7.10a). Analysis of *Chondrostoma* recovered two lineages, Iberian and Eurasiatic; polytomies were found for the relationships of Italian and Eurasiatic species and for the species groups inhabiting the Iberian Peninsula (fig. 7.10b).

Doadrio and Carmona (2003) concluded that their results indicated that the ancestor of *Squalius* species from the Iberian Peninsula inhabited the area during the Miocene, before the Lago Mare phase. Similarly, the four main Iberian species groups of *Chondrostoma* inhabited the peninsula in the Middle–Upper Miocene. Therefore, the Lago Mare phase of the Mediterranean Sea seems to have been too recent an event to have had any impact on the biogeographic evolution of both genera in the Iberian Peninsula. Furthermore, the authors wrote that the dry and wet periods during the Cenozoic were long enough to allow a dispersal and evolution in the area more gradual than the Lago Mare hypothesis predicts.

Robalo et al. (2007) reexamined the phylogeny of *Chondrostoma* based on mitochondrial and nuclear data. They obtained a highly resolved cladogram, which prompted them to reject the hypothesis of a simultaneous origin of multiple lineages, thus confirming the previous analysis.

References

Banarescu, P. 1992. *Zoogeography of freshwater: Distribution and dispersal of freshwater animal in North America and Eurasia*, Vol. 2. Wiesbaden: Aula-Verlag.

Bianco, P. G. 1990. Potential role of the paleohistory of the Mediterranean and Parathetis basins on the early dispersal of Euro-Mediterranean freshwater fishes. *Ichthyological Exploration of Freshwaters* 1:167–184.

Doadrio, I. and J. A. Carmona. 2003. Testing freshwater Lago Mare dispersal theory on the phylogeny relationships of Iberian Cyprinid genera *Chondrostoma* and *Squalius* (Cypriniformes, Cyprinidae). *Graellsia* 59(2–3):457–473.

Huelsenbeck, J. P. and F. Ronquist. 2001. MrBayes: Bayesian inference of phylogenetic trees. *Bioinformatics* 17(8):754–755.

Robalo, J. I., V. C. Almada, A. Levy, and I. Doadrio. 2007. Re-examination and phylogeny of the genus *Chondrostoma* based on mitochondrial and nuclear data and the definition of 5 new genera. *Molecular Phylogenetics and Evolution* 42:362–372.

Strimmer, K. and A. von Haeseler. 1996. Quartet puzzling: A quartet maximum-likelihood method for reconstructing tree topologies. *Molecular Biology and Evolution* 13:964–969.

CASE STUDY 7.5 The Arrival of Caviomorph Rodents and Platyrrhine Primates in South America

South America was an isolated continent after it separated from Africa in the Cretaceous (90–100 mya), until it reconnected with North America in the Pliocene (3–2.5 mya). According to the fossil record, in the Eocene–Early Oligocene (50–30 mya), South

(continued)

CASE STUDY 7.5 The Arrival of Caviomorph Rodents and Platyrrhine Primates in South America *(continued)*

America was colonized by two groups of placental mammals, rodents and primates, of which the extant caviomorphs (guinea pigs, chinchillas, agoutis, New World porcupines, and arboreal spiny rats) and platyrrhines (sakis, spider monkeys, capuchins, and marmosets) are the descendants, but the exact times and ways of colonization are debated (Arnason et al. 2000; Hoffstetter 1972; Springer et al. 2003). Poux et al. (2006) undertook a molecular clock analysis in order to demarcate the periods of possible colonization by estimating the time windows between their divergence from their respective Old World sister groups (upper or oldest bound) and the subsequent diversifications in South America (lower or most recent bound).

Poux et al. (2006) analyzed separate and combined sequences of three nuclear genes, coding for the alpha 2B adrenergic receptor, the von Willebrand factor, and the interphotoreceptor retinoid binding protein. For each gene, the authors selected sixty species using three criteria: Species should represent all placental orders and two marsupial mammals as outgroups; the sampling should reflect the diversity of primate and rodent taxa, with a broad representation of platyrrhines and caviomorphs; and species should enable the use of paleontological calibration constraints from various lineages, thus minimizing the dependence of the results on a single fossil reference. Sequences were aligned by hand with the ED editor of MUST version 2000 (Philippe 1993). The phylogenetic analyses were performed on the complete DNA data set by maximum likelihood with PAUP version 4 (Swofford 1999), and the Bayesian analyses were performed with MrBayes version 3.0 (Huelsenbeck and Ronquist 2001). Molecular dating analyses were performed according to the Bayesian relaxed molecular clock approach (Thorne et al. 1998), using the MULTIDIVTIME package (Thorne and Kishino 2002). For fossil calibrations, Poux et al. (2006) selected eight time constraints previously used in other molecular dating studies: outside the Primates and Rodentia, the diversification ages of Paenungulata (54–65 mya), Perissodactyla (54–58 mya), Cetartiodactyla (55–65 mya), and Lagomorpha (minimum age of 37 mya); within rodents, the splits *Glis–Dryomys* (minimum age of 28.5 mya) and *Aplodontia–Marmota* (minimum age of 37 mya); and within Primates, the basal primate radiation (63–90 mya) and the Cercopithecoidea–Hominoidea divergence (25–35 mya).

The maximum likelihood and Bayesian analyses yielded identical topologies (fig. 7.11). The four major mammalian clades (Afrotheria, Xenarthra, Laurasiatheria, and Euarchontoglires) were strongly supported. Within primates and rodents, the monophyly of platyrrhines and caviomorphs had maximal support. All clades relevant to the understanding of the problem analyzed were well supported, providing a reliable framework for the assessment of divergence times. According to the molecular datings (fig. 7.11), the platyrrhines arrived in South America within a time window of at most 25.5 million years, between 34.0–40.0 mya (the age of the Catarrhini–Platyrrhini divergence) and 14.5–19.1 mya (the earliest diversification of the extant platyrrhines). By contrast, the arrival of caviomorphs must have taken place in a time window of at most 16.5 million years (Middle Eocene), between 41.4–49.6 mya (the Phiomorpha–Caviomorpha split) and 33.0–4.1 mya (the earliest radiation of extant caviomorphs). There is an overlap of 7.0 million years between the latest possible arrival of caviomorphs (33.0 mya) and the earliest possible arrival of platyrrhines (40.0 mya). The authors concluded that a concomitant arrival of both groups in South America cannot be ruled out.

(continued)

CASE STUDY 7.5 The Arrival of Caviomorph Rodents and Platyrrhine Primates in South America (*continued*)

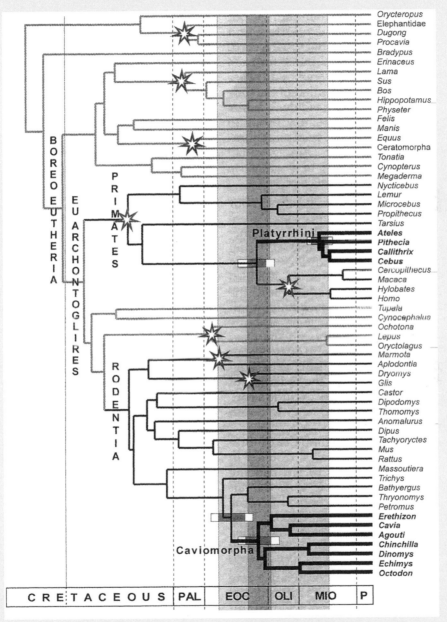

Figure 7.11 Cladogram showing the posterior divergence ages of placental taxa obtained by Poux et al. (2006; reproduced with permission of *Systematic Biology*). The vertical zones span the periods between the origin and radiation of Caviomorpha and Platyrrhini, with an overlapping period during which they could have reached South America synchronously. EOC, Eocene; MIO, Miocene; OLI, Oligocene; P, Pliocene; PAL, Paleocene.

(continued)

CASE STUDY 7.5 The Arrival of Caviomorph Rodents and Platyrrhine Primates in South America *(continued)*

With regard to the possible dispersal histories, the results are consistent with an African origin for caviomorphs and platyrrhines, from phiomorph and anthropoid stocks, respectively, followed by transatlantic migration. Despite the distance between Africa and South America in the Middle Eocene–Early Oligocene, colonization could have occurred, aided by marine currents, paleowinds, or stepping stone islands along with rafts. This transatlantic route is the preferred hypothesis for platyrrhine dispersal for two reasons: Fossils of early platyrrhines and catarrhines so far have been found only in Africa, and migration through Antarctica is unlikely for this group because at the time of the platyrrhine–catarrhine divergence (37 mya) Australia, Antarctica, and South America were no longer connected, and Antarctica was covered by ice sheets. With respect to rodents, South American caviomorphs and African phiomorphs may share an Asian ancestor, but it is not clear whether they diverged already in Asia or after dispersal of their ancestor into Africa, leaving the dispersal route to South America open to speculation. The caviomorph fossil record indicates that by 31 mya these rodents had probably started to diversify in South America and that their arrival predated the Early Oligocene, implying that extant lineages derive from early diversification events. There are still paleontological uncertainties about the time of radiation of living South American platyrrhines. However, fossils from the Late Miocene of La Venta (Colombia) are highly similar to modern taxa, consistent with the dating of the platyrrhine diversification in the Early Miocene. Poux et al. (2006) concluded that the most plausible scenario for primates was a transatlantic dispersal at the end of the Miocene, followed by the extinction of all but one of the earlier diverging lineages and the radiation of extant taxa in the Early Miocene. The arrival of rodents and primates might have been contemporaneous; however, in contrast to platyrrhines, representatives of the early diversification of caviomorphs, which occurred before the Oligocene glaciations, survived until the present.

References

Arnason, U., A. Gullberg, A. S. Burguete, and A. Janke. 2000. Molecular estimates of primate divergences and new hypotheses for primate dispersal and the origin of modern humans. *Hereditas* 133:217–228.

Hoffstetter, R. 1972. Relationships, origins, and history of the ceboid monkeys and caviomorph rodents: A modern reinterpretation. *Evolutionary Biology* 6:322–347.

Philippe, H. 1993. MUST: A computer package of management utilities for sequences and trees. *Nucleic Acids Research* 21:5264–5272.

Poux, C., P. Chevret, D. Huchon, W. W. de Jong, and E. J. P. Douzery. 2006. Arrival and diversification of caviomorph rodents and platyrrhine primates in South America. *Systematic Biology* 55(2):228–244.

Springer, M. S., W. J. Murphy, E. Eizirik, and S. J. O'Brien. 2003. Placental mammal diversification and the Cretaceous–Tertiary boundary. *Proceedings of the National Academy of Sciences* 100:1056–1061.

Swofford, D. L. 1999. *PAUP*: Phylogenetic analysis using parsimony (*and other methods)*. Version 4.0 beta. Sunderland, Mass.: Sinauer.

Thorne, J. L. and H. Kishino. 2002. Divergence time and evolutionary rate estimation with multilocus data. *Systematic Biology* 51:689–702.

For Further Reading

Avise, J. C. 2000. *Phylogeography: The history and formation of species.* Cambridge, Mass.: Harvard University Press.

Avise, J. C., J. Arnold, R. M. Ball, E. Bermingham, T. Lamb, J. E. Neigel, C. A. Reeb, and N. C. Saunders. 1987. Intraspecific phylogeography: The mitochondrial DNA bridge between population genetics and systematics. *Annual Review of Ecology and Systematics* 18:489–522.

Benton, M. J. and P. C. J. Donoghue. 2007. Paleontological evidence to date the tree of life. *Molecular Biology and Evolution* 24:26–53.

Grande, L. 1985. The use of paleontology in systematics and biogeography, and a time control refinement for historical biogeography. *Paleobiology* 11:234–243.

Hunn, C. A. and P. Upchurch. 2001. The importance of time/space in diagnosing the causality of phylogenetic events: Towards a "chronobiogeographical paradigm." *Systematic Biology* 50:391–407.

Problems

Problem 7.1

A hypothetical time-slicing analysis of the Mexican Transition Zone (fig. 7.12a) has led to the recognition of three time slices, which correspond to the Cretaceous (fig. 7.12b), Eocene (fig. 7.12c), and Pleistocene (fig. 7.12d). How would you interpret these results?

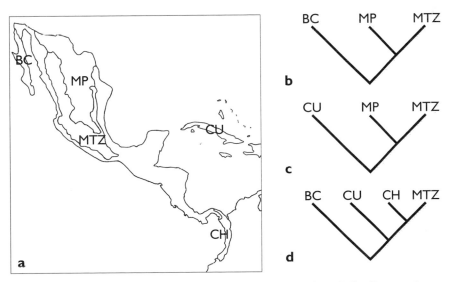

Figure 7.12   (a) Mexican Transition Zone and other areas analyzed; (b–d) general area cladograms corresponding to different time slices; (b) Cretaceous; (c) Eocene; (d) Pleistocene. BC, Baja California; CH, Chocó; CU, Cuba; MP, Mexican Plateau; MTZ, Mexican Transition Zone.

Problem 7.2

Figure 7.13 illustrates the matrilineal phylogeny of *Pronolagus rupestris*, Smith's red rock rabbit, from South Africa (Matthee and Robinson 1996). Which phylogeographic pattern would you infer for this species?

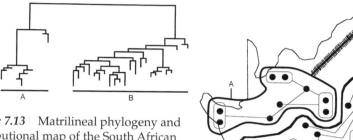

Figure 7.13   Matrilineal phylogeny and distributional map of the South African rabbit *Pronolagus rupestris*.

Problem 7.3

Figure 7.14 illustrates the matrilineal phylogenies of three different sympatric species (Avise 2000). Which phylogeographic histories would you infer from these hypotheses?

Figure 7.14   Matrilineal phylogenies of three different sympatrid species.

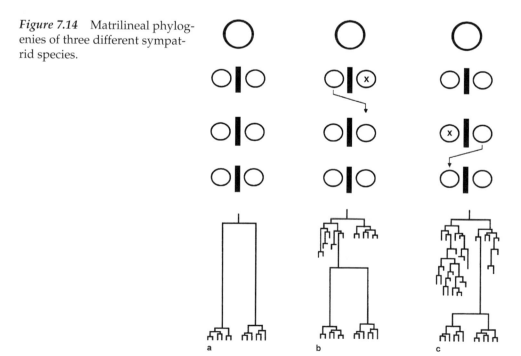

For Discussion

1. Carefully read the following article:

    Upchurch, P., C. A. Hunn, and D. B. Norman. 2002. An analysis of dinosaurian biogeography: Evidence for the existence of vicariance and dispersal patterns caused by geological events. *Proceedings of the Royal Society of London, Series B* 269:613–621.

    a. Identify and extract the main ideas.

    b. Transform each main idea into a question.

2. Carefully read the following article:

    Avise, J. C., J. Arnold, R. M. Ball, E. Bermingham, T. Lamb, J. E. Neigel, C. A. Reeb, and N. C. Saunders. 1987. Intraspecific phylogeography: The mitochondrial DNA bridge between population genetics and systematics. *Annual Review of Ecology and Systematics* 18:489–522.

    a. Identify and extract the main ideas.

    b. Transform each main idea into a question.

3. Imagine you are studying a taxon with three species distributed on one continental area and three others distributed on an island. Geological information indicates that the vicariance between the continental and insular species might date to 2 mya, when the island originated. How would you use the following elements to falsify a vicariance hypothesis?

    a. Phylogenetic analysis of the six species.

    b. Intraspecific phylogeography of the insular species.

    c. Molecular clock of the six species.

    d. Fossils from the continent and the island.

CHAPTER 8

Construction of a Geobiotic Scenario

Once we have identified the biotic components and cenocrons, we may be able to construct a geobiotic scenario. By compiling biological data (e.g., means of dispersal) and nonbiological data (e.g., past continental configurations) we can integrate a plausible scenario to help explain the episodes of vicariance or biotic divergence and dispersal or biotic convergence that have shaped the biotic evolution of the biotic components analyzed. In this chapter I discuss some basic concepts that are considered in the construction of geobiotic scenarios, with a brief account of plate tectonics.

Geographic Features

Both panbiogeographers and cladistic biogeographers have shown interest in geology, geophysics, and plate tectonics (Cooper 1989; Craw 1988a; Craw et al. 1999; Ebach and Humphries 2002; Grehan 2001b; Heads 1989; Michaux 1989). Geology and biogeography have a causal relationship; they are the independent and dependent variables, respectively (Michaux 1989). This does not imply that geological hypotheses necessarily validate biogeographic hypotheses. Geologists may not necessarily have interpreted geological history of the area adequately enough to justify validation. In fact, the relationship between geology and biogeography should be based on its capacity for "reciprocal illumination" (in the sense of Hennig 1950). In order to make this relationship more fruitful, it would be important to develop a common language that allows interconnection of the biological and geological systems. This is because evolution in space and time of taxa and biotas is a unique geobiotic phenomenon. Tracks and area cladograms are appropriate instruments to develop such a common language.

Biogeographers have classified geographic features in terms of their impact on dispersal and vicariance (Cox and Moore 1998; MacDonald 2003; Rapoport 1975; Simpson 1953, 1965; Vargas 1992b). The most important are barriers (geographic features that hinder dispersal) and corridors (geographic features that facilitate dispersal). Barriers are easily identified with geographic elements such as mountains, rivers, and seas. In the marine environment, in addition to land barriers (e.g., the Isthmus of Panama), there can be more subtle barriers, represented by changes in physicochemical properties (Cecca 2002). Corridors include such a variety of habitats that many of the organisms found at either end of them have little difficulty traversing them (Cox and Moore 1998). These terms are relative because, for example, a cordillera may act as a barrier for certain species but be a corridor for others. Instead of barriers, cladistic biogeographers usually refer to vicariance events.

In some instances, physical or biological conditions make it easier or more difficult for certain species to cross a certain barrier. Features that are not equally favorable for dispersal of all species are called filters. For example, before the rise of the Isthmus of Panama, a chain of small islands (stepping stones) on a shallow sea of about 150 m in depth occupied its place. These islands facilitated dispersal of some species (e.g., mice) but acted as a barrier for other species (e.g., flightless birds). After the Isthmus of Panama developed, during the Pleistocene, most of Central America was occupied by dense tropical forests, which allowed the dispersal of forest species but acted as a barrier for biota from the savannas (MacDonald 2003).

There are some areas completely surrounded by totally different environments, such as islands, caves, and high mountain peaks, where chances of dispersal are very low for most taxa. They are known as sweepstake routes, and they differ from filters in kind, not merely in degree, because almost all species that traverse them cannot survive (Cox and Moore 1998).

Plate Tectonics

In dealing with long-term changes (table 8.1) in the biotic distributional patterns, continental drift may be a relevant factor (Briggs 1987; Cox and Moore 1998). Not only do the splitting and collision of landmasses affect distributional patterns directly, but new mountains, oceans, and land barriers change the climatic patterns on the landmasses.

Continental drift was originally proposed by Wegener (1912) and met with enormous opposition. Plate tectonics was a mechanism that explained continental drift and made it a credible theory. Seafloor spreading is believed to be caused by great convection currents that bring material to the surface from the hot interior of the earth, inducing the movement of tectonic plates. These are the moving units at the surface of the earth and may contain continental masses or may consist of ocean floor. The movement of the plates had great relevance for organisms. The movement of the continents relative to the poles and the equator caused climatic changes. Additionally, shallow epicontinental seas covered parts of the continents or formed seas within them during the Jurassic and Cretaceous periods, forming barriers to dispersal. The splitting of continents also altered the patterns of water circulation in the oceans. Furthermore, the appearance of new mountains as a result of continental drift had dramatic consequences for biotic distributional patterns.

A major feature of the Late Paleozoic and Mesozoic was the supercontinent of Gondwana. It included the land areas that later became South America,

Table 8.1

Era	Period	Epoch	Approximate Duration (millions of years)	Approximate Date of Commencement (millions of years)
Cenozoic	Quaternary	Holocene	0.01	0.01
		Pleistocene	1.59	1.6
	Tertiary	Pliocene	3.7	5.3
		Miocene	21.9	23.5
		Oligocene	10.5	34.0
		Eocene	19.0	53.0
		Paleocene	12.0	65.0
Mesozoic	Cretaceous		70	135
	Jurassic		70	205
	Triassic		40	245
Paleozoic	Permian		50	295
	Carboniferous		65	360
	Devonian		50	410
	Silurian		25	435
	Ordovician		71	500
	Cambrian		70	570
Proterozoic			4,000	4,600

Africa, Madagascar, Antarctica, Australia, New Zealand, and India. Its northern edge broke up into a series of minor landmasses or terranes, which moved north and joined the southern edge of Eurasia to form southern Europe, Tibet, and two separate portions of China (Cox and Moore 1998). By the Silurian (435–410 mya), there were three continents: Euramerica (North America and Eurasia), Siberia, and Gondwana. In the Late Carboniferous to Early Permian (300–270 mya) Euramerica joined Gondwana, and in the Late Permian (260 mya) Siberia joined this landmass, forming the world continent of Pangaea. Pangaea soon divided into two landmasses: Laurasia in the north and Gondwana in the south. From the Jurassic to the Cretaceous, Laurasia was penetrated by epicontinental seas and Gondwana started to break up into separate continents. India separated from the rest of Gondwana in the Early Cretaceous. In the Late Cretaceous, Europe and Asia were separated by the Obisk Sea; the former was connected to eastern North America (the Euramerican landmass), and the latter was connected to western North America (the Asiamerican landmass).

Sanmartín and Ronquist (2004) provided a recent synthesis of ideas concerning the fragmentation of Gondwana, which they presented as a geological area cladogram (fig. 8.1). Gondwana started to break up in the Jurassic (165–150 mya), when rifting began between India and Australia–east Antarctica. Shortly

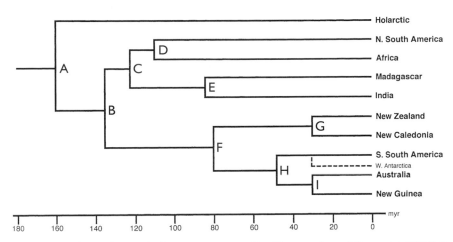

Figure 8.1   Geological area cladogram presented by Sanmartín and Ronquist (2004) that represents the relationships between the southern continents based on paleogeographic evidence. A, North Atlantic (180–160 mya); B, southern South Atlantic (135 mya); C, Somali Basin (121 mya); D, northern South Atlantic (110 mya); E, Mascarene Basin (84 mya); F, Tasman Sea (80 mya); G, New Caledonian Basin (30 mya); H, South Tasman Sea (52–35 mya); I, Coral Sea Basin (30 mya).

after, the Madagascar–India block, which was adjacent to Somalia, broke away from Africa and began moving southeast, attaining its present position in the Early Cretaceous (121 mya). India separated from Madagascar in the Late Cretaceous (88–84 mya), with the opening of the Mascarene Basin, and began drifting north, eventually to collide with Asia about 50 mya. South America began to separate from Africa in the Early Cretaceous (135 mya), with the opening of the South Atlantic Ocean at the latitude of Argentina and Chile. Northern South America and Africa remained connected until the mid-Late Cretaceous (110–95 mya), when a transform fault opened between Brazil and Guinea. As a result, Africa started drifting northeast and collided with Eurasia in the Paleocene (60 mya), whereas southern South America drifted southeast into contact with Antarctica. New Zealand, Australia, South America, and Antarctica remained connected until the Late Cretaceous: East Antarctica was adjacent to southern Australia, whereas New Zealand and southern South America were in contact with west Antarctica. About 80 mya, the Tasmantis block (New Zealand and New Caledonia) broke away from west Antarctica and moved northwest, opening the Tasman Sea. New Zealand and New Caledonia were finally separated in the mid-Tertiary (40–30 mya), when the Norfolk Ridge foundered, opening the New Caledonian Basin. Australia and South America remained in contact across Antarctica until the Eocene. Australia began to separate from Antarctica in the Late Cretaceous (90 mya), but both continents remained in contact along Tasmania, and complete separation did not occur until the Late Eocene (35 mya) with the opening of the South Tasman Sea. Southern South America and Antarctica remained in contact through the Antarctic peninsula until the Oligocene (20–28 mya), when the Drake Passage opened between these continents, allowing the establishment of the Antarctic Circumpolar Current. After its separation from Antarctica, Australia began to drift rapidly toward Asia. New Guinea was then joined to the northern margin of the Australian plate, although only the southern margin of New Guinea was emergent at that time. The collision of the Australian and Pacific plates in the Oligocene (30 mya) initiated the tectonic uplift of New Guinea, but by the Early Miocene much of southern New Guinea was again submerged. Subsequent episodes of uplift in the Miocene, after the collision of the Australian and Asian plates, led to the accretion of numerous terranes to the northern margin of New Guinea. The link between North and South America, the Isthmus of Panama, was formed in the Late Pliocene (2 mya).

There is still much discussion about some aspects of tectonics. Theories postulating a lost Pacifica continent (Kamp 1980; Nur and Ben-Avraham 1980) or an expanding Earth (McCarthy 2003, 2007; Shields 1979, 1991, 1996) have been proposed to explain certain "anomalies" in Wegener's theory, but they have not gained support by geophysicists (Humphries and Ebach 2004).

Any of these theories implies a major role for vicariance in isolating populations of plant and animal species (Cox and Moore 1998). Continents split, and their fragments carry away their cargo of living organisms ("Noah's arks") and buried fossils ("Viking funeral ships") (McKenna 1973). Hallam (1974) characterized two main historical phenomena caused by plate tectonics: convergence and divergence. Biogeographic convergence indicates intermixing of biotas through continental shifting: Two continents approach each other, and their biotas merge. Biogeographic divergence occurs when two continents move apart or two landmasses are isolated by a seaway.

For Further Reading

Briggs, J. C. 1987. *Biogeography and plate tectonics*. Amsterdam: Elsevier.
Cecca, F. 2002. *Palaeobiogeography of marine fossil invertebrates: Concepts and methods*. London: Taylor and Francis.
Lieberman, B. S. 2000. *Paleobiogeography: Using fossils to study global change, plate tectonics and evolution*. New York: Kluwer.

For Discussion

1. What is the relevance of a geobiotic scenario for the integration of biogeography with other disciplines?
2. Carefully read the following article:

    Ebach, M. C. and C. J. Humphries. 2002. Cladistic biogeography and the art of discovery. *Journal of Biogeography* 29:427–444.

    a. Identify and extract the main ideas.
    b. Transform each main idea into a question.

3. Carefully read the following articles:

Briggs, J. C. 2004. The ultimate expanding Earth hypothesis. *Journal of Biogeography* 31:855–857.

McCarthy, D. 2003. The trans-Pacific zipper effect: Disjunct sister taxa and matching geological outlines that link the Pacific margins. *Journal of Biogeography* 30:1545–1561.

Establish the arguments in favor of the expanding Earth hypothesis and those against it.

CHAPTER 9

Toward an Integrative Biogeography

The enormous methodological diversity in biogeography has led to extreme views. Keast (1991), Tassy and Deleporte (1999), and Vuilleumier (1999) suggested that the existence of multiple methods indicates that biogeography is far from coherent as a discipline. Morrone and Crisci (1995) and Riddle and Hafner (2004) found that this attests to the vitality of the field. Ebach and Humphries (2003:959) stated that "the present plethora of techniques reflect a lack of scientific debate and agreement as to what constitutes the ontology (specification of conceptualization) of biogeography." Brown (2004:32) found "this explosion both exciting and intimidating." Walter (2004:907) stated that "the excitement and challenge of biogeography lies in this extraordinary diversity of research." Many authors are not interested in theoretical aspects and are interested only in the biotic patterns exhibited by the taxa that they study. This is evident in the high number of published works where, after the phylogenetic analysis of the taxon, the author discusses its "biogeography," based on the cladogram obtained. The real losers in this complex situation are the students, who look for answers and find instead a perplexing situation, leaving them feeling alienated and powerless.

If we compare the situation of biogeography with that of systematics in recent decades, there is a remarkable difference. Although in systematics alternative approaches have existed (e.g., evolutionary taxonomy, cladistics, and phenetics), the theoretical–methodological development led to the clear preeminence of the cladistic approach. In biogeography, no unified approach has yet emerged. However, in my opinion several points indicate that a biogeographic synthesis might be developing:

- The recognition that historical and ecological factors interact to determine the global patterns of the biodiversity and the need for more effective communication between evolutionary and ecological biogeographers (Crisci et al. 2006; Davis and Scholtz 2001; Holloway 2003; Ruggiero and Ezcurra 2003). As expressed by Riddle (2005:186), "the dichotomy between an ecological vs. historical biogeography simply does not track the many patterns and processes considered relevant and worthy of attention."
- The realization that dispersal and vicariance are relevant phases of biotic evolution (Morrone 2004a). Biogeography should abandon the sterile opposition between dispersal and vicariance and accept that after biotic patterns

are elucidated, the processes (vicariance, dispersal, and extinction) that caused them can be investigated fully.

- The studies showing that biotas are mosaics created by vicariance, dispersal, and extinction and therefore show reticulate patterns, necessitating a better understanding of biotic evolution from a phylogenetic perspective (Brooks and McLennan 2001; Brown 2004; Lieberman 2004; Losos 1996; McLennan and Brooks 2002; Webb et al. 2002).
- The rediscovery of regionalization as a relevant biogeographic inquiry (Craw et al. 1999; Morrone 2001a). Biogeographic classifications are not definitive because they change as our vision of biotic evolution is modified, representing hypotheses about the delimitation of biotic components and their historical relationships that help us explore nature (Morrone 2006).
- The increasing awareness of the need to integrate temporal information into biogeographic analyses (Donoghue and Moore 2003) while accepting that fossils give minimum ages for groups, with its implications for accurate molecular clock calibration (Benton and Donoghue 2007; Magallón 2004).
- The development of phylogeography as a bridge between microevolution and macroevolution (Avise 2000; Riddle and Hafner 2006).
- The development of macroecology, the analysis of statistical patterns of ecological attributes between sets of species on large scales, which allows a bridge between ecology and biogeography (Maurer 2000).
- Reflection about the interaction between the genealogical and ecological hierarchies to help explain evolutionary processes and patterns (Eldredge 1985; Eldredge and Salthe 1984; Morrone 2004c).
- The development of biogeography as an interdisciplinary science (Brown 2004) or as an independent and pluralist discipline (Nihei 2006) that should continue expanding its realm of inquiry and extend its interfaces with other disciplines.

In this book I have revised some criteria and concepts, with the intention of prompting reflection about biogeography. I hope that the development of a new biogeographic synthesis helps eliminate archaic notions such as centers of origin. In order to end the constant debate in biogeography it would be necessary to rethink many biogeographic strategies. With greater collaboration, ecological and evolutionary biogeographers may eventually discover that their theories can be articulated as parts of a more inclusive theory. I hope that the ideas outlined here help develop such integrative biogeography.

For Further Reading

Donoghue, M. J. and B. R. Moore. 2003. Toward an integrative historical biogeography. *Integrative and Comparative Biology* 43:261–270.

Riddle, B. R. and D. J. Hafner. 2006. A step-wise approach to integrating phylogeographic and phylogenetic biogeographic perspectives on the history of a core North American warm deserts biota. *Journal of Arid Environments* 66:435–461.

For Discussion

1. Search for articles dealing with the integration of biogeographic ideas.

   a. What are the differences and similarities between them?

   b. Do you think that different authors see integration in the same way?

2. Carefully read the following articles:

   Donoghue, M. J. and B. R. Moore. 2003. Toward an integrative historical biogeography. *Integrative and Comparative Biology* 43:261–270.

   Riddle, B. R. and D. J. Hafner. 2006. A step-wise approach to integrating phylogeographic and phylogenetic biogeographic perspectives on the history of a core North American warm deserts biota. *Journal of Arid Environments* 66:435–461.

   a. Extract their main ideas concerning integration.

   b. Compare them, finding their similarities and differences.

Glossary

a posteriori methods   cladistic biogeographic methods that deal with dispersal, extinction, and duplicated lineages after the parsimony analysis of a data matrix based on the unmodified taxon–area cladograms. They include Brooks parsimony analysis and component compatibility.

a priori methods   cladistic biogeographic methods that allow modification of the area relationships in the taxon–area cladograms to deal with dispersal, extinctions, or duplicated lineages, in order to obtain resolved area cladograms and provide the optimum fit to a general area cladogram. They include component analysis, tree reconciliation analysis, and three area statement analysis.

area cladistics   cladistic biogeographic method derived from component and three area statement analyses in which one begins by replacing the names of the terminal taxa of two or more taxon–area cladograms and deriving areagrams and then resolving paralogy using the transparent method. Combining areagrams of different taxa inhabiting the same areas allows identification of patterns.

area of endemism   area of nonrandom distributional congruence between different taxa.

Asiamerica   supercontinent from the Late Cretaceous that included western North America and Asia.

assumption 0   considers areas inhabited by a widespread taxon as a monophyletic group in the resolved area cladogram, meaning that the taxon is treated as a synapomorphy of the areas.

assumption 1   considers areas inhabited by a widespread taxon as a monophyletic or paraphyletic group.

assumption 2   considers areas inhabited by a widespread taxon as a monophyletic, paraphyletic, or polyphyletic group.

barrier   geographic feature that hinders dispersal. It is easily identified with geographic elements such as mountains, rivers, and seas. In the marine environment, in addition to land barriers, there can be more subtle barriers, represented by changes in physicochemical properties.

baseline   spatial correlation between an individual track and a geographic or geological feature, such as an ocean or marine basin, a river, or a mountain chain, that is used to orient an individual track.

biogeographic convergence   biotic mixture caused by geodispersal that leads to the reticulated, nonhierarchical evolution of a biotic component.

biogeographic divergence   result of area split (due to vicariant events). Vicariance is the most common interpretation of divergence patterns.

biogeographic homology   sorting procedure used to establish meaningful comparisons within biogeography. It represents the relationship between biotic components rather than the components themselves.

biogeography   study of the geographic distribution of taxa and their attributes in space and time.

biotic component   set of spatiotemporally integrated taxa that coexist in a given area, which represents a biogeographic unit, from a synchronic or proximal perspective. Biotic components may remain the same despite the possible transformations they could undergo, may split into two or more independent biotic components as a result of vicariance, may mix into a new biotic component as a result of biogeographic convergence, and eventually may become extinct.

Brooks parsimony analysis (BPA)   cladistic biogeographic method that consists of the parsimony analysis of taxon–area cladograms that are codified as two-state variables and analyzed as characters.

Buffon's law   discovery made by Comte Georges-Louis Leclerc de Buffon (1707–1788), who found that different tropical areas of the world, even those that had some similar climatic and environmental conditions, were inhabited by completely different mammal species. It represented the first falsification of Linnaeus's explanation.

causal biogeography   subdiscipline that investigates the causes of an observed pattern, encompassing both historical and ecological biogeography.

cenocron   set of taxa that share the same biogeographic history, constituting an identifiable subset within a biotic component by their common biotic origin and evolutionary history, from a diachronic perspective. Cenocrons allow one to represent how the convergence of biotic components occurs during biotic evolution.

cenogenesis   evolution of biotic associations through time.

center of origin   restricted geographic area from which new species are constantly evolving and dispersing to other parts of the earth through different means of dispersal.

chorology   subdiscipline that uses evolutionary models to trace distributional pathways.

cladistic biogeographic analysis   a method consisting of three basic steps: (1) constructing taxon–area cladograms, from the taxonomic cladograms of two or more different taxa, by replacing their terminal taxa with the areas they inhabit; (2) obtaining resolved area cladograms from the taxon–area cladograms (when demanded by the method applied); and (3) obtaining a general area cladogram, based on the information contained in the resolved area cladograms.

cladistic biogeography   approach originated by Gareth Nelson, Donn Eric Rosen, and Norman Platnick, who associated Croizat's panbiogeography with Hennig's phylogenetic systematics. Cladistic biogeography assumes a correspondence between taxonomic relationships and area relationships. If we compare area cladograms derived from taxonomic cladograms of different groups of plant and animals inhabiting a certain region, we may recognize the general pattern of fragmentation of the areas analyzed. Other authors include Chris Humphries, Lynne Parenti, Dan Brooks, Malte Ebach, and Bruce Lieberman.

classical biogeography   inaugurated by Carl Linnaeus (1707–1778), who provided an explanation of the geographic distribution of living beings. Other classical biogeographers are Johann Reinhold Forster (1729–1798), Eberhardt August Wilhelm von Zimmermann (1743–1815), Karl Willdenow (1765–1812), Alexander von Humboldt (1769–1805), and Augustin Pyrame de Candolle (1779–1841).

comparative phylogeography   comparison of the phylogeographic structure exhibited by sympatric species to discover whether they exhibit congruent patterns, geographically structured by vicariance events. Incongruent patterns may indicate that the species colonized the area more recently, whereas congruent patterns may suggest a longer history of association of the different species. This approach is similar to cladistic biogeography.

component analysis   cladistic biogeographic method that solves the problems derived from redundant distributions, widespread taxa, and missing areas using assumptions 0, 1, and 2 and then intersects the solved sets of resolved area cladograms to obtain the general area cladograms.

constrained dispersal–vicariance analysis (DIVA)   variation of the tree reconciliation analysis that distinguishes between random dispersals (those that imply that the taxon passes through a barrier) and predictable dispersals (those that occur when a barrier disappears).

corridor   geographic feature that facilitates dispersal. Corridors include a variety of habitats so that a large number of the organisms found at either end of them have little difficulty traversing them.

Darwinian biogeography   classic dispersalist model, first articulated by Charles Lyell (1797–1875) and then developed by Charles Darwin (1809–1882) and Alfred Russel Wallace (1823–1913).

descriptive biogeography   subdiscipline that describes biogeographic patterns.

descriptive dispersal   extent to which organisms are able to move within their distributional area.

descriptive vicariance   disjunct distribution, showing where a biotic component vicariated, not when the isolation occurred. It functions as a general statement of geographic distribution.

diffusion   gradual movement of populations across adjacent suitable habitats, over several generations. It is also known as range expansion.

dispersal   expansion of the distributional area of a taxon, covering all types of geographic translocation.

dispersalism   general biogeographic approach that locates centers of origin or ancestral areas and then uses dispersal from them to explain the biogeographic history of particular taxa.

dispersal–vicariance analysis   cladistic biogeographic method that reconstructs ancestral distributions from one given phylogenetic hypothesis without assuming a particular process a priori, taking into account vicariance, dispersal, and extinction. It reconstructs the biogeographic history of individual taxa, but it can also be used to find the general relationships of an area, especially when these relationships do not conform to a hierarchical pattern.

dispersal–vicariance model   evolution of biotic distributions in two steps. (1) Dispersal: When climatic and geographic factors are favorable, organisms actively expand their geographic distribution according to their dispersal capabilities, thus acquiring their ancestral distribution or primitive cosmopolitism. (2) Vicariance: When organisms have occupied all available geographic or ecologic space, their distribution may stabilize, allowing the isolation of populations in different sectors of the area and the differentiation of new species through the appearance of geographic barriers.

dispersion   movement of an organism within its area of distribution, also known as organismic or intrarange dispersal.

district   lowest biogeographic category.

dominion   biogeographic category intermediate between region and province.

dynamic vicariance   action of climatic changes that displace biotic components gradually in a certain direction, which finally find a barrier that causes vicariance.

ecological biogeography   subdiscipline that analyzes patterns at the species or population level, at small spatial and temporal scales, accounting for distributions in terms of biotic and abiotic interactions that happen in short periods of time.

ecological hierarchy   an arrangement of entities called interactors, which are involved in the matter–energy interchange. Interactors include molecules, cells, organisms, populations, and biotas.

endemicity analysis   method to identify areas of endemism that takes into consideration the spatial position of the species in order to identify the set of grid cells that represent an optimal area of endemism, according to a score based on the number of species endemic to it.

Euramerica   supercontinent from the Silurian that included North America and Eurasia.

event-based methods   cladistic biogeographic methods that derive explicit models of particular biogeographic processes. They include tree reconciliation analysis, vicariance events analysis, and dispersal–vicariance analysis.

evolutionary biogeography   approach that integrates distributional, phylogenetic, molecular, and paleontological data in order to discover biogeographic patterns and assess the historical changes that shaped them.

extensionism   belief that long-distance dispersal is an unlikely process to explain disjunct distributions, so ancient land bridges and continents now submerged in the oceans, which once linked the surviving continents, must be postulated. Extensionists include Joseph Dalton Hooker (1817–1911), Hermann von Ihering (1850–1930), Florentino Ameghino (1854–1911), Daniele Rosa (1857–1944), John Christopher Willis (1868–1958), René Jeannel (1879–1965), and Alfred Wegener (1880–1930).

extinction   local extirpation or total disappearance of a taxon.

filter   physical or biological feature that is not equally favorable for dispersal of all species.

genealogical hierarchy   an arrangement of entities called replicators that contain information, reproduce in similar entities, and evolve. Replicators include genes, chromosomes, organisms, and clades.

general area cladogram   area cladogram based on the information from the different resolved area cladograms.

generalized track   result of the significant superposition of different individual tracks, which indicate the preexistence of an ancestral biotic component that became fragmented by geological or tectonic events.

geobiotic scenario   plausible explanation of the episodes of vicariance or biotic divergence and dispersal or biotic convergence that have shaped the evolution of the biotic components analyzed.

geodispersal   simultaneous movement of several taxa caused by the effacement of a barrier, followed by the emergence of a new barrier that produces subsequent vicariance. It is also known as mass coherent dispersal, biotic dispersal, concerted dispersal, or predicted dispersal.

geological area cladogram   area cladogram based on geological or tectonic data.

Gondwana   supercontinent from the Late Paleozoic and Mesozoic, which included the land areas that later became South America, Africa, Madagascar, Antarctica, Australia, New Zealand, and India.

hierarchical thinking   the idea that nature is structured in entities that are ordered hierarchically, with smaller entities nested within larger ones. Each level of a hierarchy has its emergent properties and some autonomy; it is not a mere assemblage of smaller entities.

historical biogeography   subdiscipline that analyzes patterns of species and supraspecific taxa, at large spatial and temporal scales, being more interested in processes that happen over long periods of time.

horofauna   event of adaptive evolution of different lineages that interact, maximize the exploitation of available resources, and reach stability, covering all possible ecological niches.

individual track   primary spatial coordinates of a species or supraspecific taxon. It is a line graph drawn on a map that connects the different localities or distributional areas of the taxon according to their geographic proximity. From the topological viewpoint, an individual track is a minimum-spanning tree that for n localities contains $n - 1$ connections.

integrative biogeography   combination of ecological and evolutionary biogeography into a single general approach.

intraspecific phylogeography   study of the principles and processes governing the geographic distribution of genealogical lineages, especially those within and between closely related species, based on molecular data.

items of error   number of nodes and areas that are necessary to add to a taxon–area cladogram so that it agrees with the general area cladogram, that is, to map one cladogram onto the other to determine their congruence.

jump dispersal   random movement of organisms through barriers, which allows successful establishment of the species in very distant areas. It is also known as long-distance dispersal, waif dispersal, founder effect dispersal, or random dispersal.

Laurasia   supercontinent that included the northern landmasses. From the Jurassic to the Cretaceous, it was penetrated by epicontinental seas. In the Late Cretaceous, Europe and Asia were separated by the Obisk Sea; the former was connected to eastern North America (Euramerica), and the latter was connected to western North America (Asiamerica).

main massing   greatest concentration of numerical, genetic, or morphological diversity within the range of a taxon, which can be used to orient an individual track.

minimum-spanning tree method   a method of delineating in maps the individual tracks of different taxa and then superimposing them in order to find the generalized tracks. Nodes are identified in the areas where two or more generalized tracks superimpose.

missing areas   when no terminal taxon is distributed in one of the areas analyzed, this area will not be represented in the taxon–area cladogram.

modified Brooks parsimony analysis   variation of Brooks parsimony analysis, designed to deal with geodispersal, which involves two separate steps: retrieving congruent episodes of vicariance and retrieving congruent episodes of geodispersal.

molecular clock   the assumption that the rate of molecular evolution is approximately constant over time for proteins in all lineages, allowing inference of a clock-like accumulation of molecular changes. The "ticks" of the molecular clock, which correspond to mutations, occur not at regular intervals but rather at random points in time. This time is measured in arbitrary units and then calibrated in millions of years by reference to the fossil record or geological data, giving minimum estimates of the age of a clade, which in turn may help elucidate the relative minimum ages of the cenocron to which it belongs.

New York school of zoogeography   zoogeographic approach originated in the first decades of the twentieth century, founded on the dispersal of organisms over a static Earth. Authors include William Diller Matthew (1871–1930), George Gaylord Simpson (1902–1984), Philip J. Darlington Jr. (1904–1983), George Sprague Myers (1905–1985), and Ernst Mayr (1904–2005).

node   complex area where two or more generalized tracks superimpose, which is usually interpreted as a tectonic and biotic convergence zone.

orientation   formulating a hypothesis on the sequence of the disjunctions implied in an individual track, based on a baseline, main massing, or phylogenetic information.

panbiogeographic analysis   analysis consisting of three basic steps: (1) constructing individual tracks for two or more different taxa, (2) obtaining generalized tracks based on the comparison of the individual tracks, and (3) identifying nodes in the areas where two or more generalized tracks intersect.

panbiogeography   biogeographic approach originally developed by Léon Croizat (1894–1982) that emphasizes the spatial or geographic dimension of biodiversity to allow a better understanding of evolutionary patterns and processes. It is based on four assumptions: Distributional patterns constitute an empirical database for biogeographic analyses; distributional patterns provide information about where, when, and how plants and animals evolved; the spatial and temporal component of these distributional patterns can be represented graphically; and testable hypotheses about historical relationships between the evolution of distributions and Earth history can be derived from geographic correlations between distribution graphs and geological or geomorphic features. Authors include Robin Craw, Michael Heads, John Grehan, Ian Henderson, and Rod Page.

Pangaea   supercontinent that in the Late Carboniferous to Early Permian included all the continents we know in the present. Pangaea soon became divided into two landmasses: Laurasia in the north and Gondwana in the south.

paralogy-free subtree analysis   cladistic biogeographic method that reduces complex cladograms to paralogy-free subtrees, then inputs them in a component or three-item matrix, which is analyzed with a parsimony algorithm.

parsimony analysis of endemicity (PAE)   biogeographic method that constructs cladograms based on the parsimony analysis of presence–absence data of species and supraspecific taxa. Units analyzed include localities, areas of endemism, grid cells, hydrological basins, real and virtual islands, transects, communities, and political entities. PAE is used for panbiogeographic analysis and for obtaining areas of endemism.

pattern   nonrandom, repetitive arrangement or distribution of organisms and clades in geographic space.

pattern-based methods   cladistic biogeographic methods that search for general patterns of relationships between areas, without initial assumptions about particular biogeographic processes. They include component analysis, Brooks parsimony analysis, component compatibility, and paralogy-free subtree analysis.

permanentism   idea that postulates the permanent position of the continents and holds that long-distance dispersal is the fundamental process that caused distributional patterns.

phylogenetic analysis for comparing trees (PACT)   cladistic biogeographic method that compares Venn diagrams, obtained from taxon–area cladograms, looking for common elements.

phylogenetic biogeography   approach developed by Willi Hennig (1913–1976), combining the phylogeny of a monophyletic group with the distribution of its species to trace the progression in space. Another author is Lars Brundin (1907–1993).

phylogenetic orientation   use of cladistic information to orient the individual track of a supraspecific taxon.

phylogeography   analysis of the geographic distribution of intraspecific genealogical lineages.

plate tectonics   mechanism that explains continental drift. Seafloor spreading is believed to be caused by great convection currents that bring material to the surface from the hot interior of the earth, inducing the movement of the tectonic plates. These constitute the moving units at the surface of the earth and may contain continental masses or may consist of ocean floor.

prescientific biogeography   presence of biogeographic concepts in Saint Augustine (350–430), Abu al-Rayhan Mohamed ben Ahmad al-Biruni (973–1050), Avicenna (980–1037), Saint Thomas Aquinas (1225–1274), Joseph d'Acosta (1540–1600), Athanasius Kircher (1602–1680), and Matthew Hale (1609–1676).

primary biogeographic homology   conjecture on a common biogeographic history, which means that different taxa, even those with completely different means of dispersal, are spatiotemporally integrated in a biotic component.

process   cause of the geographic distribution of a taxon.

province   biogeographic category intermediate between dominion and district.

pseudo-congruence   appearance of the same area relationships in different area cladograms, although the taxa diversified at different times, presumably from different underlying causes.

pseudo-incongruence   appearance of conflict in different area cladograms when the taxa evolved at the same time but diversified in response to different events.

realm   highest biogeographic category. The biota of the world is classified into three realms: Holarctic, Holotropical, and Austral.

redundant distribution   distribution that occurs when an area appears more than once in a taxon–area cladogram because two or more terminal species are distributed in it.

refuge theory   hypothesis developed by Jürgen Haffer to explain the biogeographic patterns of the Amazon forest, which postulates that during the glacial periods of the Northern Hemisphere, some centers of endemism were islands of rain forest surrounded by a sea of grasslands, allowing many forest species to survive therein and eventually evolve in isolation as separate species.

region   biogeographic category intermediate between realm and dominion. The biota of the world is classified into twelve regions: Nearctic and Palearctic (Holarctic realm); Neotropical, Afrotropical or Ethiopian, Oriental, and Australian Tropical (Holotropical realm); and Andean, Antarctic, Cape or Afrotemperate, Neoguinean, Australian Temperate, and Neozelandic (Austral realm).

regionalization   biogeographic classification based on successively nested biotic components. It takes place before cenocrons are elucidated and a geobiotic scenario is proposed.

resolved area cladogram   taxonomic cladogram in which widespread taxa, redundant distributions, and missing areas are resolved before generalized tracks are obtained. Also known as areagram.

retrodiction   "prediction" of past events.

scientific research program   basic unit of scientific investigation. It consists of three parts   the hard core, a set of assumptions that define the program, which are irrefutable and which scientists protect as part of their research tradition; the positive heuristic, a set of methodological rules that specify the research policy of the adherents of the research program; and the protective belt, a flexible set of auxiliary hypotheses that are constructed, readjusted, or discarded as directed by the positive heuristic. Research programs are progressive if each modification leads to novel predictions, which are confirmed by subsequent research or by the chance discovery of new facts; they are degenerating if they are characterized by ad hoc modifications or by the persistent failure of their predictions to be corroborated empirically.

secondary biogeographic homology   cladistic test of the biotic components.

secondary Brooks parsimony analysis   variation of Brooks parsimony analysis in which areas involved in parallelisms in the general area cladogram are duplicated and dealt with separately to determine whether they were a unique area or whether they were different areas incorrectly treated as a single one.

secular migration   movement involving a short distance that occurs so slowly that the species evolves in the meantime.

sweepstake route   area completely surrounded by totally different environments, such as islands, caves, or high mountain peaks, where chances of dispersal are very low for most taxa. They differ from filters in kind, not merely in degree, because almost all the species that traverse them cannot survive.

systematic biogeography   subdiscipline that describes, compares, and classifies biotas.

taxon–area cladogram   a cladogram obtained by replacement of the name of each terminal taxon in the cladogram of the taxon analyzed with the area where it is distributed.

taxon pulse   model that accounts for habitat shifts along certain pathways, according to a unidirectional progression along a sequence of habitats.

three-area statement analysis   cladistic biogeographic method based on the three-item statement analysis in which each node is considered a relationship between branches, where some branches are related more closely than other branches of the tree, and is expressed minimally as a three-item statement. All area statements are recorded in a data matrix and analyzed with a parsimony or compatibility algorithm.

time slicing   partitioning of the taxa analyzed in a cladistic biogeographic analysis according to different time planes.

track compatibility   quantitative panbiogeographic method based on character compatibility, in which individual tracks are coded in an area × track matrix that is analyzed for track compatibility, where two individual tracks are compatible or congruent with each other if one is a subset of the other or they are the same in a pairwise comparison.

transition zone   complex area of mixed biota between two realms or regions. In a panbiogeographic analysis, it is indicated by the presence of one or more nodes, whereas in a cladistic biogeographic analysis it may give conflicting results because it appears to be sister area to different areas.

transparent method   procedure that treats taxon–area cladograms as individual points that may be part of a common pattern (the general area cladogram), viewing them in terms of proximal relationships and resolving them so each area is represented once.

tree reconciliation analysis   cladistic biogeographic method that maximizes the amount of codivergence or shared history between area cladograms from different taxa, minimizing losses (due to extinctions or lack of collection) and duplications (independent vicariance events) when different area cladograms are combined to obtain a general area cladogram. The reconstruction is based on a cost matrix.

vicariance   appearance of a barrier that allows fragmentation of the distribution of an ancestral species, after which the descendant species may evolve in isolation. The appearance of the barrier causes the disjunction, so they both have the same age.

vicariance biogeography   general biogeographic approach that looks for coincidence in the distributional patterns of different unrelated taxa. It includes panbiogeography and cladistic biogeography.

widespread taxon   a terminal taxon of a taxon–area cladogram that inhabits two or more of the studied areas. It is also known as mast (for "multiple areas on a single terminal").

References

Abogast, B. S. and G. J. Kenagy. 2001. Comparative phylogeography as an integrative approach to historical biogeography. *Journal of Biogeography* 28:819–825.

Abrahamovich, A., N. B. Díaz, and J. J. Morrone. 2004. Distributional patterns of the Neotropical and Andean species of the genus *Bombus* (Hymenoptera: Apidae). *Acta Zoológica Mexicana (nueva serie)* 20:99–117.

Aguilar-Aguilar, R. and R. Contreras-Medina. 2001. La distribución de los mamíferos marinos de México: Un enfoque panbiogeográfico. In *Introducción a la biogeografía en Latinoamérica: Teorías, conceptos, métodos y aplicaciones,* ed. J. Llorente Bousquets and J. J. Morrone, 213–219. Mexico, D.F.: Las Prensas de Ciencias, UNAM.

Aguilar-Aguilar, R., R. Contreras-Medina, A. Martínez-Aquino, G. Salgado-Maldonado, and A. González-Zamora. 2005. Aplicación del análisis de parsimonia de endemismos (PAE) en los sistemas hidrológicos de México: Un ejemplo con helmintos parásitos de peces dulceacuícolas. In *Regionalización biogeográfica en Iberoamérica y tópicos afines: Primeras Jornadas Biogeográficas de la Red Iberoamericana de Biogeografía y Entomología Sistemática (RIBES XII.I–CYTED),* ed. J. Llorente Bousquets and J. J. Morrone, 227–239. Mexico, D.F.: Las Prensas de Ciencias, UNAM.

Aguilar-Aguilar, R., R. Contreras-Medina, and G. Salgado Maldonado. 2003. Parsimony analysis of endemicity (PAE) of Mexican hydrological basins based on helminth parasites of freshwater fishes. *Journal of Biogeography* 30:1861–1872.

Ahrens, D. 2004. Orogenese und evolutive Radiation: Verbreitungs- und Speziationsmuster der Sericini (Coleoptera: Scarabaeidae) im Himalaya. *Mitteilungen der Deutschen Gesellschaft fur Allgemeine und Angewandte Entomologie* 14:107–110.

Allen, J. A. 1871. On the mammals and winter birds of East Florida. *Bulletin of the Museum of Comparative Zoology* 2:161–450.

Álvarez Mondragón, E. and J. J. Morrone. 2004. Propuesta de áreas para la conservación de aves de México, empleando herramientas panbiogeográficas e índices de complementariedad. *Interciencia* 29:112–120.

Amorim, D. de S. 2001. Dos Amazonias. In *Introducción a la biogeografía en Latinoamérica: Teorías, conceptos, métodos y aplicaciones,* ed. J. Llorente Bousquets and J. J. Morrone, 245–255. Mexico, D.F.: Las Prensas de Ciencias, UNAM.

Amorim, D. de S. and M. R. S. Pires. 1996. Neotropical biogeography and a method for maximum biodiversity estimation. In *Biodiversity in Brazil: A first approach,* ed. C. E. M. Bicudo and N. A. Menezes, 183–219. São Paulo: CNPq.

Amorim, D. S. and S. H. S. Tozoni. 1994. Phylogenetic and biogeographic analysis of the Anisopodoidea (Diptera, Bibionomorpha), with an area cladogram for intercontinental relationships. *Revista Brasileira de Entomologia* 38:517–543.

Andersen, N. M. 1982. *The semiaquatic bugs (Hemiptera, Gerromorpha): Phylogeny, adaptations, biogeography and classification.* Klampenborg: Scandinavian Science Press Limited.

Andersen, N. M. 1991. Cladistic biogeography of marine water striders (Insecta, Hemiptera) in the Indo-Pacific. *Australian Systematic Botany* 4:151–163.

Andersson, L. 1996. An ontological dilemma: Epistemology and methodology of historical biogeography. *Journal of Biogeography* 23:269–277.

Andrés Hernández, A. R., J. J. Morrone, T. Terrazas, and L. López Mata. 2006. Análisis de trazos de las especies mexicanas de *Rhus* subgénero *Lobadium* (Angiospermae: Ancardiaceae). *Interciencia* 31:900–904.

Andrews, P. 2007. The biogeography of hominid evolution. *Journal of Biogeography* 34:381–382.

Aris-Brosou, S. and Z. Yang. 2001. *PHYBAYES: A program for phylogenetic analyses in a Bayesian framework*. London: Department of Biology, University College London.

Arnaiz-Villena, A., J. Guillén, V. Ruiz-del-Valle, E. Lowy, J. Zamora, P. Varela, D. Stefani, and L. M. Allende. 2001. Phylogeography of crossbills, bullfinches, grosbeaks, and rosefinches. *Cellular and Molecular Life Sciences* 58:1–8.

Arnason, U., A. Gullberg, A. S. Burguete, and A. Janke. 2000. Molecular estimates of primate divergences and new hypotheses for primate dispersal and the origin of modern humans. *Hereditas* 133:217–228.

Aubréville, A. 1969. Essais sur la distribution et l'histoire des angiospermes tropicales dans le monde. *Adansonia* 2:189–247.

Aubréville, A. 1970. A propos de l'introduction raisonée à la biogéographie de l'Áfrique de Léon Croizat. *Adansonia* 2:489–497.

Aubréville, A. 1974a. Les origines des angiospermes. *Adansonia* 2:5–27.

Aubréville, A. 1974b. Origines polytopiques des angiospermes tropicales. *Adansonia* 2:145–198.

Aubréville, A. 1975. Essais de géophylétique des bombacacées. *Adansonia* 2:57–64.

Aurioles, G. D. 1993. Biodiversidad y estado actual de los mamíferos marinos de México. *Revista de la Sociedad Mexicana de Historia Natural* 44:397–412.

Ávila, L. J., M. Morando, and J. W. Sites Jr. 2006. Congeneric phylogeography: Hypothesizing species limits and evolutionary processes in Patagonian lizards of the *Liolaemus boulengeri* group (Squamata: Liolaemini). *Biological Journal of the Linnean Society* 89:241–275.

Avise, J. C. 1989. A role for molecular genetics in the recognition and conservation of endangered species. *Trends in Ecology and Evolution* 4:279–281.

Avise, J. C. 1992. Molecular population structure and the biogeographic history of a regional fauna: A case-history with lessons for conservation biology. *Oikos* 63:62–76.

Avise, J. C. 1996. Toward a regional conservation genetics perspective: Phylogeography of faunas in the southeastern United States. In *Conservation genetics: Case histories from nature*, ed. J. C. Avise and J. L. Hamrick, 431–470. New York: Chapman & Hall.

Avise, J. C. 2000. *Phylogeography: The history and formation of species*. Cambridge, Mass.: Harvard University Press.

Avise, J. C. 2004. What is the field of biogeography, and where is it going? *Taxon* 53:893–898.

Avise, J. C., J. Arnold, R. M. Ball, E. Bermingham, T. Lamb, J. E. Neigel, C. A. Reeb, and N. C. Saunders. 1987. Intraspecific phylogeography: The mitochondrial DNA bridge between population genetics and systematics. *Annual Review of Ecology and Systematics* 18:489–522.

Avise, J. C., B. W. Bowen, T. Lamb, A. B. Meylan, and E. Bermingham. 1992. Mitochondrial-DNA evolution at a turtle's pace: Evidence for low genetic variability and reduced microevolutionary rate in the Testudines. *Molecular Biology and Evolution* 9:457–473.

Avise, J. C. and D. Walker. 1998. Pleistocene phylogeographic effects of avian populations and the speciation process. *Proceedings of the Royal Society of London, Series B: Biological Sciences* 265:457–463.

Axelrod, D. I. 1963. Fossil floras suggest stable not drifting continents. *Journal of Geophysical Research* 68:3257–3263.

Axelrod, D. I. 1979. Age and origin of Sonoran Desert vegetation. *Occasional Papers of the California Academy of Sciences* 132:1–74.

Axelrod, D. I. 1983. Paleobotanical history of the western deserts. In *Origin and evolution of deserts*, ed. S. G. Wells and D. R. Haragan, 113–129. Albuquerque: University of New Mexico Press.

Báez, J. C., R. Real, J. M. Vargas, and A. Flores-Moya. 2004. A biogeographical analysis of the genera *Audouinella* (Rhodophyta), *Cystoseira* (Phaeophyceae) and *Cladophora* (Chlorophyta) in the western Mediterranean Sea and the Adriatic Sea. *Phicologia* 43:404–415.

Ball, I. R. 1976. Nature and formulation of biogeographical hypotheses. *Systematic Zoology* 24:407–430.

Ball, I. R. and J. C. Avise. 1992. Mitochondrial-DNA phylogeographic differentiation among avian populations and the evolutionary significance of subspecies. *The Auk* 109:626–636.

Banarescu, P. 1992. *Zoogeography of freshwater: Distribution and dispersal of freshwater animal in North America and Eurasia*, Vol. 2. Wiesbaden: Aula-Verlag.

Baroni-Urbani, C. 1977. Hologenesis, phylogenetic systematics, and evolution. *Systematic Zoology* 26:343–346.

Baroni-Urbani, C., S. Ruffo, and A. Vigna Taglianti. 1978. Materiali per una biogeografia italiana fondata su alcuni generi di coleotteri cicindelidi, carabidi e crisomelidi. *Estrato dalle Memorie della Societa Entomologica Italiana* 56:35–92.

Bartholomew, J. G., W. Eagle Clark, and P. H. Grimshaw. 1911. *Atlas of zoogeography*. London: Royal Geographical Society.

Bates, J. M. and T. C. Demos. 2001. Do we need to devalue Amazonia and other large tropical forests? *Diversity and Distributions* 7:249–255.

Bates, J. M., S. J. Hackett, and J. Cracraft. 1998. Area-relationships in the Neotropical lowlands: An hypothesis based on raw distributions of passerine birds. *Journal of Biogeography* 25:783–793.

Bellan, G. and D. Bellan Santini. 1997. Utilizzazione delle analisi di parsimonia (cladistica) in sinecologia bentonica: Esempi in una zona inquinata. *Societa Italiana di Ecologia Atti* 18:247–250.

Benton, M. J. and P. C. J. Donoghue. 2007. Paleontological evidence to date the tree of life. *Molecular Biology and Evolution* 24:26–53.

Bermingham, E. and A. P. Martin. 1998. Comparative mtDNA phylogeography of Neotropical fishes: Testing shared history to infer the evolutionary landscape of lower Central America. *Molecular Ecology* 7:499–517.

Bermingham, E. and C. Moritz. 1998. Comparative phylogeography: Concepts and applications. *Molecular Ecology* 7:367–369.

Bernatchez, L. 2001. The evolutionary history of brown trout (*Salmo trutta* L.) inferred from phylogeographic, nested clade, and mismatch analyses of mitochondrial DNA variation. *Evolution* 55:351–379.

Bernatchez, L. and J. J. Dodson. 1990. Allopatric origin of sympatric populations of lake whitefish (*Coregonus clupeaformis*) as revealed by mitochondrial-DNA restriction analysis. *Evolution* 44:1263–1271.

Berry, P. E. 1982. The systematics and evolution of *Fuchsia* sect. *Fuchsia* (Onagraceae). *Annals of the Missouri Botanical Garden* 69:1–198.

Bessega, C., J. C. Vilardi, and B. O. Saidman. 2006. Genetic relationships among American species of the genus *Prosopis* (Mimosoidae, Leguminosae) inferred from ITS sequences: Evidence for long-distance dispersal. *Journal of Biogeography* 33:1905–1915.

Betancourt, J. L. 2004. Arid lands in paleobiogeography: The rodent midden record in the Americas. In *Frontiers of biogeography: New directions in the geography of nature,* ed. M. V. Lomolino and L. R. Heaney, 27–46. Sunderland, Mass.: Sinauer.

Bianco, P. G. 1990. Potential role of the paleohistory of the Mediterranean and Parathetis basins on the early dispersal of Euro-Mediterranean freshwater fishes. *Ichthyological Exploration of Freshwaters* 1:167–184.

Biondi, M. 1998. Comparison of some methods for a cladistically founded biogeographical analysis. *Memorie del Museo Civico di Storia Naturale di Verona, 2. serie, Sezione Scienze della Vita* 13:9–31.

Biondi, M. and P. D'Alessandro. 2006. Biogeographical analysis of the flea beetle genus *Chaetocnema* in the Afrotropical region: Distribution patterns and areas of endemism. *Journal of Biogeography* 33:720–730.

Bisconti, M., W. Landini, G. Bianucci, G. Cantalamessa, G. Carnevale, L. Ragiani, and G. Valleri. 2001. Biogeographic relationships of the Galapagos terrestrial biota: Parsimony analyses of endemicity based on reptiles, land birds and *Scalesia* land plants. *Journal of Biogeography* 28:495–510.

Biswas, S. and S. S. Pawar. 2006. Phylogenetic tests of distribution patterns in South Asia: Towards an integrative approach. *Journal of Biosciences* 31:95–113.

Blondel, J. 1986. *Biogéographie évolutive.* Paris: Collection d'Écologie, Masson.

Bohme, M. U., U. Fritz, T. Kotenko, G. Dzukic, K. Ljubisavljevic, N. Tzankov, and T. U. Berendonk. 2007. Phylogeography and cryptic variation within the *Lacerta viridis* complex (Lacertidae, Reptilia). *Zoologica Scripta* 36:119–131.

Boniolo, G. and M. Carrara. 2004. On biological identity. *Biology and Philosophy* 19:443–457.

Bossuyt, F., R. M. Brown, D. M. Hillis, D. C. Cannatella, and M. C. Milinkovitch. 2006. Phylogeny and biogeography of a cosmopolitan frog radiation: Late Cretaceous diversification resulted in continent-scale endemism in the family Ranidae. *Systematic Biology* 55:579–594.

Bowler, P. S. 1989. *Evolution: The history of an idea.* Berkeley: University of California Press.

Bowler, P. S. 1996. *Life's splendid drama.* Chicago: University of Chicago Press.

Braby, M. F. and N. E. Pierce. 2007. Systematics, biogeography and diversification of the Indo-Australian genus *Delias* Hübner (Lepidoptera: Pieridae): Phylogenetic evidence supports an "out-of-Africa" origin. *Systematic Entomology* 32:2–25.

Bremer, K. 1992. Ancestral areas: A cladistic reinterpretation of the center of origin concept. *Systematic Biology* 41:436–445.

Bremer, K. 1993. Intercontinental relationships of African and South American Asteraceae: A cladistic biogeographic analysis. In *Biological relationships between Africa and South America,* ed. P. Goldbatt, 105–135. New Haven, Conn.: Yale University Press.

Bremer, K. 1995. Ancestral areas: Optimization and probability. *Systematic Biology* 44:255–259.

Briggs, J. C. 1974. *Marine zoogeography.* New York: McGraw-Hill.

Briggs, J. C. 1984. *Centres of origin in biogeography.* Leeds: University of Leeds.

Briggs, J. C. 1987. *Biogeography and plate tectonics.* Amsterdam: Elsevier.

Briggs, J. C. 1995. *Global biogeography.* Amsterdam: Elsevier.

Briggs, J. C. 2004. The ultimate expanding Earth hypothesis. *Journal of Biogeography* 31:855–857.

Briggs, J. C. and C. J. Humphries. 2004. Early classics. In *Foundations of biogeography: Classic papers with commentaries*, ed. M. V. Lomolino, D. F. Sax, and J. H. Brown, 5–13. Chicago: University of Chicago Press.

Britton, T., B. Oxelman, A. Vinnersten, and K. Bremer. 2002. Phylogenetic dating with confidence intervals using mean path lengths. *Molecular Phylogenetics and Evolution* 24:58–65.

Bromham, L. and M. Woolfit. 2004. Explosive radiations and the reliability of molecular clocks: Island endemic radiations as a test case. *Systematic Biology* 53:758–766.

Brooks, D. R. 1981. Hennig's parasitological method: A proposed solution. *Systematic Zoology* 30:229–249.

Brooks, D. R. 1985. Historical ecology: A new approach to studying the evolution of ecological associations. *Annals of the Missouri Botanical Garden* 72:60–680.

Brooks, D. R. 1990. Parsimony analysis in historical biogeography and coevolution: Methodological and theoretical update. *Systematic Zoology* 39:14–30.

Brooks, D. R. 2004. Reticulations in historical biogeography: The triumph of time over space in evolution. In *Frontiers of biogeography: New directions in the geography of nature*, ed. M. V. Lomolino and L. R. Heaney, 125–144. Sunderland, Mass.: Sinauer.

Brooks, D. R. 2005. Historical biogeography in the age of complexity: Expansion and integration. *Revista Mexicana de Biodiversidad* 76:79–94.

Brooks, D. R., P. G. Dowling, M. G. P. van Veller, and E. P. Hoberg. 2003. Ending a decade of deception: A valiant failure, a not-so valiant failure, and a success story. *Cladistics* 20:32–46.

Brooks, D. R. and D. A. McLennan. 1991. *Phylogeny, ecology and behavior: A research program in comparative biology.* Chicago: University of Chicago Press.

Brooks, D. R. and D. A. McLennan. 2001. A comparison of a discovery-based and an event-based method of historical biogeography. *Journal of Biogeography* 28:757–767.

Brooks, D. R. and D. A. McLennan. 2003. Extending phylogenetic studies of coevolution: Secondary Brooks parsimony analysis, parasites, and the great apes. *Cladistics* 19:104–119.

Brooks, D. R. and M. G. P. Van Veller. 2003. Critique of parsimony analysis of endemicity as a method of historical biogeography. *Journal of Biogeography* 30:819–825.

Brooks, D. R., M. G. P. van Veller, and D. A. McLennan. 2001. How to do BPA, really. *Journal of Biogeography* 28:345–358.

Brown, G. K., G. Nelson, and P. Y. Ladiges. 2006. Historical biogeography of *Rhododendron* section *Vireya* and the Malesian Archipelago. *Journal of Biogeography* 33:1929–1944.

Brown, J. H. 1995. *Macroecology.* Chicago: University of Chicago Press.

Brown, J. H. 2004. Concluding remarks. In *Frontiers of biogeography: New directions in the geography of nature*, ed. M. V. Lomolino and L. R. Heaney, 361–368. Sunderland, Mass.: Sinauer.

Brown, J. H. and M. V. Lomolino. 1998. *Biogeography*, 2nd ed. Sunderland, Mass.: Sinauer.

Browne, J. 1983. *The secular ark: Studies in the history of biogeography.* New Haven, Conn.: Yale University Press.

Brundin, L. 1966. Transantarctic relationships and their significance as evidenced by midges. *Kungliga Svenska Vetenskaps Akademien Handlingar, Series 4* 11:1–472.

Brundin, L. 1972a. Evolution, causal biology, and classification. *Zoologica Scripta* 1:107–120.

Brundin, L. 1972b. Phylogenetics and biogeography. *Systematic Zoology* 21:68–79.

Brundin, L. 1981. Croizat's panbiogeography versus phylogenetic biogeography. In *Vicariance biogeography: A critique*, ed. G. Nelson and D. E. Rosen, 94–158. New York: Columbia University Press.

Brundin, L. 1988. Phylogenetic biogeography. In *Analytical biogeography: An integrated approach to the study of animal and plant distributions,* ed. A. A. Myers and P. S. Giller, 343–369. London: Chapman and Hall.

Bueno-Hernández, A. and J. Llorente Bousquets. 2003. *El pensamiento biogeográfico de Alfred Russel Wallace.* Santafé de Bogotá: Colección Luis Duque Gómez no. 1, Academia Colombiana de Ciencias Exactas, Físicas y Naturales.

Bueno-Hernández, A. and J. E. Llorente Bousquets. 2006. The other face of Lyell: Historical biogeography in his *Principles of Geology. Journal of Biogeography* 33:549–559.

Bueno-Hernández, A. A., J. J. Morrone, M. de las M. Luna-Reyes, and C. Pérez-Malváez. 1999. Raíces históricas del concepto de centro de origen en la biogeografía dispersionista: Del Edén bíblico al modelo de Darwin–Wallace. *Sciences et Techniques en Perspective* (Paris) 3:27–45.

Buffon, G. L. C. de. 1749–1788. *Histoire naturelle, générale et particulière, avec la description du Cabinet du Roy.* 36 vols. Paris: L'Imprimerie Royale.

Buhay, J. E., G. Moni, N. Mann, and K. A. Crandall. 2007. Molecular taxonomy in the dark: Evolutionary history, phylogeography, and diversity of cave crayfish in the subgenus *Aviticambarus,* genus *Cambarus. Molecular Phylogenetics and Evolution* 42:435–448.

Burridge, C. P., D. Craw, and J. M. Waters. 2007. An empirical test of freshwater vicariance via river capture. *Molecular Ecology* 16:1883–1895.

Bush, M. B. 1994. Amazonian speciation: A necessarily complex model. *Journal of Biogeography* 21:5–17.

Cabrera, A. L. and A. Willink. 1973. *Biogeografía de América Latina.* Washington, D.C.: Organización de Estados Americanos.

Caccone, A., M. C. Milinkovitch, V. Sbordoni, and J. R. Powell. 1997. Molecular biogeography: The Corsica–Sardinia microplate disjunction to calibrate mitochondrial rDNA evolutionary rates in mountain newts (*Euproctus*). *Journal of Evolutionary Biology* 7:227–245.

Cain, J. A. 1993. Common problems and cooperative solutions: Organizational activity in evolutionary studies, 1936–1947. *Isis* 84:1–25.

Cain, S. A. 1944. *Foundations of plant geography.* New York: Harper.

Camerini, J. R. 1993. Evolution, geography, and maps: An early history of Wallace's line. *Isis* 84:700–727.

Candela, A. and J. J. Morrone. 2003. Biogeografía de puercoespines neotropicales (Rodentia, Hystricognathi): Integrando datos fósiles y actuales a través de un enfoque panbiogeográfico. *Ameghiniana* 40:361–378.

Cano, J. M. and P. Gurrea. 2003. La distribución de las zigenas (Lepidoptera, Zygaenidae) ibéricas: Una consecuencia del efecto península. *Graellsia* 59:273–285.

Cao, N. and J. Ducasse. 2005. *Nelson05: A program for cladistics and biogeography.* Paris: Authors.

Carlquist, S. 1974. *Island biology.* New York: Columbia University Press.

Carlquist, S. 1995. Introduction. In *Hawaiian biogeography: Evolution on a hotspot archipelago,* ed. W. L. Wagner and V. A. Funk, 1–13. Washington, D.C.: Smithsonian Institution Press.

Carpenter, J. M. 1992. Incidit in Scyllam qui vult vitare Charybdim. *Cladistics* 8:100–102.

Carpenter, J. M. 1993. Biogeographic patterns in the Vespidae (Hymenoptera): Two views of Africa and South America. In *Biological relationships between Africa and South America,* ed. P. Goldblatt, 139–155. New Haven, Conn.: Yale University Press.

Carracedo, J. C. and S. Day. 2002. *Canary Islands.* Harpenden, Hertfordshire: Terra.

Carrillo-Ruiz, H. and M. A. Morón. 2003. Fauna de Coleoptera Scarabaeoidea de Cuetzalan del Progreso, Puebla, Mexico. *Acta Zoológica Mexicana* 88:87–121.

Carvalho, C. J. B. de, M. Bortolanza, M. C. Cardoso da Silva, and E. D. Giustina Soares. 2003. Distributional patterns of the Neotropical Muscidae (Diptera). In *Una perspectiva latinoamericana de la biogeografía*, ed. J. J. Morrone and J. Llorente Bousquets, 263–274. Mexico, D.F.: Las Prensas de Ciencias, UNAM.

Carvalho, D. J. B. de and M. S. Couri. 2002. A cladistic and biogeographic analysis of *Apsil* Malloch and *Reynoldsia* Malloch (Diptera, Muscidae) of southern South America. *Proceedings of the Entomological Society of Washington* 104:309–317.

Cavieres, L. A., M. T. K. Arroyo, P. Posadas, C. Marticorena, O. Matthei, R. Rodríguez, F. A. Squeo, and G. Arancio. 2002. Identification of priority areas for conservation in an arid zone: Application of parsimony analysis of endemicity in the vascular flora of the Antofagasta region, northern Chile. *Biodiversity and Conservation* 11:1301–1311.

Cavieres, L. A., M. Mihoc, A. Marticorena, C. Marticorena, O. Matthei, and F. A. Squeo. 2001. Determinación de áreas prioritarias para la conservación: Análisis de parsimonia de endemismos (PAE) en la flora de la IV Región de Coquimbo. In *Libro Rojo de la flora nativa de Coquimbo y de los sitios prioritarios para su conservación*, ed. F. A. Squeo and J. R. Gutiérrez, 119–132. La Serena, Chile: Ediciones Universidad de La Serena.

Cecca, F. 2002. *Palaeobiogeography of marine fossil invertebrates: Concepts and methods*. London: Taylor and Francis.

Charleston, M. A. 1998. Jungles: A new solution to the host/parasite phylogeny reconciliation problem. *Mathematical Biosciences* 149:191–223.

Chen, X. L., T. Y. Chiang, H. D. Lin, H. S. Zheng, K. T. Shao, Q. Zhang, and K. C. Hsu. 2007. Mitochondrial DNA phylogeography of *Glyptothorax fokiensis* and *Glyptothorax hainanensis* in Asia. *Journal of Fish Biology* 70:75–93.

Chen, Y. and J. Bi. 2007. Biogeography and hotspots of amphibian species of China: Implications to reserve selection and conservation. *Current Science* 92:480–489.

Choudhury, A. and G. Pérez-Ponce de León. 2005. The roots of historical biogeography in Latin American parasitology: The legacy of Hermann von Ihering and Lothar Szidat. In *Regionalización biogeográfica en Iberoamérica y tópicos afines: Primeras Jornadas Biogeográficas de la Red Iberoamericana de Biogeografía y Entomología Sistemática (RIBES XII.I–CYTED)*, ed. J. Llorente Bousquets and J. J. Morrone, 45–53. Mexico, D.F.: Las Prensas de Ciencias, UNAM.

Christiansen, K. and D. Culver. 1987. Biogeography and the distribution of cave Collembola. *Journal of Biogeography* 14:459–477.

Clement, M., D. Posada, and K. A. Crandall. 2000. TCS: A computer program to estimate gene genealogies. *Molecular Ecology* 9:1557–1659.

Colacino, C. 1997. Léon Croizat's biogeography and macroevolution, or "out of nothing, nothing comes." *Philippines Scientist* 34:73–88.

Colacino, C. and J. R. Grehan. 2003. Ostracismo alle frontiere della biologia evoluzionistica: Il caso Léon Croizat. In *Scienza e democrazia*, ed. M. Mamone Capria, 195–220. Naples: Instituto Italiano per gli Studi Filosofici, Liguori Editori.

Colinvaux, P. A. 1997. Amazonian diversity in light of the paleoecological record. *Quaternary Research* 34:330–345.

Colinvaux, P. A. 1998. A new vicariance model for Amazonic endemics. *Global Ecology and Biogeography Letters* 7:95–96.

Comes, H. P. and J. W. Kadereit. 1998. The effect of Quaternary climatic changes on plant distribution and evolution. *Trends in Plant Science* 3:432–438.

Confalonieri, V. A., A. S. Sequeira, L. Todaro, and J. C. Vilardi. 1998. Mitochondrial DNA and phylogeography of the grasshopper *Trimerotropis pallidipennis* in relation to clinal distribution of chromosome polymorphisms. *Heredity* 81:444–452.

Conran, J. G. 1995. Family distributions in the Liliiflorae and their biogeographical implications. *Journal of Biogeography* 22:1023–1034.

Conroy, V. C. J., Y. Hortelano, F. A. Cervantes, and J. A. Cook. 2001. The phylogenetic position of southern relictual species of *Microtus* (Muridae: Rodentia) in North America. *Mammalian Biology* 66:332–344.

Contreras-Medina, R. and H. Eliosa León. 2001. Una visión panbiogeográfica preliminar de México. In *Introducción a la biogeografía en Latinoamérica: Conceptos, teorías, métodos y aplicaciones,* ed. J. Llorente Bousquets and J. J. Morrone, 197–211. Mexico, D.F.: Las Prensas de Ciencias, UNAM.

Contreras-Medina, R. and I. Luna-Vega. 2002. On the distribution of gymnosperm genera, their areas of endemism and cladistic biogeography. *Australian Systematic Botany* 15:193–203.

Contreras-Medina, R., I. Luna-Vega, and J. J. Morrone. 1999. Biogeographic analysis of the genera of Cycadales and Coniferales (Gymnospermae): A panbiogeographic approach. *Biogeographica* 75:163–176.

Contreras-Medina, R., I. Luna Vega, and J. J. Morrone. 2007a. Application of parsimony analysis of endemicity to Mexican gymnosperm distributions: Grid-cells, biogeographical provinces and track analysis. *Biological Journal of the Linnean Society* 92:405–417.

Contreras-Medina, R., I. Luna-Vega, and J. J. Morrone. 2007b. Gymnosperms and cladistic biogeography of the Mexican Transition Zone. *Taxon* 56: 905–915.

Contreras-Medina, R., J. J. Morrone, and I. Luna-Vega. 2001. Biogeographic methods identify gymnosperm biodiversity hotspots. *Naturwissenschaften* 88:427–430.

Cooper, R. A. 1989. New Zealand tectonostratigraphic terranes and panbiogeography. *New Zealand Journal of Zoology* 16:699–712.

Corner, E. J. H. 1959. Panbiogeography. *New Phytologist* 58:237–238.

Corona, A. M., R. Acosta, and J. J. Morrone. 2005. Estudios biogeográficos de la Zona de Transición Mexicana. In *Regionalización biogeográfica en Iberoamérica y tópicos afines: Primeras Jornadas Biogeográficas de la Red Iberoamericana de Biogeografía y Entomología Sistemática (RIBES XII.I–CYTED),* ed. J. Llorente Bousquets and J. J. Morrone, 241–255. Mexico, D.F.: Las Prensas de Ciencias, UNAM.

Corona, A. and J. J. Morrone. 2005. Track analysis of the species of *Lampetis (Spinthoptera)* Casey, 1909 (Coleoptera: Buprestidae) in North America, Central America, and the West Indies. *Caribbean Journal of Science* 41:37–41.

Corona, A. M., V. H. Toledo, and J. J. Morrone. 2007. Does the Trans-Mexican Volcanic Belt represent a natural biogeographic unit?: An analysis of the distributional patterns of Coleoptera. *Journal of Biogeography* 34:1008–1015.

Cortés, B. R. and P. Franco. 1997. Análisis panbiogeográfico de la flora de Chiribiquete, Colombia. *Caldasia* 19:465–478.

Coscarón, M. del C. and J. J. Morrone. 1995. Systematics, cladistics, and biogeography of the *Peirates collarti* and *P. lepturoides* species groups (Heteroptera: Reduviidae, Peiratinae). *Entomologica Scandinavica* 26:191–228.

Coscarón, M. del C. and J. J. Morrone. 1997. Cladistics and biogeography of the assassin bug genus *Melanolestes* Stål (Heteroptera: Reduviidae). *Proceedings of the Entomological Society of Washington* 99:55–59.

Costa, L. P. 2003. The historical bridge between the Amazon and the Atlantic forest of Brazil: A study of molecular phylogeography with small mammals. *Journal of Biogeography* 30:71–86.

Costa, L. P., Y. L. R. Leite, G. A. B. da Fonseca, and M. T. da Fonseca. 2000. Biogeography of South American forest mammals: Endemism and diversity in the Atlantic forest. *Biotropica* 32:872–881.

Cowie, R. W. and B. S. Holland. 2006. Dispersal is fundamental to biogeography and the evolution of biodiversity on oceanic islands. *Journal of Biogeography* 33: 193–198.

Cox, C. B. 1998. From generalised tracks to ocean basins: How useful is panbiogeography? *Journal of Biogeography* 25:813–828.

Cox, C. B. 2001. The biogeographic regions reconsidered. *Journal of Biogeography* 28: 511–523.

Cox, C. B. and P. D. Moore. 1998. *Biogeography: An ecological and evolutionary approach.* Oxford: Blackwell Science.

Cracraft, J. 1982. Geographic differentiation, cladistics, and vicariance biogeography: Reconstructing the tempo and mode of evolution. *American Zoologist* 22:411–424.

Cracraft, J. C. 1985. Species selection, macroevolutionary analysis and the "hierarchical theory" of evolution. *Systematic Zoology* 34:222–229.

Cracraft, J. 1988. Deep-history biogeography: Retrieving the historical pattern of evolving continental biotas. *Systematic Zoology* 37:221–236.

Cracraft, J. 1991. Patterns of diversification within continental biotas: Hierarchical congruence among the areas of endemism of Australian vertebrates. *Australian Systematic Botany* 4:211–227.

Cracraft, J. 1994. Species diversity, biogeography and the evolution of biotas. *American Zoologist* 34:33–47.

Cracraft, J. and R. O. Prum. 1988. Patterns and processes of diversification: Speciation and historical congruence in some Neotropical birds. *Journal of Biogeography* 12:603–620.

Cranwell, L. 1962. Endemism and isolation in the Three Kings islands, New Zealand: With notes on pollen and spore types of the endemics. *Records of the Auckland Institute and Museum* 5:215–232.

Craw, R. C. 1979. Generalized tracks and dispersal in biogeography: A response to R. M. McDowall. *Systematic Zoology* 28:99–107.

Craw, R. C. 1982. Phylogenetics, areas, geology and the biogeography of Croizat: A radical view. *Systematic Zoology* 31:304–316.

Craw, R. C. 1983. Panbiogeography and vicariance cladistics: Are they truly different? *Systematic Zoology* 32:431–438.

Craw, R. C. 1984a. Léon Croizat's biogeographic work: A personal appreciation. *Tuatara* 27:8–13.

Craw, R. C. 1984b. Never a serious scientist: The life of Léon Croizat. *Tuatara* 27:5–7.

Craw, R. C. 1985. Classic problems of Southern Hemisphere biogeography re-examined: Panbiogeographic analysis of the New Zealand frog *Leiopelma*, the ratite birds and *Nothofagus*. *Zeitschrift für Zoologische Systematik und Evolutionsforschung* 23:1–10.

Craw, R. C. 1988a. Continuing the synthesis between panbiogeography, phylogenetic systematics and geology as illustrated by empirical studies on the biogeography of New Zealand and the Chatham Islands. *Systematic Zoology* 37:291–310.

Craw, R. C. 1988b. Panbiogeography: Method and synthesis in biogeography. In *Analytical biogeography: An integrated approach to the study of animal and plant distributions*, ed. A. A. Myers and P. S. Giller, 405–435. London: Chapman and Hall.

Craw, R. C. 1989a. New Zealand biogeography: A panbiogeographic approach. *New Zealand Journal of Zoology* 16:527–547.

Craw, R. C. 1989b. Quantitative panbiogeography: Introduction to methods. *New Zealand Journal of Zoology* 16:485–494.

Craw, R. C. and G. W. Gibbs. 1984. Croizat's panbiogeography and *Principia Botanica:* Search for a novel biological synthesis. *Tuatara* 27:1–75.

Craw, R. C., J. R. Grehan, and M. J. Heads. 1999. *Panbiogeography: Tracking the history of life.* Oxford: Oxford Biogeography.

Craw, R. C. and M. J. Heads. 1988. Reading Croizat: On the edge of biology. *Rivista di Biologia—Biology Forum* 81:499–532.

Craw, R. C. and R. Page. 1988. Panbiogeography: Method and metaphor in the new biogeography. In *Evolutionary processes and metaphors,* ed. M.-W. Ho and S. W. Fox, 163–189. Chichester: Wiley.

Craw, R. C. and G. Sermonti, eds. 1988. Special issue on panbiogeography. *Rivista di Biologia—Biology Forum* 81:457–615.

Craw, R. C. and P. Weston. 1984. Panbiogeography: A progressive research program? *Systematic Zoology* 33:1–33.

Crisci, J. V. 2001. The voice of historical biogeography. *Journal of Biogeography* 28:157–168.

Crisci, J. V., M. M. Cigliano, J. J. Morrone, and S. Roig-Juñent. 1991a. A comparative review of cladistic biogeography approaches to historical biogeography of southern South America. *Australian Systematic Botany* 4:117–126.

Crisci, J. V., M. M. Cigliano, J. J. Morrone, and S. Roig-Juñent. 1991b. Historical biogeography of southern South America. *Systematic Zoology* 40:152–171.

Crisci, J. V., M. S. de la Fuente, A. A. Lanteri, J. J. Morrone, E. Ortiz Jaureguizar, R. Pascual, and J. L. Prado. 1993. Patagonia, Gondwana Occidental (GW) y Oriental (GE), un modelo de biogeografía histórica. *Ameghiniana* 30:104.

Crisci, J. V., S. Freire, G. Sancho, and L. Katinas. 2001. Historical biogeography of the Asteraceae from Tandilia and Ventania mountain ranges (Buenos Aires, Argentina). *Caldasia* 23:21–41.

Crisci, J. V., L. Katinas, and P. Posadas. 2000. *Introducción a la teoría y práctica de la biogeografía histórica.* Buenos Aires: Sociedad Argentina de Botánica. (English translation: 2003, *Historical biogeography: An introduction,* Cambridge, Mass.: Harvard University Press.)

Crisci, J. V. and J. J. Morrone. 1992. A comparison of biogeographic models: A response to Bastow Wilson. *Global Ecology and Biogeography Letters* 2:174–176.

Crisci, J. V., O. E. Sala, L. Katinas, and P. Posadas. 2006. Bridging historical and ecological approaches in biogeography. *Australian Systematic Botany* 19:1–10.

Crisp, M. D. and L. G. Cook. 2005. Do early branching lineages signify ancestral traits? *Trends in Ecology and Evolution* 20:122–128.

Crisp, M. D., H. P. Linder, and P. H. Weston. 1995. Cladistic biogeography of plants in Australia and New Guinea: Congruent pattern revealed two endemic tropical tracks. *Systematic Zoology* 44:457–473.

Croizat, L. 1952. *Manual of phytogeography.* The Hague: Junk.

Croizat, L. 1954. La faja xerófila del Estado Mérida. *Universitas Emeritensis* (Mérida) 1:100–106.

Croizat, L. 1958a. An essay on the biogeographic thinking of J. C. Willis. *Archivo Botanico y Biogeografico Italiano* 34:90–116.

Croizat, L. 1958b. *Panbiogeography,* Vols. 1 and 2. Caracas, Venezuela: Author.

Croizat, L. 1961. Entre lo viejo y lo nuevo. *Revista Shell* (Caracas) 10:29–36.

Croizat, L. 1964. *Space, time, form: The biological synthesis.* Caracas, Venezuela: Author.

Croizat, L. 1971. De la "pseudovicariance" et de la "disjonction illusoire." *Anuário da Socie-dade Broteriana* 37:113–140.

Croizat, L. 1973. La "panbiogeographia" in breve. *Webbia* 28:189–226.

Croizat, L. 1976. *Biogeografía analítica y sintética ("panbiogeografía") de las Américas.* Caracas, Venezuela: Biblioteca de la Academia de Ciencias Físicas, Matemáticas y Naturales.

Croizat, L. 1978. Hennig (1966) entre Rosa (1918) y Lovtrup (1977): Medio siglo de sistemática filogenética. *Boletín de la Academia de Ciencias Físicas, Matemáticas y Naturales* (Caracas) 38:59–147.

Croizat, L. 1982. Vicariance/vicariism, panbiogeography, "vicariance biogeography," etc.: A clarification. *Systematic Zoology* 31:291–304.

Croizat, L. 1983. La biogeografía desde mi punto de vista. *Zoología Neotropical, Actas del VIII Congreso Latinoamericano de Zoología (Caracas)* 1:165–175.

Croizat, L. 1984a. Charles Darwin and his theories. In R. C. Craw and G. W. Gibbs, eds., Croizat's panbiogeography and *Principia Botanica:* Search for a novel biological synthesis. *Tuatara* 27:21–25.

Croizat, L. 1984b. Mayr vs. Croizat: Croizat vs. Mayr: An enquiry. In R. C. Craw and G. W. Gibbs, eds., Croizat's panbiogeography and *Principia Botanica:* Search for a novel biological synthesis. *Tuatara* 27:49–66.

Croizat, L., G. Nelson, and D. E. Rosen. 1974. Centers of origin and related concepts. *Systematic Zoology* 23:265–287.

Crother, B. I. and C. Guyer. 1996. Caribbean historical biogeography. Was the dispersal–vicariance debate eliminated by an extraterrestrial bolide? *Herpetologica* 52:440–465.

Crovello, T. J. 1981. Quantitative biogeography: An overview. *Taxon* 30:563–575.

Cué-Bär, E. M., J. L. Villaseñor, J. J. Morrone, and G. Ibarra-Manríquez. 2006. Identifying priority areas for conservation in Mexican tropical deciduous forest based on tree species. *Interciencia* 31:712–719.

Cunningham, C. W. and T. M. Collins. 1998. Beyond area relationships: Extinction and recolonization in molecular marine biogeography. In *Molecular approaches to ecology and evolution,* ed. R. DeSalle and B. Schierwater, 297–321. Basel: Birkhäuser.

Cusset, G. and C. Cusset. 1988a. Etudes sur les Podostemales. 11. Répartition et evolution des Tristichaceae. *Bulletin du Museum National d'Histoire Naturelle (Paris),* Ser. 4, 10:223–262.

Cusset, G. and C. Cusset. 1988b. Etudes sur les Podostemales. 12. Biogéographie évolutive des *Tristicha trifaria* (Bory ex Wild.) Sprengel. *Bulletin du Museum National d'Histoire Naturelle (Paris),* Ser. 4, 10:39–70.

D'Acosta, J. 1590. *Historia natural y moral de las Indias, en que se tratan de cosas notables del cielo, y elementos, metales, plantas y animales dellas y de los ritos y ceremonias, leyes y gobiernos, y guerras de los indios.* Seville: Juan de León.

Dapporto, L., H. Wolf, and F. Strumia. 2007. Recent geography determines the distribution of flying Hymenoptera in the Tuscan archipelago. *Journal of Zoology* 272:37–44.

Darlington, P. J. Jr. 1943. Carabidae of mountains and islands: Data on the evolution of isolated faunas, and on atrophy of winds. *Ecological Monographs* 13:37–61.

Darlington, P. J. Jr. 1957. *Zoogeography: The geographical distribution of animals.* New York: Wiley.

Darlington, P. J. Jr. 1965. *Biogeography of the southern end of the world: Distribution and history of far-southern life and land, with an assessment of continental drift.* Cambridge, Mass.: Harvard University Press.

Darwin, C. R. 1844. Essay. In *The foundations of the origin of species,* ed. F. Darwin, 1909, Cambridge: Cambridge University Press.

Darwin, C. R. 1859. *The origin of species by means of natural selection, or, the preservation of favoured races in the struggle for life.* London: John Murray.

Da Silva, J. M. C. and D. C. Oren. 1996. Application of parsimony analysis of endemicity in Amazonian biogeography: An example with primates. *Biological Journal of the Linnean Society* 39:427–437.

Da Silva, J. M. C., M. C. Sousa, and C. H. M. Castelletti. 2004. Areas of endemism for passerine birds in the Atlantic forest, South America. *Global Ecology and Biogeography* 13:85–92.

Dávalos, L. M. 2004. Phylogeny and biogeography of Caribbean mammals. *Biological Journal of the Linnean Society* 81:373–394.

Dávalos, L. M. 2006. The geography of diversification in the mormoopids (Chiroptera: Mormoopidae). *Biological Journal of the Linnean Society* 88:101–118.

Davis, A. L. and C. H. Scholtz. 2001. Historical vs. ecological patterns of scarabaeine dung beetle diversity. *Diversity and Distributions* 7:161–174.

Davis, A. L., C. H. Scholtz, and T. K. Philips. 2002. Historical biogeography of scarabaeine dung beetles. *Journal of Biogeography* 29:1217–1256.

Dawson, M. N., J. L. Staton, and D. K. Jacobs. 2001. Phylogeography of the tidewater goby, *Eucyclogobius newberryi* (Teleostei, Gobidae), in coastal California. *Evolution* 55:1167–1179.

De Candolle, A. P. 1820. Géographie botanique. In *Dictionnaire des sciences naturelles,* 359–422. Strasbourg: Treuttel and Würtz.

De Candolle, A. P. 1838. *Statistique de la famille des Composées.* Strasbourg: Treuttel and Würtz.

De Grave, S. 2001. Biogeography of Indo-Pacific Pontoniinae (Crustacea, Decapoda): A PAE analysis. *Journal of Biogeography* 28:1239–1253.

Deleporte, P. and M. Colyn. 1999. Biogeographie et dynamique de la biodiversite: Application de la "PAE" aux forêts planitiaires d'Afrique centrale. *Biosystema* 17:37–43.

De Marmels, J. 2000. The larva of *Allopetalia pustulosa* Selys, 1873 (Anisoptera: Aeshnidae), with notes on Aeshnoid evolution and biogeography. *Odonatologica* 29:113–128.

Demastes, J. W., T. A. Spradling, M. S. Hafner, D. J. Hafner, and D. L. Reed. 2002. Systematics and phylogeography of pocket gophers in the genera *Cratogeomys* and *Pappogeomys. Molecular Phylogenetics and Evolution* 22:144–154.

Demesure, B., B. Comps, and R. J. Petit. 1996. Chloroplast DNA phylogeography of the common beach (*Fagus sylvatica* L.) in Europe. *Evolution* 50:2515–2520.

De Meyer, M. 1996. Cladistic and biogeographic analysis of Hawaiian Pipunculidae (Diptera) revisited. *Cladistics* 12:291–303.

Deo, A. J. and R. DeSalle. 2006. Nested areas of endemism analysis. *Journal of Biogeography* 33:1511–1526.

de Pinna, M. C. C. 1991. Concepts and tests of homology in the cladistic paradigm. *Cladistics* 7:367–394.

de Pinna, M. C. C. 1996. Comparative biology and systematics: Some controversies in retrospective. *Journal of Comparative Biology* 2:3–15.

De Queiroz, A. 2005. The resurrection of oceanic dispersal in historical biogeography. *Trends in Ecology and Evolution* 20:68–73.

DeSalle, R. and J. A. Hunt. 1987. Molecular evolution in Hawaiian drosophiloids. *Trends in Ecology and Evolution* 2:212–216.

Devitt, T. J. 2006. Phylogeography of the western lyresnake (*Trimorphodon biscilatus*): Testing aridland biogeographical hypotheses across the Nearctic–Neotropical transition. *Molecular Ecology* 15:4387–4407.

de Weerdt, W. H. 1989. Phylogeny and biogeography of North Atlantic Chalinidae (Haplosclerida, Demospongiae). *Beaufortia* 39:55–58.

Doadrio, I. and J. A. Carmona. 2003. Testing freshwater Lago Mare dispersal theory on the phylogeny relationships of Iberian Cyprinid genera *Chondrostoma* and *Squalius* (Cypriniformes, Cyprinidae). *Graellsia* 59:457–473.

Domínguez, M. C., S. Roig-Juñent, J. J. Tassin, F. C. Ocampo, and G. E. Flores. 2006. Areas of endemism of the Patagonian steppe: An approach based on insect distributional patterns using endemicity analysis. *Journal of Biogeography* 33:1527–1537.

Domínguez-Domínguez, O., L. Boto, F. Alda, and G. Pérez Ponce de León. 2007. Human impacts on drainages of the Mesa Central, Mexico, and its genetic effects on an endangered fish, *Zoogoneticus quitzeoensis*. *Conservation Biology* 21:168–180.

Donato, M. 2006. Historical biogeography of the family Tristiridae (Orthoptera: Acridomorpha) applying dispersal–vicariance analysis. *Journal of Arid Environments* 66:421–434.

Donato, M., P. Posadas, D. R. Miranda Esquivel, E. Ortiz Jaureguizar, and G. Cladera. 2003. Historical biogeography of the Andean region: Evidence from Listroderina (Coleoptera: Curculionidae: Rhytirrhinini) in the context of the South American geobiotic scenario. *Biological Journal of the Linnean Society* 80:339–352.

Donoghue, M. J., C. D. Bell, and J. Li. 2001. Phylogenetic patterns in Northern Hemisphere plant geography. *International Journal of Plant Science* 162:S41–S52.

Donoghue, M. J. and B. R. Moore. 2003. Toward an integrative historical biogeography. *Integrative and Comparative Biology* 43:261–270.

Dowling, A. P. G. 2002. Testing the accuracy of TreeMap and Brooks parsimony analyses of coevolutionary patterns using artificial associations. *Cladistics* 18:416–435.

Drummond, A. J. and A. Rambaut. 2003. *BEAST*. Version 1.0. Retrieved May 25, 2008, from http://evolve.zoo.ox.ac.uk/beast/.

Dumolin-Lapegue, S., B. Demesure, S. Fineschi, B. LeCorre, and R. J. Petit. 1997. Structure of white oaks throughout the European continent. *Genetics* 146:1475–1487.

Dutton, P. H., B. W. Bowen, D. W. Owens, A. Barragan, and S. K. Davis. 1999. Global phylogeography of the leatherback turtle (*Dermochelys coriacea*). *Journal of Zoology* 248:397–409.

Ebach, M. C. 1999. Paralogy and the centre of origin concept. *Cladistics* 15:387–391.

Ebach, M. C. 2001. Extrapolating cladistic biogeography: A brief comment on van Veller et al. (1999, 2000, 2001). *Cladistics* 17:383–388.

Ebach, M. C. 2003. Area cladistics. *Biologist* 50:169–172.

Ebach, M. C. and G. D. Edgecombe. 2001. Cladistic biogeography: Component-based methods and paleontological application. In *Fossils, phylogeny, and form: An analytical approach*, ed. J. M. Adrain, G. D. Edgecombe, and B. S. Lieberman, 235–289. New York: Kluwer/Plenum.

Ebach, M. C. and D. F. Goujet. 2006. The first biogeographical map. *Journal of Biogeography* 33:761–769.

Ebach, M. C. and C. J. Humphries. 2002. Cladistic biogeography and the art of discovery. *Journal of Biogeography* 29:427–444.

Ebach, M. C. and C. J. Humphries. 2003. Ontology of biogeography. *Journal of Biogeography* 30:959–962.

Ebach, M. C., C. J. Humphries, R. A. Newman, D. Williams, and S. A. Walsh. 2005a. Assumption 2: Opaque to intuition? *Journal of Biogeography* 32:781–787.

Ebach, M. C., C. J. Humphries, and D. M. Williams. 2003. Phylogenetic biogeography deconstructed. *Journal of Biogeography* 30:1285–1296.

Ebach, M. and J. J. Morrone. 2005. Forum on historical biogeography: What is cladistic biogeography? *Journal of Biogeography* 32:2179–2183.

Ebach, M. C., R. A. Newman, C. J. Humphries, and D. M. Williams. 2005b. *3item version 2.0: Three-item analysis for cladistics and area cladistics.* Oxford: Authors.

Ebach, M. C. and D. M. Williams. 2004. Congruence and language. *Taxon* 53:113–118.

Edmunds, G. F. Jr. 1972. Biogeography and evolution of Ephemeroptera. *Annual Review of Entomology* 17:21–42.

Ekman, S. 1935. *Tiergeographie des Meeres.* Leipzig: Akademische Verlagsgesellschaft.

Eldredge, N. 1981. Discussion. In *Vicariance biogeography: A critique,* ed. G. Nelson and D. E. Rosen, 34–38. New York: Columbia University Press.

Eldredge, N. 1985. *Unfinished synthesis: Biological hierarchies and modern evolutionary thought.* New York: Oxford University Press. (Spanish translation: 1997, *Síntesis inacabada: Jerarquías biológicas y pensamiento evolutivo moderno,* Mexico, D.F.: Ciencia y Tecnología, Fondo de Cultura Económica.)

Eldredge, N. and S. N. Salthe. 1984. Hierarchy and evolution. In *Oxford surveys in evolutionary biology,* Vol. 1, ed. R. Dawkins and M. Ridley, 182–206. Oxford: Oxford University Press.

Emerson, B. C., S. Forgie, S. Goodacre, and P. Oromí. 2006. Testing phylogeographic predictions on an active volcanic island: *Brachyderes rugatus* (Coleoptera: Curculionidae) on La Palma (Canary Islands). *Molecular Ecology* 15:449–458.

Enghoff, H. 1995. Historical biogeography of the Holarctic: Area relationships, ancestral areas, and dispersal of non-marine animals. *Cladistics* 11:223–263.

Enghoff, H. 1996. Widespread taxa, sympatry, dispersal, and an algorithm for resolved area cladograms. *Cladistics* 12:349–364.

Enghoff, H. 1998. Widespread taxa and Component 2.0. *Cladistics* 14:383–386.

Enghoff, H. 2000. Reversals as branch support in biogeographical parsimony analysis. *Vie et Milieu* 50:255–260.

Erwin, T. L. 1979. Thoughts on the evolutionary history of ground beetles: Hypotheses generated from comparative faunal analyses of lowland forest sites in temperate and tropical regions. In *Carabid beetles: Their evolution, natural history, and classification,* ed. T. L. Erwin, G. E. Ball, and D. R. Whitehead, 539–592. The Hague: Junk.

Erwin, T. L. 1981. Taxon pulses, vicariance, and dispersal: An evolutionary synthesis illustrated by carabid beetles. In *Vicariance biogeography: A critique,* ed. G. Nelson and D. E. Rosen, 371–391. New York: Columbia University Press.

Escalante, T. 2003. Avances en el atlas biogeográfico de los mamíferos terrestres de México. In *Una perspectiva latinoamericana de la biogeografía,* ed. J. J. Morrone and J. Llorente Bousquets, 297–302. Mexico, D.F.: Las Prensas de Ciencias, UNAM.

Escalante, T., D. Espinosa, and J. J. Morrone. 2002. Patrones de distribución geográfica de los mamíferos terrestres de México. *Acta Zoológica Mexicana* 87:47–65.

Escalante, T., D. Espinosa, and J. J. Morrone. 2003. Using parsimony analysis of endemicity to analyze the distribution of Mexican land mammals. *Southwestern Naturalist* 48:563–578.

Escalante, T. and J. J. Morrone. 2003. ¿Para qué sirve el análisis de parsimonia de endemismos? In *Una perspectiva latinoamericana de la biogeografía,* ed. J. J. Morrone and J. Llorente Bousquets, 167–172. Mexico, D.F.: Las Prensas de Ciencias, UNAM.

Escalante, T., G. Rodríguez, N. Cao, M. C. Ebach, and J. J. Morrone. 2007a. Cladistic biogeographic analysis suggests an early Caribbean diversification in Mexico. *Naturwissenschaften* 94:561–565.

Escalante, T., G. Rodríguez, and J. J. Morrone. 2004. The diversification of the Nearctic mammals in the Mexican transition zone. *Biological Journal of the Linnean Society* 83:327–339.

Escalante, T., G. Rodríguez, and J. J. Morrone. 2005. Las provincias biogeográficas del componente Mexicano de Montaña desde la perspectiva de los mamíferos continentales. *Revista Mexicana de Biodiversidad* 76:199–205.

Escalante, T., V. Sánchez-Cordero, J. J. Morrone, and M. Linaje. 2007b. Areas of endemism of Mexican terrestrial mammals: A case study using species' ecological niche modeling, parsimony analysis of endemicity and Goloboff fit. *Interciencia* 32:151–159.

Escalante, T., V. Sánchez-Cordero, J. J. Morrone, and M. Linaje. 2007c. Deforestation affects biogeographical regionalization: A case study contrasting potential and extant distributions of Mexican terrestrial mammals. *Journal of Natural History* 41:965–984.

Espadas-Manrique, C., R. Durán, and J. Argáez. 2003. Phytogeographic analysis of taxa endemic to the Yucatán Peninsula using geographic information systems, the domain heuristic method and parsimony analysis of endemicity. *Diversity and Distributions* 9:313–330.

Espinosa Organista, D., J. Llorente, and J. J. Morrone. 2006. Historical biogeographic patterns of the species of *Bursera* (Burseraceae) and their taxonomical implications. *Journal of Biogeography* 33:1945–1958.

Espinosa Organista, D. and J. Llorente Bousquets. 1993. *Fundamentos de biogeografías filogenéticas*. México, D.F.: Facultad de Ciencias, UNAM.

Espinosa Organista, D., J. J. Morrone, C. Aguilar, and J. Llorente Bousquets. 2000. Regionalización biogeográfica de México: Provincias bióticas. In *Biodiversidad, taxonomía y biogeografía de artrópodos de México: Hacia una síntesis de su conocimiento*, Vol. II, ed. J. Llorente Bousquets, E. González, and N. Papavero, 61–94. Mexico, D.F.: Conabio.

Espinosa Organista, D., J. J. Morrone, J. Llorente Bousquets, and O. Flores Villela. 2002. *Análisis de patrones biogeográficos históricos*. México, D.F.: Las Prensas de Ciencias, UNAM.

Espinosa Pérez, H. and L. Huidobro Campos. 2005. Ictiogeografía de los peces dulceacuícolas de la vertiente del Golfo de México. In *Regionalización biogeográfica en Iberoamérica y tópicos afines: Primeras Jornadas Biogeográficas de la Red Iberoamericana de Biogeografía y Entomología Sistemática (RIBES XII.I–CYTED)*, ed. J. Llorente Bousquets and J. J. Morrone, 295–318. Mexico, D.F.: Las Prensas de Ciencias, UNAM.

Farris, J. S. 1988. *Hennig86 reference*. Version 1.5. Port Jefferson, N.Y.: Author.

Farris, J. S. 2000. Diagnostic efficiency of three-taxon analysis. *Cladistics* 16:403–410.

Farris, J. S. and A. G. Kluge. 1998. A/the brief history of three-taxon analysis. *Cladistics* 14:349–362.

Fattorini, S. 2002. Biogeography of tenebrionid beetles (Coleoptera, Tenebrionidae) on the Aegean islands (Greece). *Journal of Biogeography* 29:49–67.

Fattorini, S. and A. P. Fowles. 2005. A biogeographical analysis of the tenebrionid beetles (Coleoptera, Tenebrionidae) of the island of Thasos in the context of the Aegean islands (Greece). *Journal of Natural History* 39:3919–3949.

Fedorov, V. B., A. V. Goropashnaya, M. Jaarola, and J. A. Cook. 2003. Phylogeography of lemmings (*Lemmus*): No evidence for postglacial colonization of Arctic from the Beringian refugium. *Molecular Ecology* 12:725–731.

Felsenstein, J. 1986. *Phylogenetic inference package (PHYLIP)*. Seattle: University of Washington.

Felsenstein, J. 1993. *Phylogenetic inference package (PHYLIP)*. Version 3.5. Seattle: University of Washington.

Fernandes, M. E. B., J. M. Da Silva, and J. Silva Jr. 1995. The monkeys of the islands of the Amazon estuary, Brazil: A biogeographic analysis. *Mammalia* 59:213–221.

Ferrusquía-Villafranca, I. 1998. Geología de México: Una sinopsis. In *Diversidad biológica de México*, ed. T. P. Ramamoorthy, R. Bye, A. Lot, and J. Fa, 3–108. Mexico, D.F.: Instituto de Biología, UNAM.

Fiala, K. 1984. *CLINCH: Cladistic inference using character compatibility 6.2*. Seattle: University of Washington.

Fittkau, E. J. 1969. The fauna of South America. In *Biogeography and ecology in South America*, Vol. 2, ed. E. J. Fittkau, J. Illies, H. Klinge, G. H. Schwabe, and H. Sioli, 624–650. The Hague: Junk.

Fjeldså, J. 1992. Biogeographic patterns and evolution of the avifauna of relict high-altitude woodlands of the Andes. *Steenstrupia* 18:9–62.

Flores Villela, O. and I. Goyenechea. 2001. A comparison of hypotheses of historical biogeography for Mexico and Central America, or in search for the lost pattern. In *Mesoamerican herpetology: Systematics, zoogeography, and conservation*, ed. J. D. Johnson, R. G. Webb, and O. Flores Villela, 171–181. El Paso: The University of Texas.

Folinsbee, K. E. and D. R. Brooks. 2007. Miocene hominoid biogeography: Pulses of dispersal and differentiation. *Journal of Biogeography* 34:383–397.

Fontenla, J. L. 2003. Biogeography of Antillean butterflies (Lepidoptera: Rhopalocera): Patterns of association among areas of endemism. *Transactions of the American Entomological Society* 129:399–410.

Fontenla, J. L. 2005. Relaciones biogeográficas de las hormigas (Hymenoptera: Formicidae) de las Antillas. In *Regionalización biogeográfica en Iberoamérica y tópicos afines: Primeras Jornadas Biogeográficas de la Red Iberoamericana de Biogeografía y Entomología Sistemática (RIBES XII.I–CYTED)*, ed. J. Llorente Bousquets and J. J. Morrone, 395–416. Mexico, D.F.: Las Prensas de Ciencias, UNAM.

Fordyce, J. A. and C. C. Nice. 2003. Contemporary patterns in a historical context: Phylogeographic history of the pipevine swallowtail, *Battus philenor* (Papilionidae). *Evolution* 57:1089–1099.

Forster, J. R. 1778. *Observations made during a voyage round the world, or physical geography, natural history and ethic philosophy*. London: G. Robinson.

Fortino, A. D. and J. J. Morrone. 1997. Signos gráficos para la representación de análisis panbiogeográficos. *Biogeographica* 73:49–56.

Foucault, M. 1966. *Les mots et les choses: Une archéologie des sciences humaines*. Paris: Gallimard. (English translation: 1994, *The order of things: An archaeology of the human sciences*, New York: Vintage.)

Foucault, M. 1969. *L'archéologie du savoir*. Paris: Gallimard. (English translation: 1972, *The archaeology of knowledge*, New York: Harper and Row.)

Franco Rosselli, P. 2001. Estudios panbiogeográficos en Colombia. In *Introducción a la biogeografía en Latinoamérica: Teorías, conceptos, métodos y aplicaciones*, ed. J. Llorente Bousquets and J. J. Morrone, 221–224. Mexico, D.F.: Las Prensas de Ciencias, UNAM.

Franco Rosselli, P. and C. C. Berg. 1997. Distributional patterns of *Cecropia* (Cecropiaceae): A panbiogeographic analysis. *Caldasia* 19:285–296.

Fritsch, P. 2001. Phylogeny and biogeography of the flowering plant genus *Styrax* (Styracaceae) based on chloroplast DNA restriction sites and DNA sequences of the internal transcribed space region. *Molecular Phylogenetics and Evolution* 19:387–408.

Funk, V. A. 2004. Revolutions in historical biogeography. In *Foundations of biogeography: Classic papers with commentaries*, ed. M. V. Lomolino, D. F. Sax, and J. H. Brown, 647–657. Chicago: University of Chicago Press.

Funk, V. A. and W. L. Wagner. 1995. Biogeographic patterns in the Hawaiian islands. In *Hawaiian biogeography: Evolution on a hotspot archipelago,* ed. W. L. Wagner and V. A. Funk, 379–419. Washington, D.C.: Smithsonian Institution Press.

García-Barros, E. 2003. Mariposas diurnas endémicas de la región Paleártica Occidental: Patrones de distribución y su análisis mediante parsimonia (Lepidoptera, Papilionoidea). *Graellsia* 59:233–258.

García-Barros, E., P. Gurrea, M. J. Luciáñez, J. M. Cano, M. L. Munguira, J. C. Moreno, H. Sainz, M. J. Sanza, and J. C. Simón. 2002. Parsimony analysis of endemicity and its application to animal and plant geographical distributions in the Ibero-Balearic region (western Mediterranean). *Journal of Biogeography* 29:109–124.

García-Trejo, E. A. and A. G. Navarro. 2004. Patrones biogeográficos de la riqueza de especies y el endemismo de la avifauna en el oeste de México. *Acta Zoológica Mexicana* 20:167–185.

Garraffoni, A. R. S., S. S. Nihei, and P. C. Lana. 2006. Distribution patterns of Terebellidae (Annelida: Polychaeta): An application of parsimony analysis of endemicity (PAE). *Scientia Marina* 70:269–276.

Gaston, J. J. 2003. *The structure and dynamics of geographic ranges.* Oxford: Oxford University Press.

George, W. 1964. *Biologist philosopher.* London: Abelard-Schuman.

Geraads, D. 1998. Biogeography of circum-Mediterranean Miocene–Pliocene rodents: A revision using factor analysis and parsimony analysis of endemicity. *Palaeogeography, Palaeoclimatology, Palaeoecology* 137:273–288.

Gesundheit, P. and C. Macías García. 2005. Biogeografía cladística de la familia Goodeidae (Cyprinodontiformes). In *Regionalización biogeográfica en Iberoamérica y tópicos afines: Primeras Jornadas Biogeográficas de la Red Iberoamericana de Biogeografía y Entomología Sistemática (RIBES XII.I–CYTED),* ed. J. Llorente Bousquets and J. J. Morrone, 319–338. Mexico, D.F.: Las Prensas de Ciencias, UNAM.

Gillespie, J. H. 1991. *The causes of molecular evolution.* New York: Oxford University Press.

Giribet, G. and G. D. Edgecombe. 2006. The importance of looking at small-scale patterns when inferring Gondwanan biogeography: A case study of the centipede *Paralamyctes* (Chilopoda, Lithobiomorpha, Henicopidae). *Biological Journal of the Linnean Society* 89:65–78.

Glasby, C. J. 2005. Polychaete distribution patterns revisited: An historical explanation. *Marine Ecology* 26:235–245.

Glasby, C. J. and B. Álvarez. 1999. Distribution patterns and biogeographic analysis of Austral Polychaeta (Annelida). *Journal of Biogeography* 26:507–533.

Goldani, A. and G. S. Carvalho. 2003. Análise de parcimônia de endemismo de cercopídeos neotropicais (Hemiptera, Cercopidae). *Revista Brasileira de Entomologia* 47:437–442.

Goldani, A., A. Ferrari, G. S. Carvalho, and A. J. Creão-Duarte. 2002. Análise de parcimonia de endemimso de membracídeos neotropicais (Hemiptera, Membracidae, Hoplophorionini). *Revista Brasileira de Zoologia* 19:187–193.

Goloboff, P. 1993. NONA. *Cladistics* 9:83–91.

Goloboff, P. 1998. *NONA.* Version 2.0. Retrieved May 25, 2008, from http://www.cladistics.com/about_nona.htm.

Goloboff, P. 2004. *NDM/VNDM programs.* Version 1.5. Retrieved May 25, 2008, from http://www.zmuc.dk-/public/phylogeny/Endemism/.

Goloboff, P., J. Farris, and K. Nixon. 2008. *TNT.* Willi Hennig Society. Retrieved May 25, 2008, from http://www.zmuc.dk/public/phylogeny/TNT/.

Gómez-González, S., L. A. Cavieres, E. A. Teneb, and J. Arroyo. 2004. Biogeographical analysis of species of the tribe Cytiseae (Fabaceae) in the Iberian Peninsula and Balearic Islands. *Journal of Biogeography* 31:1659–1671.

González-Rodríguez, A., J. F. Bain, J. L. Golden, and K. Oyama. 2004. Chloroplast DNA variation in the *Quercus affinis–Q. laurina* complex in Mexico: Geographical structure and associations with nuclear and morphological variation. *Molecular Ecology* 13:3467–3476.

González-Zamora, A., I. Luna-Vega, J. L. Villaseñor, and C. A. Ruiz-Jiménez. 2007. Distributional patterns and conservation of species of Asteraceae (asters etc.) endemic to eastern Mexico: A panbiogeographical approach. *Systematics and Biodiversity* 5:135–144.

Good, R. 1974. *The geography of the flowering plants.* London: Longman.

Goodman, M., J. Czelusniak, G. W. Moore, A. E. Romero-Herrera, and G. Matsuda. 1979. Fitting the gene lineage into its species lineage: A parsimony strategy illustrated by cladograms constructed from globin sequences. *Systematic Zoology* 28:132–168.

Gorog, A. J., M. H. Sinaga, and M. D. Engstrom. 2004. Vicariance or dispersal? Historical biogeography of three Sunda shelf murine rodents (*Maxomys surifer, Leopoldamys sabanus* and *Maxomys whiteheadi*). *Biological Journal of the Linnean Society* 81:91–109.

Gows, G., B. A. Stewart, and S. R. Daniels. 2006. Phylogeographic structure of a freshwater crayfish (Decapoda: Parastacidae: *Cherax preissii*) in south-western Australia. *Marine and Freshwater Research* 57:837–848.

Goyenechea, I., Ó. Flores Villela, and J. J. Morrone. 2001. Introducción a los fundamentos y métodos de la biogeografía cladística. In *Introducción a la biogeografía en Latinoamérica: Conceptos, teorías, métodos y aplicaciones,* ed. J. Llorente Bousquets and J. J. Morrone, 225–232. Mexico, D.F.: Las Prensas de Ciencias, UNAM.

Grande, L. 1985. The use of paleontology in systematics and biogeography, and a time control refinement for historical biogeography. *Paleobiology* 11:234–243.

Grande, L. 1990. Vicariance biogeography. In *Paleobiology: A synthesis,* ed. D. E. G. Briggs and P. R. Crowther, 448–451. Oxford: Blackwell Scientific.

Grant, L. J., R. Sluys, and D. Blair. 2006. Biodiversity of Australian freshwater planarians (Platyhelminthes: Tricladida: Paludicola): New species and localities, and a review of paludicolan distribution in Australia. *Systematics and Biodiversity* 4:435–471.

Gray, R. D. 1989. Oppositions in panbiogeography: Can the conflict between selection, constraint, ecology, and history be resolved? *New Zealand Journal of Zoology* 16:787–806.

Grehan, J. R. 1984. Evolution by law: Croizat's "orthogeny" and Darwin's "laws of growth." *Tuatara* 27:14–19.

Grehan, J. R. 1988a. Biogeographic homology: Ratites and the southern beeches. *Rivista di Biologia—Biology Forum* 81:577–587.

Grehan, J. R. 1988b. Panbiogeography: Evolution in space and time. *Rivista di Biologia—Biology Forum* 81:469–498.

Grehan, J. R. 1989. New Zealand panbiogeography: Past, present, and future. *New Zealand Journal of Zoology* 16:513–525.

Grehan, J. R. 1991. Panbiogeography 1981–91: Development of an Earth/life synthesis. *Progress in Physical Geography* 15:331–363.

Grehan, J. R. 1993. Conservation biogeography and the biodiversity crisis: A global problem in space/time. *Biodiversity Letters* 1:134–140.

Grehan, J. R. 1994. The beginning and end of dispersal: The representation of "panbiogeography." *Journal of Biogeography* 21:451–462.

Grehan, J. R. 2001a. Biogeography and evolution of the Galapagos: Integration of the biological and geological evidence. *Biological Journal of the Linnean Society* 74:267–287.

Grehan, J. R. 2001b. Islas Galápagos: Biogeografía, tectónica y evolución en un archipiélago oceánico. In *Introducción a la biogeografía en Latinoamérica: Teorías, conceptos, métodos y aplicaciones,* ed. J. Llorente Bousquets and J. J. Morrone, 153–160. Mexico, D.F.: Las Prensas de Ciencias, UNAM.

Grehan, J. R. 2001c. Panbiogeografía y la geografía de la vida. In *Introducción a la biogeografía en Latinoamérica: Teorías, conceptos, métodos y aplicaciones,* ed. J. Llorente Bousquets and J. J. Morrone, 181–195. Mexico, D.F.: Las Prensas de Ciencias, UNAM.

Grehan, J. R. 2001d. Panbiogeography from tracks to ocean basins: Evolving perspectives. *Journal of Biogeography* 28:413–429.

Grehan, J. R. 2007. A brief look at Pacific biogeography: The trans-oceanic travels of *Microseris* (Angiosperms: Asteraceae). In *Biogeography in a changing world,* ed. M. C. Ebach and R. S. Tangney, 83–94. Boca Raton, Fla.: CRC Press.

Grehan, J. R. and J. E. Rawlings. 2003. Larval description of a New World ghost moth, *Phassus* sp., and the evolutionary biogeography of wood-boring Hepialidae (Lepidoptera: Exoporia: Hepialoidea). *Proceedings of the Entomological Society of Washington* 105(3):748–755.

Grene, M. 1990. Is evolution at crossroads? *Evolutionary Biology.* 24:51–81.

Grismer, L. L. 1994. The origin and evolution of the peninsular herpetofauna of Baja California, Mexico. *Herpetological Natural History* 2:51–106.

Griswold, C. E. 1991. Cladistic biogeography of afromontane spiders. *Australian Systematic Botany* 4:73–89.

Haeckel, E. 1868. *Natürliche Schöpfungsgeschichte. Gemeinverständliche wissenschaftliche Vorträge über die Entwicklungslehre im Allgemeinen und diejenige von Darwin, Goethe und Lamarck im Besonderen über die Anwendung derselben auf den Ursprung des Menschen und andere damit zusammenhängende Grundfragen der Naturwissenschaft.* Berlin: Georg Reimer.

Haffer, J. 1969. Speciation in Amazonian forest birds. *Science* 165:131–137.

Haffer, J. 1974. *Avian speciation in tropical South America, with a systematic survey of the toucans (Rhamphastidae) and jacamars (Galbulidae).* Cambridge, Mass.: Nuttall Ornithological Club.

Haffer, J. 1978. Distribution of Amazon forest birds. *Bonner Zoologische Beitrage* 29:38–78.

Haffer, J. 1981. Aspects of Neotropical bird speciation during the Cenozoic. In *Vicariance biogeography: A critique,* ed. G. Nelson and D. E. Rosen, 371–391. New York: Columbia University Press.

Haffer, J., E. Rutschke, and K. Wunderlich. 2000. Erwin Stresemann 1889–1972: Leben und Werk eines Pioniers der wissenschaftlichen Ornithologie. *Acta Historica Leopoldiana* 34:1–465.

Hafner, M. S., J. E. Light, D. J. Hafner, S. V. Brant, T. A. Spradling, and J. W. Demastes. 2005. Cryptic species in the Mexican pocket gopher *Cratogeomys merriami. Journal of Mammalogy* 86(6):1095–1108.

Halas, D., D. Zamparo, and D. R. Brooks. 2005. A protocol for studying biotic diversification by taxon pulses. *Journal of Biogeography* 32:249–260.

Halffter, G. 1962. Explicación preliminar de la distribución geográfica de los Scarabaeidae mexicanos. *Acta Zoológica Mexicana* 5:1–17.

Halffter, G. 1964. La entomofauna americana, ideas acerca de su origen y distribución. *Folia Entomológica Mexicana* 6:1–108.

Halffter, G. 1965. Algunas ideas acerca de la zoogeografía de América. *Revista de la Sociedad Mexicana de Historia Natural* 26:1–16.

Halffter, G. 1972. Eléments anciens de l'entomofaune Neotropicale: Ses implications biogéographiques. *Biogeographie et Liaisons Intercontinentales au cours du Mésozoique, 17me Congr. Int. Zool., Monte Carlo* 1:1–40.

Halffter, G. 1974. Eléments anciens de l'entomofaune Neotropicale: Ses implications biogéographiques. *Quaestiones Entomologicae* 10:223–262.

Halffter, G. 1976. Distribución de los insectos en la zona de transición mexicana: Relaciones con la entomofauna de Norteamérica. *Folia Entomológica Mexicana* 35:1–64.

Halffter, G. 1978. Un nuevo patrón de dispersión en la zona de transición mexicana: El mesoamericano de montaña. *Folia Entomológica Mexicana* 39–40:219–222.

Halffter, G. 1987. Biogeography of the montane entomofauna of Mexico and Central America. *Annual Review of Entomology* 32:95–114.

Halffter, G., M. E. Favila, and L. Arellano. 1995. Spatial distribution of three groups of Coleoptera along an altitudinal transect in the Mexican Transition Zone and its biogeographical implications. *Elytron* 9:151–185.

Hall, J. P. W. and D. J. Harvey. 2002. The phylogeography of Amazonia revisited: New evidence from Riodinid butterflies. *Evolution* 56:1489–1497.

Hallam, A. 1973. *Atlas of palaeobiogeography.* Amsterdam: Elsevier.

Hallam, A. 1974. Changing patterns of provinciality and diversity of fossil animals in relation to plate tectonics. *Journal of Biogeography* 1:213–225.

Harold, A. S. and R. D. Mooi. 1994. Areas of endemism: Definition and recognition criteria. *Systematic Biology* 43:261–266.

Harvey, A. W. 1992. Three-taxon statements: More precisely, an abuse of parsimony? *Cladistics* 8:345–354.

Hausdorf, B. 1998. Weighted ancestral area analysis and a solution of the redundant distribution problem. *Systematic Biology* 47:445–456.

Hausdorf, B. 2002. Units in biogeography. *Systematic Biology* 51:648–652.

Hausdorf, B. and C. Hennig. 2003. Biotic element analysis in biogeography. *Systematic Biology* 52:717–723.

Hausdorf, B. and C. Hennig. 2004. Does vicariance shape biotas? Biogeographical tests of the vicariance model in the north-west European land snail fauna. *Journal of Biogeography* 31:1751–1757.

Hausdorf, B. and C. Hennig. 2007. Biotic element analysis and vicariance biogeography. In *Biogeography in a changing world,* ed. M. C. Ebach and R. S. Tangney, 95–115. Boca Raton, Fla.: CRC Press.

Haydon, D. T., B. I. Crother, and E. R. Pianka. 1994. New directions on biogeography? *Trends in Ecology and Evolution* 9:403–406.

Heads, M. J. 1984. *Principia Botanica:* Croizat's contribution to botany. *Tuatara* 27:26–48.

Heads, M. J. 1985a. Biogeographic analysis of *Nothofagus* (Fagaceae). *Taxon* 34:474–480.

Heads, M. J. 1985b. On the nature of ancestors. *Systematic Zoology* 34:205–215.

Heads, M. J. 1986. A panbiogeographic analysis of Auckland islands archipelago. In *The Lepidoptera, bryophytes, and panbiogeography of Auckland islands,* ed. R. D. Archibald, 30–44. Dunedin: New Zealand Entomological Society.

Heads, M. J. 1989. Integrating earth and life sciences in New Zealand natural history: The parallel arcs model. *New Zealand Journal of Zoology* 16:549–585.

Heads, M. J. 1994. A biogeographic review of *Parahebe* (Scrophulariaceae). *Botanical Journal of the Linnean Society of London* 115:65–89.

Heads, M. J. 1996. Biogeography, taxonomy and evolution in the Pacific genus *Coprosma* (Rubiaceae). *Candollea* 51:381–405.

Heads, M. J. 1999. Vicariance biogeography and terrane tectonics in the South Pacific: Analysis of the genus *Abrotanella* (Compositae). *Biological Journal of the Linnean Society* 67:391–432.

Heads, M. 2001. Birds of paradise (Paradiseidae) and bowerbirds (Ptilonorhynchidae): Regional levels of biodiversity and terrane tectonics in New Guinea. *Journal of Zoology, London* 255:331–339.

Heads, M. J. 2004. What is a node? *Journal of Biogeography* 31:1883–1891.

Heads, M. J. 2005a. The history and philosophy of panbiogeography. In *Regionalización biogeográfica en Iberoamérica y tópicos afines: Primeras Jornadas Biogeográficas de la Red Iberoamericana de Biogeografía y Entomología Sistemática (RIBES XII.I–CYTED)*, ed. J. Llorente Bousquets and J. J. Morrone, 67–123. Mexico, D.F.: Las Prensas de Ciencias, UNAM.

Heads, M. J. 2005b. Toward a panbiogeography of the seas. *Biological Journal of the Linnean Society* 84:675–723.

Henderson, I. M. 1989. Quantitative panbiogeography: An investigation into concepts and methods. *New Zealand Journal of Zoology* 16:495–510.

Henderson, I. M. 1991. Biogeography without area? *Australian Systematic Botany* 4:59–71.

Hengeveld, R. 1990. *Dynamic biogeography*. Cambridge: Cambridge University Press.

Hennig, W. 1950. *Grundzüge einer Theorie der phylogenetischen Systematik*. Berlin: Deutscher Zentralverlag. (English translation: 1966, *Phylogenetic systematics*, Urbana: University of Illinois Press; Spanish translation: 1968, *Elementos de una sistemática filogenética*, Buenos Aires: Eudeba.)

Hess, H. H. 1962. History of ocean basins. In *Petrologic studies: A volume in honor of A. F. Buddington*, ed. A. E. Engel, J. H. L. James, and B. F. Leonard, 599–620. Boulder, Colo.: Geological Society of America.

Hewitt, G. M. 1996. Some genetic consequences of ice ages, and their role in divergence and speciation. *Biological Journal of the Linnean Society* 58:247–276.

Hewitt, G. M. 1999. Post-glacial re-colonization of European biota. *Biological Journal of the Linnean Society* 68(1–2):87–112.

Hewitt, G. M. 2000. The genetic legacy of the Quaternary ice ages. *Nature* 405:907–913.

Hewitt, G. M. 2004. The structure of biodiversity from molecular phylogeography. *Frontiers in Zoology* 1:1–4.

Hibbett, D. S. 2001. Shiitake mushrooms and molecular clock: Historical biogeography of *Lentinula*. *Journal of Biogeography* 28:231–241.

Hoffstetter, R. 1972. Relationships, origins, and history of the ceboid monkeys and caviomorph rodents: A modern reinterpretation. *Evolutionary Biology* 6:322–347.

Hofsten, N. G. E. von. 1916. Zur älteren Geschichte des Diskontinuitätsproblems in Biogeographie. *Zoologische Annalen Zeitschrift für Geschichte der Zoologie* 7:197–353.

Holden, J. C. and R. S. Dietz. 1972. Galapagos Gore, NazCoPac triple junction and Carnegie/Cocos ridge. *Nature* 235:266–269.

Holloway, J. D. 1992. Croizat's panbiogeography: A New Zealand perspective. *Journal of Biogeography* 19:233–238.

Holloway, J. D. 2003. Biological images of geological history: Through a glass darkly or brightly face to face? *Journal of Biogeography* 30:165–179.

Holsinger, K. E. 2006. *Nested clade analysis*. Retrieved May 25, 2008, from http://darwin.eeb.uconn.edu/eeb348/lecture-notes/nca.pdf.

Hooker, J. D. 1844–1860. *The botany of the Antarctic voyage of* H. M. Discovery *ships* Erebus *and* Terror *in the years 1839–1843. I. Flora Antarctica (1844–47)*. London: Reeve.

Hovenkamp, P. 1997. Vicariance events, not areas, should be used in biogeographic analysis. *Cladistics* 13:67–79.

Hovenkamp, P. 2001. A direct method for the analysis of vicariance patterns. *Cladistics* 17:260–265.

Hubbard, J. P. 1973. Avian evolution in the aridlands of North America. *Living Bird* 12:155–196.

Hubbel, S. P. 2001. *The unified neutral theory of biodiversity and biogeography*. Princeton, N.J.: Princeton University Press.

Huelsenbeck, J. P., B. Rannala, and B. Larget. 2000. A Bayesian framework for the analysis of cospeciation. *Evolution* 54:352–364.

Huelsenbeck, J. P. and F. Ronquist. 2001. MrBayes: Bayesian inference of phylogenetic trees. *Bioinformatics* 17(8):754–755.

Huidobro, L., J. J. Morrone, J. L. Villalobos, and F. Álvarez. 2006. Distributional patterns of freshwater taxa (fishes, crustaceans and plants) from the Mexican transition zone. *Journal of Biogeography* 33:731–741.

Hull, D. L. 1983. Popper and Plato's metaphor. In *Advances in cladistics, volume 2: Proceedings of the second meeting of the Willi Hennig Society,* ed. N. I. Platnick and V. A. Funk, 177–189. New York: Columbia University Press.

Hull, D. L. 1988. *Science as a process: An evolutionary account of the social and conceptual development of science*. Chicago: University of Chicago Press.

Hulsey, C. D., F. J. García de León, Y. Sánchez Johnson, D. A. Hendrickson, and T. J. Near. 2004. Temporal diversification of Mesoamerican cichlid fishes across a major biogeographic boundary. *Molecular Phylogenetics and Evolution* 31:754–764.

Humboldt, A. von. 1805. *Essai sur la géographie des plantes, accompagné d'un tableau physique des régions équinoxiales, fondé sur les mesures exécutées, depuis le dixième degré de latitude boréale jusqu'au dixième degré de latitude australe, pendant les années 1799, 1800, 1801, 1802, et 1903,* ed. A. de Humboldt and A. Bonpland. Paris: Levrault, Schoelle.

Humboldt, A. von. 1815. *Personal narrative of travels to the equinoctial regions of America, during the years 1799–1804*. London: H. H. Bohn.

Humphries, C. J. 1981. Biogeographical methods and the southern beeches (Fagaceae: *Nothofagus*). In *Advances in cladistics, 1, Proceedings of the first meeting of the Willi Hennig Society,* ed. V. A. Funk and D. R. Brooks, 117–207. Bronx: New York Botanical Garden.

Humphries, C. J. 1982. Vicariance biogeography in Mesoamerica. *Annals of the Missouri Botanical Garden* 69:444–463.

Humphries, C. J. 1985. Temperate biogeography and an intemperate botanist. *Taxon* 34:480–492.

Humphries, C. J. 1989. Any advance on assumption 2? *Journal of Biogeography* 16:101–102.

Humphries, C. J. 1992. Cladistic biogeography. In *Cladistics: A practical course in systematics,* ed. P. L. Forey, C. J. Humphries, I. J. Kitching, R. W. Scotland, D. J. Siebert, and D. M. Williams, 137–159. Oxford: Clarendon.

Humphries, C. J. 2000. Form, space and time; which comes first? *Journal of Biogeography* 27:11–15.

Humphries, C. J. 2004. From dispersal to geographic congruence: Comments on cladistic biogeography in the twentieth century. In *Milestones in systematics,* ed. D. M. Williams and P. L. Forey, 225–260. Boca Raton, Fla.: CRC Press.

Humphries, C. J. and M. C. Ebach. 2004. Biogeography on a dynamic Earth. In *Frontiers of biogeography: New directions in the geography of nature*, ed. M. V. Lomolino and L. R. Heaney, 67–86. Sunderland, Mass.: Sinauer.

Humphries, C. J., P. Y. Ladiges, M. Roos, and M. Zandee. 1988. Cladistic biogeography. In *Analytical biogeography: An integrated approach to the study of animal and plant distributions*, ed. A. A. Myers and P. S. Giller, 371–404. London: Chapman and Hall.

Humphries, C. J. and L. R. Parenti. 1986. *Cladistic biogeography.* Oxford: Clarendon.

Humphries, C. J. and L. R. Parenti. 1999. *Cladistic biogeography,* 2nd ed.: *Interpreting patterns of plant and animal distributions.* Oxford: Oxford University Press.

Humphries, C. J. and O. Seberg. 1989. Graphs and generalized tracks: Some comments on method. *Systematic Zoology* 38:69–76.

Hunn, C. A. and P. Upchurch. 2001. The importance of time/space in diagnosing the causality of phylogenetic events: Towards a "chronobiogeographical paradigm." *Systematic Biology* 50:391–407.

Ippi, S. and V. Flores. 2001. Las tortugas neotropicales y sus áreas de endemismo. *Acta Zoológica Mexicana (nueva serie)* 84:49–63.

Irwin, D. E. 2002. Phylogeographic breaks without geographic barriers to gene flow. *Evolution* 56:2383–2394.

Jansa, S. A., F. K. Barber, and L. R. Heaney. 2006. The pattern and timing of diversification of Philippine endemic rodents: Evidence from mitochondrial and nuclear gene sequences. *Systematic Biology* 55:73–88.

Janvier, P. 1984. Cladistics: Theory, purpose and evolutionary implications. In *Evolutionary theory: Paths into the future*, ed. J. W. Pollard, 39–75. Chichester: Wiley.

Jeannel, R. 1938. Les Migadopides (Coleoptera, Adephaga), une lignée subantarctique. *Revue Française d'Entomologie* 5:1–55.

Jeannel, R. 1942. *La genèse des faunes terrestres: Élements de biogéographie.* Paris: Presses Universitaires de France.

Juan, C., B. C. Emerson, P. Oromí, and G. M. Hewitt. 2000. Colonization and diversification: Towards a phylogeographic synthesis for the Canary Islands. *Trends in Ecology and Evolution* 15:104–109.

Kamp, P. J. J. 1980. Pacifica and New Zealand, proposed Eastern elements in Gondwanaland's history. *Nature* 288:659–664.

Katinas, L., J. V. Crisci, W. L. Wagner, and P. C. Hoch. 2004. Geographical diversification of tribes Epilobieae, Gongylocarpeae, and Onagreae (Onagraceae) in North America, based on parsimony analysis of endemicity and track compatibility analysis. *Annals of the Missouri Botanical Garden* 91:159–185.

Katinas, L., J. J. Morrone, and J. V. Crisci. 1999. Track analysis reveals the composite nature of the Andean biota. *Australian Journal of Botany* 47:111–130.

Keast, A. 1991. Panbiogeography: Then and now. *Quarterly Review of Biology* 66:467–472.

Kishino, H., J. L. Thorne, and W. J. Bruno. 2001. Performance of a divergence time estimation method under a probabilistic model of rate evolution. *Molecular Biology and Evolution* 18:352–361.

Kluge, A. G. 1988. Parsimony in vicariance biogeography: A quantitative method and a greater Antillean example. *Systematic Zoology* 37:315–328.

Kluge, A. G. 1989. A concern for evidence, and a phylogenetic hypothesis of relationships among Epicrates (Boidae, Serpentes). *Systematic Zoology* 38:7–25.

Kluge, A. G. 1993. Three-taxon transformation in phylogenetic inference: Ambiguity and distortion as regards explanatory power. *Cladistics* 9:246–259.

Knowles, L. L. and W. P. Maddison. 2002. Statistical phylogeography. *Molecular Ecology* 11:2623–2635.

Koehler, J. 2000. Amphibian diversity in Bolivia: A study with special reference to montane forest regions. *Bonner Zoologische Monographien* 48:5–243.

Kolibác, J. 1998. New Australian Thanerocleridae, with notes on the biogeography of the subtribe Isoclerina Kolibác (Coleoptera: Cleroidea). *Invertebrate Taxonomy* 12:951–975.

Kozac, K. H., R. A. Blaine, and A. Larson. 2006. Gene lineages and eastern North American paleodrainage basins: Phylogeography and speciation in salamanders of the *Eurycea bislineata* species complex. *Molecular Ecology* 15:191–207.

Krzywinski, J., R. C. Wilkerson, and N. J. Besansky. 2001. Toward understanding Anophelinae (Diptera, Culicidae) phylogeny: Insights from nuclear single-copy genes and the weight of evidence. *Systematic Biology* 50:540–556.

Kuch, M., N. Rohland, J. L. Betancourt, C. L. Latorre, S. Steppan, and H. N. Poinar. 2002. Molecular analysis of an 11,700-year old rodent midden from the Atacama Desert, Chile. *Molecular Ecology* 11:913–924.

Kuhn, T. S. 1971. *La estructura de las revoluciones científicas.* Mexico, D.F.: Breviarios, no. 213, Fondo de Cultura Económica.

Kumar, S., K. Tamura, I. B. Jakobsen, and M. Nei. 2001. MEGA2: Molecular evolutionary genetics analysis software. *Bioinformatics* 17:1244–1245.

Kuschel, G. 1969. Biogeography and ecology of South American Coleoptera. In *Biogeography and ecology in South America,* Vol. 2, ed. E. Fittkau, J. J. Illies, H. Klinge, G. H. Schwabe, and H. Sioli, 709-722. The Hague: Junk.

Ladiges, P. Y., J. Kellermann, G. Nelson, C. J. Humphries, and F. Udovicic. 2005. Historical biogeography of Australian Rhamnaceae, tribe Pomaderreae. *Journal of Biogeography* 32:1909–1919.

Ladiges, P. Y., G. Nelson, and J. Grimes. 1997. Subtree analysis, *Nothofagus* and Pacific biogeography. *Cladistics* 13:125–129.

Ladiges, P. Y., S. M. Prober, and G. Nelson. 1992. Cladistic and biogeographic analysis of the "blue ash" eucalypts. *Cladistics* 8:103–124.

Lakatos, I. 1970. Falsification and the methodology of scientific research programmes. In *Criticism and the growth of knowledge,* ed. I. Lakatos and A. Musgrave, 91–196. Cambridge: Cambridge University Press.

Lakatos, I. 1978. *The methodology of scientific research programmes.* Cambridge: Cambridge University Press.

Lalueza-Fox, C., J. Castresana, L. Sampietro, T. Marques-Bonet, J. A. Alcover, and J. Bertranpetit. 2005. Molecular dating of caprines using ancient DNA sequences of *Myotragus balearicus,* an extinct endemic Balearic mammal. *BMC Evolutionary Biology* 5:1–11.

Lamarck, J. B. P. A. de M. and A. P. de Candolle. 1805. *Flore française, ou descriptions succintes de toutes les plantes qui croissent naturellement en France, disposées selon une nouvelle méthode d'analyse, et precedées par une exposé des principes élémentaires de la botanique.* Paris: Desray.

Lankester, E. R. 1905. *Extinct animals.* London: Constable.

Lanteri, A. A. and V. A. Confalonieri. 2003. Filogeografía: Objetivos, métodos y ejemplos. In *Una perspectiva latinoamericana de la biogeografía,* ed. J. J. Morrone and J. Llorente Bousquets, 185–193. Mexico, D.F.: Las Prensas de Ciencias, UNAM.

Lapointe, F. J. and L. J. Rissler. 2005. Congruence, consensus, and the comparative phylogeography of codistributed species in California. *The American Naturalist* 166: 290–299.

Larson, J. 1986. Not without a plan: Geography and natural history in the late eighteenth century. *Journal of the History of Biology* 19:447–488.

Lattke, J. E. 2003. Biogeographic analysis of the ant genus *Gnamptogenys* Roger in southeast Asia–Australasia (Hymenoptera: Formicidae: Ponerinae). *Journal of Natural History* 37:1879–1897.

Laudan, L. 1977. *Progress and its problems: Toward a theory of scientific growth.* Berkeley: University of California Press.

León-Paniagua, L., E. García, J. Arroyo-Cabrales, and S. Castañeda-Rico. 2004. Patrones biogeográficos de la mastofauna. In *Biodiversidad de la Sierra Madre Oriental,* ed. I. Luna, J. J. Morrone, and D. Espinosa, 469–486. Mexico D.F.: Las Prensas de Ciencias, UNAM.

León-Paniagua, L., A. G. Navarro-Sigüenza, B. E. Hernández-Baños, and J. C. Morales. 2007. Diversification of the arboreal mice of the genus *Habromys* (Rodentia: Cricetidae: Neotominae) in the Mesoamerican highlands. *Molecular Phylogenetics and Evolution* 42:653–664.

Lequesne, W. J. 1982. Compatibility analysis and its applications. *Zoological Journal of the Linnean Society* 74:267–275.

Lévi-Strauss, C. 1961. *Tristes tropiques.* New York: Criterion.

Lieberman, B. S. 1997. Early Cambrian paleogeography and tectonic history: A biogeographic approach. *Geology* 25:1039–1042.

Lieberman, B. S. 2000. *Paleobiogeography: Using fossils to study global change, plate tectonics and evolution.* New York: Kluwer.

Lieberman, B. S. 2002. Phylogenetic biogeography with and without the fossil record: Gauging the effects of extinction and paleontological incompleteness. *Palaeogeography, Palaeoclimatology, Palaeoecology* 178:39–52.

Lieberman, B. S. 2003a. Paleobiogeography: The relevance of fossils to biogeography. *Annual Review of Ecology and Systematics* 34:51–69.

Lieberman, B. S. 2003b. Unifying theory and methodology in biogeography. *Evolutionary Biology* 33:1–25.

Lieberman, B. S. 2004. Range expansion, extinction, and biogeographic congruence: A deep time perspective. In *Frontiers of biogeography: New directions in the geography of nature,* ed. M. V. Lomolino and L. R. Heaney, 111–124. Sunderland, Mass.: Sinauer.

Lieberman, B. S. 2005. Geobiology and paleobiogeography: Tracking the coevolution of the earth and its biota. *Palaeobiogeography, Palaeoclimatology, Palaeoecology* 219:23–33.

Lieberman, B. S. and N. Eldredge. 1996. Trilobite biogeography in the Middle Devonian: Geological processes and analytical methods. *Paleobiology* 22:66–79.

Liebherr, J. K. 1988. General patterns in West Indian insects, and graphical biogeographic analysis of some circum-Caribbean *Platynus* beetles (Carabidae). *Systematic Zoology* 37:385–409.

Liebherr, J. K. 1991. A general area cladogram for montane Mexico based on distributions in the platynine genera *Elliptoleus* and *Calathus* (Coleoptera: Carabidae). *Proceedings of the Entomological Society of Washington* 93:390–406.

Liebherr, J. K. 1994. Biogeographic patterns of montane Mexican and Central American Carabidae (Coleoptera). *Canadian Entomologist* 126:841–860.

Liebherr, J. K. 1997. Dispersal and vicariance in Hawaiian platynine carabid beetles (Coleoptera). *Pacific Science* 51:424–439.

Liebherr, J. K. and A. E. Hajek. 1990. A cladistic test of the taxon cycle and taxon pulse hypotheses. *Cladistics* 5:39–59.

Liebherr, J. K. and E. C. Zimmermann. 1998. Cladistic analysis, phylogeny and biogeography of the Hawaiian Platynini (Coleoptera: Carabidae). *Systematic Entomology* 23:137–162.

Linder, H. P. 1999. *Rytidosperma vickeryae*—a new danthonioid grass from Kosciuszko (New South Wales, Australia): Morphology, phylogeny and biogeography. *Australian Systematic Botany* 12:743–755.

Linder, H. P. 2001. On areas of endemism, with an example of the African Restionaceae. *Systematic Biology* 50:892–912.

Linder, H. P. and M. D. Crisp. 1995. *Nothofagus* and Pacific biogeography. *Cladistics* 11:5–32.

Linder, H. P. and D. M. Mann. 1998. The phylogeny and biogeography of *Thamnochortus* (Restionaceae). *Botanical Journal of the Linnean Society of London* 128:319–357.

Linnaeus, C. 1744. *Oratio de Telluris habitabilis incremento (habita cum . . . Johannem Westermannum Medicinae Doctorem in Academia Regia Upsaliensi anno MDCCXLIIII Aprilis 12., renunciarte etc.).* Leiden: Lugduni Batavorum.

Linz, B., F. Balloux, Y. Moodley, A. Manica, H. Liu, P. Roumagnac, D. Falush, C. Stamer, F. Prugnolle, S. W. van der Merwe, Y. Yamaoka, D. Y. Graham, E. Pérez-Trallero, T. Wadstrom, S. Suerbaum, and M. Achtman. 2007. An African origin for the intimate association between humans and *Helicobacter pylori*. *Nature* 445:915–918.

Liu, P. S. and R. Hershler. 2007. A test of the vicariance hypothesis of western North American freshwater biogeography. *Journal of Biogeography* 34:534–548.

Llorente Bousquets, J., J. J. Morrone, A. Bueno, R. Pérez, Á. Viloria, and D. Espinosa Organista. 2000. Historia del desarrollo y la recepción de las ideas panbiogeográficas de Léon Croizat. *Revista de la Academia Colombiana de Ciencias* 24:549–577.

Lomolino, M. V., B. R. Riddle, and J. H. Brown. 2006. *Biogeography,* 3rd ed. Sunderland, Mass.: Sinauer.

Lomolino, M. V., D. F. Sax, and J. H. Brown, eds. 2004. *Foundations of biogeography: Classic papers with commentaries.* Chicago: University of Chicago Press.

López Almirall, A. 2005. Nueva perspectiva para la regionalización de Cuba: Definición de los sectores. In *Regionalización biogeográfica en Iberoamérica y tópicos afines: Primeras Jornadas Biogeográficas de la Red Iberoamericana de Biogeografía y Entomología Sistemática (RIBES XII.I–CYTED),* ed. J. Llorente Bousquets and J. J. Morrone, 417–428. Mexico, D.F.: Las Prensas de Ciencias, UNAM.

López Ruf, M., J. J. Morrone, and E. P. Hernández. 2006. Patrones de distribución de las Naucoridae argentinas (Insecta: Heteroptera). *Revista de la Sociedad Entomológica Argentina* 65:111–121.

Lopretto, E. C. and J. J. Morrone. 1998. Anaspidacea, Bathynellacea (Syncarida), generalised tracks, and the biogeographical relationships of South America. *Zoologica Scripta* 27:311–318.

Losos, J. B. 1996. Phylogenetic perspectives on community ecology. *Ecology* 77:1344–1354.

Lourenço, W. R. 1998. Panbiogeographie, les distributions disjointes et le concept de famille relictuelle chez les scorpions. *Biogeographica* 74:133–144.

Löve, A. 1967. A remarkable biological synthesis. *Ecology* 48:704–705.

Lovejoy, N. R. 1997. Stingrays, parasites, and Neotropical biogeography: A closer look at Brooks et al.'s hypotheses concerning the origins of Neotropical freshwater rays (Potamotrygonidae). *Systematic Biology* 46:218–230.

Löwenberg-Neto, P. and C. J. B. de Carvalho. 2004. Análise Parcimoniosa de Endemicidade (PAE) na delimitação de áreas de endemismos: Inferências para conservação da biodiversidade na região sul do Brasil. *Natureza & Conservaçao* 2:58–65.

Ludt, C. J., W. Schröder, O. Rottmann, and R. Kühn. 2004. Mitochondrial DNA phylogeography of red deer (*Cervus elaphus*). *Molecular Phylogenetics and Evolution* 31: 1064–1083.

Luis Martínez, A., J. Llorente Bousquets, and I. Vargas Fernández. 2005. Una megabase de datos de mariposas y la regionalización biogeográfica de México. In *Regionalización biogeográfica en Iberoamérica y tópicos afines: Primeras Jornadas Biogeográficas de la Red Iberoamericana de Biogeografía y Entomología Sistemática (RIBES XII.I–CYTED)*, ed. J. Llorente Bousquets and J. J. Morrone, 269–294. Mexico, D.F.: Las Prensas de Ciencias, UNAM.

Luna, I., L. Almeida, and J. Llorente. 1989. Florística y aspectos fitogeográficos del bosque mesófilo de montaña de las cañadas de Ocuilan, estados de Morelos y México. *Anales del Instituto de Biología de la UNAM, Serie Botánica* 59:63–87.

Luna-Vega, I. and O. Alcántara. 2002. Placing the Mexican cloud forests in a global context: A track analysis based on vascular plant genera. *Biogeographica* 78:1–14.

Luna-Vega, I., O. Alcántara, D. Espinosa Organista, and J. J. Morrone. 1999. Historical relationships of the Mexican cloud forests: A preliminary vicariance model applying parsimony analysis of endemicity to vascular plant taxa. *Journal of Biogeography* 26:1299–1305.

Luna-Vega, I., O. Alcántara, J. J. Morrone, and D. Espinosa Organista. 2000. Track analysis and conservation priorities in the cloud forests of Hidalgo, Mexico. *Diversity and Distributions* 6:137–143.

Luna-Vega, I. and R. Contreras-Medina. 2000. Distribution of the genera of Theaceae (Angiospermae: Theales): A panbiogeographic analysis. *Biogeographica* 76:79–88.

Luna-Vega, I., J. J. Morrone, O. Alcántara Ayala, and D. Espinosa Organista. 2001. Biogeographical affinities among Neotropical cloud forests. *Plant Systematics and Evolution* 228:229–239.

Luzzatto, M., C. Palestrini, and P. P. D'Entrèves. 2000. Hologenesis: The last and lost theory of evolutionary change. *Italian Journal of Zoology* 67:129–138.

Lyell, C. 1830–1833. *Principles of geology*, 3 vols. London: John Murray.

MacArthur, R. H. 1972. *Geographical ecology: Patterns in the distribution of species*. New York: Harper and Row.

MacArthur, R. H. and E. O. Wilson. 1967. *The theory of island biogeography*. Princeton, N.J.: Princeton University Press.

MacDonald, G. M. 2003. *Biogeography: Space, time, and life*. New York: Wiley.

MacHugh, D. E., M. D. Shriver, R. T. Loftus, P. Cunningham, and D. G. Bradley. 1997. Microsatellite DNA variation and the evolution, domestication and phylogeography of taurine and zebu cattle (*Bos taurus* and *Bos indicus*). *Genetics* 146:1071–1086.

Magallón, S. A. 2004. Dating lineages: Molecular and paleontological approaches to the temporal framework of clades. *International Journal of Plant Sciences* 165:S7–S21.

Magallón, S. A. and M. J. Sanderson. 2005. Angiosperm divergence time: The effect of genes, codon positions, and time constraints. *Evolution* 59:1653–1670.

Mahner, M. and M. Bunge. 1997. *Foundations of biophilosophy*. Berlin: Springer Verlag.

Maldonado, M. and M. J. Uriz. 1995. Biotic affinities in a transitional zone between the Atlantic and the Mediterranean: A biogeographical approach based on sponges. *Journal of Biogeography* 22:89–110.

Marino, P. I., G. R. Spinelli, and P. Posadas. 2001. Distributional patterns of species of Ceratopogonidae (Diptera) in southern South America. *Biogeographica* 77:113–122.

Marks, B. D., S. J. Hackett, and A. P. Capparella. 2002. Historical relationships among Neotropical lowland forest areas of endemism as determined by mitochondrial DNA sequence variation within the wedge-billed woodcreeper (Aves: Dendrocolaptidae: *Glyphorynchus spirurus*). *Molecular Phylogenetics and Evolution* 24:153–167.

Marques, A. C. 2005. Three-taxon statement analysis and its relation with primary data: Implications for cladistics and biogeography. In *Regionalización biogeográfica en Iberoamérica y tópicos afines: Primeras Jornadas Biogeográficas de la Red Iberoamericana de Biogeografía y Entomología Sistemática (RIBES XII.I–CYTED)*, ed. J. Llorente Bousquets and J. J. Morrone, 171–180. Mexico, D.F.: Las Prensas de Ciencias, UNAM.

Márquez, J. and J. J. Morrone. 2003. Análisis panbiogeográfico de las especies de *Heterolinus* y *Homalolinus* (Coleoptera: Staphylinidae: Xantholinini). *Acta Zoológica Mexicana (nueva series)* 90:15–25.

Marshall, C. J. and J. K. Liebherr. 2000. Cladistic biogeography of the Mexican transition zone. *Journal of Biogeography* 27:203–216.

Marshall, L. G. and J. G. Lundberg. 1996. Miocene deposits in the Amazonian foreland basin. *Science* 273:123–124.

Martínez Gordillo, M. and J. J. Morrone. 2005. Patrones de endemismo y disyunción de los géneros de Euphorbiaceae *sensu lato:* Un análisis panbiogeográfico. *Boletín de la Sociedad Mexicana de Botánica* 77:21–34.

Mateos, M. 2005. Comparative phylogeography of livebearing fishes in the genera *Poeciliopsis* and *Poecilia* (Poeciliidae: Cyprinodontiformes) in central Mexico. *Journal of Biogeography* 32:775–780.

Matthee, C. A. and T. J. Robinson. 1996. Mitochondrial DNA differentiation among geographical populations of *Pronolagus rupestris,* Smith's red rock rabbit (Mammalia, Lagomorpha). *Heredity* 76:514–523.

Matthee, C. A., B. J. van Vuuren, D. Bell, and T. J. Robinson. 2004. Molecular supermatrix of the rabbits and hares (Leporidae) allows for the identification of five intercontinental exchanges during the Miocene. *Systematic Biology* 53:433–447.

Matthew, W. D. 1915. Climate and evolution. *Annals of the New York Academy of Sciences* 24:171–318.

Maurer, B. A. 2000. Macroecology and consilience. *Global Ecology and Biogeography* 9:275–280.

Mayden, R. L. 1988. Vicariance biogeography, parsimony, and evolution in North American freshwater fishes. *Systematic Zoology* 37:329–355.

Mayden, R. L. 1992. The wilderness of panbiogeography: A synthesis of space, time and form? *Systematic Zoology* 40:503–519.

Mayeda, W. 1972. *Graph theory.* New York: Wiley-Interscience.

Mayr, E. 1942. *Systematics and the origin of species.* New York: Columbia University Press.

Mayr, E. 1946. History of the North American bird fauna. *Wilson Bulletin* 58:3–41.

Mayr, E. 1961. Cause and effect in biology. *Science* 134:1501–1506.

Mayr, E. 1982. *The growth of biological thought: Diversity, evolution, and inheritance.* Cambridge, Mass.: The Belknap Press of Harvard University Press.

McCarthy, D. 2003. The trans-Pacific zipper effect: Disjunct sister taxa and matching geological outlines that link the Pacific margins. *Journal of Biogeography* 30:1545–1561.

McCarthy, D. 2007. Are plate tectonic explanations for trans-Pacific disjunctions plausible? Empirical tests of radical dispersalist theories. In *Biogeography in a changing world,* ed. M. C. Ebach and R. S. Tangney, 177–198. Boca Raton, Fla.: CRC Press.

McDowall, R. M. 1978. Generalized tracks and dispersal in biogeography. *Systematic Zoology* 27:88–104.

McKenna, M. C. 1973. Sweepstakes, filters, corridors, Noah's ark and beached Viking funeral ships in paleogeography. In *Implications of continental drift to the earth sciences,* Vol. I, ed. D. H. Tarling and S. K. Runcorn, 295–308. New York: Academic Press.

McLennan, D. A. and D. R. Brooks. 2002. Complex histories of speciation and dispersal in communities: A re-analysis of some Australian bird data using BPA. *Journal of Biogeography* 29:1055–1066.

Mejía-Madrid, H. H., E. Vázquez-Domínguez, and G. Pérez-Ponce de León. 2007. Phylogeography and freshwater basins in central Mexico: Recent history as revealed by the fish parasite *Rhabdoclona lichtenfelsi* (Nematoda). *Journal of Biogeography* 34:787–801.

Melo Santos, A. M., D. Rodrigues Cavalcanti, J. M. Cardoso da Silva, and M. Tabarelli. 2007. Biogeographical relationships among tropical forests in north-eastern Brazil. *Journal of Biogeography* 34:437–446.

Méndez-Larios, I., J. L. Villaseñor, R. Lira, J. J. Morrone, P. Dávila, and E. Ortiz. 2005. Toward the identification of a core zone in the Tehuacán–Cuicatlán biosphere reserve, Mexico, based on parsimony analysis of endemicity of flowering plant species. *Interciencia* 30:264–274.

Menu-Marque, S., J. J. Morrone, and C. Locascio. 2000. Distributional patterns of the South American species of *Boeckella* (Copepoda: Centropagidae): A track analysis. *Journal of Crustacean Biology* 20:262–272.

Meschede, M. and W. Frisch. 1998. A plate tectonic model for the Mesozoic and early Cenozoic history of the Caribbean plate. *Tectonophysics* 296:269–291.

Michaux, B. 1989. Generalized tracks and geology. *Systematic Zoology* 38:390–398.

Michaux, B. 1991. Distributional patterns and tectonic development in Indonesia: Wallace reinterpreted. *Australian Systematic Botany* 4:25–36.

Michaux, B. and R. A. B. Leschen. 2005. East meets west: Biogeology of the Campbell Plateau. *Journal of the Linnean Society* 86:95–115.

Mickevich, M. F. 1981. Quantitative phylogenetic biogeography. In *Advances in cladistics: Proceedings of the first meeting of the Willi Hennig Society*, ed. V. A. Funk and D. R. Brooks, 202–222. New York: New York Botanical Garden.

Mihoc, M. A. K., J. J. Morrone, M. A. Negritto, and L. A. Cavieres. 2006. Evolución de la serie Microphyllae (Adesmia, Fabaceae) en la Cordillera de los Andes: Una perspectiva biogeográfica. *Revista Chilena de Historia Natural* 79:389–404.

Miranda-Esquivel, D. R. 2001. Efectos de la dispersión sobre la reconstrucción por árboles reconciliados y el patrón de distribución de los subgéneros neotropicales de *Simulium* (Diptera: Simuliidae). *Caldasia* 23:3–20.

Miranda-Esquivel, D. R., M. Donato, and P. Posadas. 2003. La dispersión ha muerto, larga vida a la dispersion. In *Una perspectiva latinoamericana de la biogeografía*, ed. J. J. Morrone and J. Llorente Bousquets, 179–184. Mexico, D.F.: Las Prensas de Ciencias, UNAM.

Mitchell, S. D. 2002. Integrative pluralism. *Biology and Philosophy* 17:55–70.

Mitter, C. and D. R. Brooks. 1983. Phylogenetic aspects of coevolution. In *Coevolution*, ed. D. J. Futuyma and M. Slatkin, 65–98. Sunderland, Mass.: Sinauer.

Mock, K. E., B. J. Bentz, E. M. O'Neill, J. P. Chong, J. Orwin, and M. E. Pfrender. 2007. Landscape-scale genetic variation in a forest outbreak species, the mountain pine beetle (*Dendroctonus ponderosae*). *Molecular Ecology* 16:553–568.

Moline, P. M. and H. P. Linder. 2006. Input data, analytical methods and biogeography of *Elegia* (Restionaceae). *Journal of Biogeography* 33:47–62.

Moore, M. J., A. Tye, and R. K. Jansen. 2006. Patterns of long-distance dispersal in *Tiquilia* subg. *Tiquilia* (Boraginaceae): Implications for the origins of amphitropical disjuncts and Galápagos islands endemics. *American Journal of Botany* 93:1163–1177.

Morafka, D. J. 1977. *A biogeographical analysis of the Chihuahuan desert through its herpeto-fauna*. The Hague: Junk.

Morafka, D. J., G. A. Adest, and L. M. Reyes. 1992. Differentiation of North American deserts: A phylogenetic evaluation of a vicariance model. *Tulane Studies in Zoology and Botany, Supplementary Publications* 1:195–226.

Morain, S. A. 1984. *Systematic and regional biogeography*. New York: Van Nostrand Reinhold.

Morales-Barros, N., J. A. B. Silva, C. Y. Miyaki, and J. S. Morgante. 2006. Comparative phylogeography of the Atlantic forest endemic sloth (*Bradypus torquatus*) and the widespread three-toed sloth (*Bradypus variegatus*) (Bradypodidae, Xenarthra). *Genetica* 126:189–198.

Moreira-Muñoz, A. 2007. The Austral floristic realm revisited. *Journal of Biogeography* 34:1649–1660.

Morell, P. L., J. M. Porter, and E. A. Friar. 2000. Intercontinental dispersal: The origin of the widespread South American plant species *Gilia laciniata* (Polemoniaceae) from a rare California and Oregon coastal endemic. *Plant Systematics and Evolution* 224:13–32.

Moreno, R. A., C. H. Hernández, M. M. Rivadeneira, M. A. Vidal, and N. Rozbaczylo. 2006. Patterns of endemism in south-eastern Pacific benthic polychaetes of the Chilean coast. *Journal of Biogeography* 33:750–759.

Moritz, C. C. 1994. Applications of mitochondrial DNA analysis in conservation: A critical review. *Molecular Ecology* 3:401–411.

Moritz, C. C. 1995. Uses of molecular phylogenies for conservation. *Philosophical Transactions of the Royal Society of London B*, 349:113–118.

Morrone, J. J. 1992. Revisión sistemática, análisis cladístico y biogeografía histórica de los géneros *Falklandius* Enderlein y *Lanteriella* gen. nov. (Coleoptera: Curculionidae). *Acta Entomológica Chilena* 17:157–174.

Morrone, J. J. 1993a. Beyond binary oppositions. *Cladistics* 9:437–438.

Morrone, J. J. 1993b. Cladistic and biogeographic analyses of the weevil genus *Listroderes* Schoenherr (Coleoptera: Curculionidae). *Cladistics* 9:397–411.

Morrone, J. J. 1993c. Revisión sistemática de un nuevo género de Rhytirrhinini (Coleoptera: Curculionidae), con un análisis biogeográfico del dominio Subantártico. *Boletín de la Sociedad de Biología de Concepción* 64:121–145.

Morrone, J. J. 1994a. Distributional patterns of species of Rhytirrhinini (Coleoptera: Curculionidae) and the historical relationships of the Andean provinces. *Global Ecology and Biogeography Letters* 4:188–194.

Morrone, J. J. 1994b. On the identification of areas of endemism. *Systematic Biology* 43:438–441.

Morrone, J. J. 1994c. Systematics, cladistics, and biogeography of the Andean weevil genera *Macrostyphlus, Adioristidius, Puranius,* and *Amathynetoides*, new genus (Coleoptera: Curculionidae). *American Museum Novitates* 3104:1–63.

Morrone, J. J. 1995. Asociaciones históricas en biología comparada. *Ciencia* 46:229–235.

Morrone, J. J. 1996a. Austral biogeography and relict weevil taxa (Coleoptera: Nemonychidae, Belidae, Brentidae, and Caridae). *Journal of Comparative Biology* 1:123–127.

Morrone, J. J. 1996b. The biogeographical Andean subregion: A proposal exemplified by Arthropod taxa (Arachnida, Crustacea, and Hexapoda). *Neotropica* 42:103–114.

Morrone, J. J. 1996c. Distributional patterns of the South American Aterpini (Coleoptera: Curculionidae). *Revista de la Sociedad Entomológica Argentina* 55:131–141.

Morrone, J. J. 1997. Biogeografía cladística: Conceptos básicos. *Arbor* 158:373–388.

Morrone, J. J. 1998. On Udvardy's Insulantarctica province: A test from the weevils (Coleoptera: Curculionoidea). *Journal of Biogeography* 25:947–955.

Morrone J. J. 1999a. Biodiversidad en el espacio: La importancia de los atlas biogeográficos. *Physis (Buenos Aires)* 55:47–48.

Morrone, J. J. 1999b. Presentación preliminar de un nuevo esquema biogeográfico de América del Sur. *Biogeographica* 75:1–16.

Morrone, J. J. 2000a. Biogeographic delimitation of the Subantarctic subregion and its provinces. *Revista del Museo Argentino de Ciencias Naturales, nueva serie* 2:1–15.

Morrone, J. J. 2000b. Delimitation of the Central Chilean subregion and its provinces, based mainly on Arthropod taxa. *Biogeographica* 76:97–106.

Morrone, J. J. 2000c. Entre el escarnio y el encomio: Léon Croizat y la panbiogeografía. *Interciencia* 5:41–47.

Morrone, J. J. 2000d. La importancia de los atlas biogeográficos para la conservación de la Biodiversidad. In *Hacia un proyecto CYTED para el Inventario y Estimación de la Diversidad Entomológica en Iberoamérica: PrIBES*, ed. F. Martín-Piera, J. J. Morrone, and A. Melic, 69–78. Saragossa, Spain: Monografías Tercer Milenio, Sociedad Entomológica Aragonesa.

Morrone, J. J. 2000e. *El lenguaje de la cladística*. Mexico, D.F.: Programa Libro de Texto Universitario, Dirección General de Publicaciones y Fomento Editorial, UNAM.

Morrone, J. J. 2000f. A new regional biogeography of the Amazonian subregion, mainly based on animal taxa. *Anales del Instituto de Biología de la UNAM, Serie Zoología* 71:99–123.

Morrone, J. J. 2000g. El tiempo de Darwin y el espacio de Croizat: Rupturas epistémicas en los estudios evolutivos. *Ciencia* 51:39–46.

Morrone, J. J. 2000h. What is the Chacoan subregion? *Neotropica* 46:51–68.

Morrone, J. J. 2001a. *Biogeografía de América Latina y el Caribe*. Saragossa, Spain: Manuales y Tesis SEA, no. 3.

Morrone, J. J. 2001b. A formal definition of the Paramo–Punan biogeographic subregion and its provinces, based mainly on animal taxa. *Revista del Museo Argentino de Ciencias Naturales, nueva serie* 3:1–12.

Morrone, J. J. 2001c. Homology, biogeography and areas of endemism. *Diversity and Distributions* 7:297–300.

Morrone, J. J. 2001d. The Parana subregion and its provinces. *Physis* (Buenos Aires) 58:1–7.

Morrone, J. J. 2001e. A proposal concerning formal definitions of the Neotropical and Andean regions. *Biogeographica* 77:65–82.

Morrone, J. J. 2001f. Review of the biogeographic provinces of the Patagonian subregion. *Revista de la Sociedad Entomológica Argentina* 60:1–8.

Morrone, J. J. 2001g. *Sistemática, biogeografía, evolución: Los patrones de la biodiversidad en tiempo-espacio*. Mexico, D.F.: Las Prensas de Ciencias, UNAM.

Morrone, J. J. 2001h. Toward a cladistic model for the Caribbean subregion: Delimitation of areas of endemism. *Caldasia* 23:43–76.

Morrone, J. J. 2002a. Biogeographic regions under track and cladistic scrutiny. *Journal of Biogeography* 29:149–152.

Morrone, J. J. 2002b. El espectro del dispersalismo: De los centros de origen a las áreas ancestrales. *Revista de la Sociedad Entomológica Argentina* 61:1–14.

Morrone, J. J. 2002c. The Neotropical weevil genus *Entimus* (Coleoptera: Curculionidae: Entiminae): Cladistics, biogeography, and modes of speciation. *Coleopterists Bulletin* 56: 501-513.

Morrone, J. J. 2003a. Las ideas biogeográficas de Osvaldo Reig y el desarrollo del "dispersalismo" en América Latina. In *Una perspectiva latinoamericana de la biogeografía*, ed. J. J. Morrone and J. Llorente Bousquets, 69–74. Mexico, D.F.: Las Prensas de Ciencias, UNAM.

Morrone, J. J. 2003b. ¿Quién le teme al darwinismo? *Ciencia* 54:78–88.

Morrone, J. J. 2004a. *Homología biogeográfica: Las coordenadas espaciales de la vida.* Mexico, D.F.: Cuadernos del Instituto de Biología 37, Instituto de Biología, UNAM.

Morrone, J. J. 2004b. Panbiogeografía, componentes bióticos y zonas de transición. *Revista Brasileira di Entomologia* 48:149–162.

Morrone, J. J. 2004c. La Zona de Transición Sudamericana: Caracterización y relevancia evolutiva. *Acta Entomológica Chilena* 28:41–50.

Morrone, J. J. 2005a. Cladistic biogeography: Identity and place. *Journal of Biogeography* 32:1281–1284.

Morrone, J. J. 2005b. Hacia una síntesis biogeográfica de México. *Revista Mexicana de Biodiversidad* 76:207–252.

Morrone, J. J. 2006. Biogeographic areas and transition zones of Latin America and the Caribbean Islands, based on panbiogeographic and cladistic analyses of the entomofauna. *Annual Review of Entomology* 51:467–494.

Morrone, J. J. and J. M. Carpenter. 1994. In search of a method for cladistic biogeography: An empirical comparison of component analysis, Brooks parsimony analysis, and three-area statements. *Cladistics* 10:99–153.

Morrone, J. J. and M. del C. Coscarón. 1996. Distributional patterns of the American Peiratinae (Heteroptera: Reduviidae). *Zoologische Medeligen Leiden* 70:1–15.

Morrone, J. J. and M. del C. Coscarón. 1998. Cladistics and biogeography of the assassin bug genus *Rasahus* Amyot and Serville (Heteroptera: Reduviidae: Peiratinae). *Zoologische Medeligen Leiden* 72:73–87.

Morrone, J. J. and J. V. Crisci. 1990. Panbiogeografía: Fundamentos y métodos. *Evolución Biológica* (Bogotá) 4:119–140.

Morrone, J. J. and J. V. Crisci. 1992. Aplicación de métodos filogenéticos y panbiogeográficos en la conservación de la diversidad biológica. *Evolución Biológica* (Bogotá) 6:53–66.

Morrone, J. J. and J. V. Crisci. 1993. El retorno a la historia y la conservación de la diversidad biológica. In *Elementos de política ambiental,* ed. F. Goin and R. Goñi, 361–365. La Plata: Cámara de Diputados de la Provincia de Buenos Aires.

Morrone, J. J. and J. V. Crisci. 1995. Historical biogeography: Introduction to methods. *Annual Review of Ecology and Systematics* 26:373–401.

Morrone, J. J. and T. Escalante. 2002. Parsimony analysis of endemicity (PAE) of Mexican terrestrial mammals at different area units: When size matters. *Journal of Biogeography* 29:1095–1104.

Morrone, J. J. and D. Espinosa Organista. 1998. La relevancia de los atlas biogeográficos para la conservación de la biodiversidad mexicana. *Ciencia* 49:12–16.

Morrone, J. J., D. Espinosa Organista, C. Aguilar-Zúñiga, and J. Llorente Bousquets. 1999. Preliminary classification of the Mexican biogeographic provinces: A parsimony analysis of endemicity based on plant, insect, and bird taxa. *Southwestern Naturalist* 44:508–515.

Morrone, J. J., D. Espinosa Organista, and J. Llorente Bousquets. 1996. *Manual de biogeografía histórica.* Mexico, D.F.: Universidad Nacional Autónoma de México.

Morrone, J. J., D. Espinosa Organista, and J. Llorente Bousquets. 2002. Mexican biogeographic provinces: Preliminary scheme, general characterizations, and synonymies. *Acta Zoológica Mexicana (nueva serie)* 85:83–108.

Morrone, J. J. and A. Gutiérrez. 2005. Do fleas (Insecta: Siphonaptera) parallel their mammal host diversification in the Mexican transition zone? *Journal of Biogeography* 32:1315–1325.

Morrone, J. J., L. Katinas, and J. V. Crisci. 1996. On temperate areas, basal clades, and bio-diversity conservation. *Oryx* 30:187–194.

Morrone, J. J., L. Katinas, and J. V. Crisci. 1997. A cladistic biogeographic analysis of Central Chile. *Journal of Comparative Biology* 2:25–42.

Morrone, J. J. and E. C. Lopretto. 1994. Distributional patterns of freshwater Decapoda (Crustacea: Malacostraca) in southern South America: A panbiogeographic approach. *Journal of Biogeography* 21:97–109.

Morrone, J. J. and E. C. Lopretto. 1995. Parsimony analysis of endemicity of freshwater Decapoda (Crustacea: Malacostraca) from southern South America. *Neotropica* 41:3–8.

Morrone, J. J. and E. C. Lopretto. 2001. Trichodactylid biogeographic patterns (Crustacea: Decapoda) and the Neotropical region. *Neotropica* 47:49–55.

Morrone, J. J. and J. Márquez. 2001. Halffter's Mexican Transition Zone, beetle generalised tracks, and geographical homology. *Journal of Biogeography* 28:635–650.

Morrone, J. J. and J. Márquez. 2003. Aproximación a un Atlas Biogeográfico Mexicano: Componentes bióticos principales y provincias bigeográficas. In *Una perspectiva latinoamericana de la biogeografía*, ed. J. J. Morrone and J. Llorente Bousquets, 217–220. Mexico, D.F.: Las Prensas de Ciencias, UNAM.

Morrone, J. J., S. Mazzucconi, and A. Bachmann. 2004. Distributional patterns of Chacoan water bugs (Heteroptera: Belostomatidae, Corixidae, Micronectidae, and Gerridae). *Hydrobiologica* 523:159–173.

Morrone, J. J., G. Osella, and A. M. Zuppa. 2001. Distributional patterns of the relictual subfamily Raymondionyminae (Coleoptera: Erirhinidae): A track analysis. *Folia Entomológica Mexicana* 40:381–388.

Morrone, J. J. and L. A. Pereira. 1999. On the geographical distribution of the Neotropical and Andean species of *Schendylops* (Chilopoda: Geophilomorpha: Schendylidae). *Revista de la Sociedad Entomológica Argentina* 58:165–171.

Morrone, J. J., S. Roig-Juñent, and J. V. Crisci. 1994. Cladistic biogeography of terrestrial subantarctic beetles (Insecta: Coleoptera) from South America. *National Geographic Research and Exploration* 10:104–115.

Morrone, J. J., S. Roig-Juñent, and G. E. Flores. 2002. Delimitation of biogeographic districts in central Patagonia (southern South America), based on beetle distributional patterns (Coleoptera: Carabidae and Tenebrionidae). *Revista del Museo Argentino de Ciencias Naturales, nueva serie* 4:1–6.

Morrone, J. J. and E. Urtubey. 1997. Historical biogeography of the northern Andes: A cladistic analysis based on five genera of Rhytirrhinini (Coleoptera: Curculionidae) and *Barnadesia* (Asteraceae). *Biogeographica* 73:115–121.

Mortimer, E. and J. Van Vuuren. 2007. Phylogeography of *Eupodes minutus* (Acari: Porstigmata) on sub-Antarctic Marion Island reflects the impact of historical events. *Polar Biology* 30:471–476.

Mota, J. F., F. J. Pérez-García, M. L. Jiménez, J. J. Amate, and J. Peñas. 2002. Phytogeographical relationships among high mountain areas in the Baetic Ranges (South Spain). *Global Ecology and Biogeography* 11:497–504.

Müller, P. 1973. *The dispersal centres of terrestrial vertebrates in the Neotropical realm: A study in the evolution of the Neotropical biota and its native landscapes.* The Hague: Junk.

Müller, P. 1979. *Introducción a la zoogeografía.* Barcelona: Blume.

Murphy, R. W. and G. Aguirre-León. 2002. Nonavian reptiles; origins and evolution. In *A new island biogeography of the Sea of Cortés*, ed. T. J. Case, M. L. Cody, and E. Ezcurra, 181–220. New York: Oxford University Press.

Myers, A. A. 1991. How did Hawaii accumulate its biota?: A test from the Amphipoda. *Global Ecology and Biogeography Letters* 1:24–29.

Myers, A. A. and P. S. Giller, eds. 1988a. *Analytical biogeography: An integrated approach to the study of animal and plant distributions.* London: Chapman and Hall.

Myers, A. A. and P. S. Giller. 1988b. Biogeographic patterns. In *Analytical biogeography: An integrated approach to the study of animal and plant distributions,* ed. A. A. Myers and P. S. Giller, 15–21. London: Chapman and Hall.

Myers, G. S. 1938. Fresh-water fishes and West Indian zoogeography. *Annual Reports of the Board Regents of the Smithsonian Institution* 1947:339–364.

Myers, G. S. 1963. The freshwater fishes of North America. *Proceedings of the XVI International Congress of Zoology* 4:15–20.

Navarro, A. G., H. A. Garza-Torres, S. López de Aquino, O. R. Rojas-Soto, and L. A. Sánchez-González. 2004. Patrones biogeográficos de la avifauna. *Biodiversidad de la Sierra Madre Oriental,* ed. I. Luna, J. J. Morrone, and D. Espinosa, 439–467. Mexico, D.F.: Las Prensas de Ciencias, UNAM.

Neiman, M. and C. M. Lively. 2004. Pleistocene glaciation is implicated in the phylogeographical structure of *Potamopyrgus antipodarum,* a New Zealand snail. *Molecular Ecology* 13:3085–3098.

Nelson, G. 1969. The problem of historical biogeography. *Systematic Zoology* 18:243–246.

Nelson, G. 1973. Comments on Léon Croizat's biogeography. *Systematic Zoology* 22:312–320.

Nelson, G. 1974. Historical biogeography: An alternative formalization. *Systematic Zoology* 23:555–558.

Nelson, G. 1977. Biogeografía analítica y sintética ("Panbiogeografía") de las Américas by L. Croizat (1976). *Systematic Zoology* 26:449–452.

Nelson, G. 1978a. From Candolle to Croizat: Comments on the history of biogeography. *Journal of the History of Biology* 11:269–305.

Nelson, G. 1978b. Ontogeny, phylogeny, paleontology, and the biogenetic law. *Systematic Zoology* 27:324–345.

Nelson, G. 1983. Vicariance and cladistics: Historical perspectives with implications for the future. In *Evolution, time and space: The emergence of the biosphere,* ed. R. W. Sims, J. H. Price, and P. E. S. Whalley, 469–472. San Diego: Academic Press.

Nelson, G. 1984. Cladistics and biogeography. In *Cladistics: Perspectives on the reconstruction of evolutionary history,* ed. T. Duncan and T. F. Stuessy, 273–293. New York: Columbia University Press.

Nelson, G. 1985. A decade of challenge the future of biogeography. *Journal of the History of Earth Sciences Society* 4:187–196.

Nelson, G. 1989. Cladistics and evolutionary models. *Cladistics* 5:275–289.

Nelson, G. 1994. Homology and systematics. In *Homology: The hierarchical basis of comparative biology,* ed. B. K. Hall, 101–149. San Diego: Academic Press.

Nelson, G. and P. Y. Ladiges. 1990. Biodiversity and biogeography. *Journal of Biogeography* 17:559–560.

Nelson, G. and P. Y. Ladiges. 1991a. Standard assumptions for biogeographic analysis. *Australian Systematic Botany* 4:41–58.

Nelson, G. and P. Y. Ladiges. 1991b. *TAS (MSDos computer program).* New York: Author.

Nelson, G. and P. Y. Ladiges. 1991c. Three-area statements: Standard assumptions for biogeographic analysis. *Systematic Zoology* 40:470–485.

Nelson, G. and P. Y. Ladiges. 1993. Missing data and three-item analysis. *Systematic Zoology* 40:470–485.

Nelson, G. and P. Y. Ladiges. 1995. *TASS*. New York: Author.

Nelson, G. and P. Y. Ladiges. 1996. Paralogy in cladistic biogeography and analysis of paralogy-free subtrees. *American Museum Novitates* 3167:1–58.

Nelson, G. and P. Y. Ladiges. 2001. Gondwana, vicariance biogeography and the New York school revisited. *Australian Journal of Botany* 49:389–409.

Nelson, G. and P. Y. Ladiges. 2003. Geographic paralogy. In *Una perspectiva latinoamericana de la biogeografía*, ed. J. J. Morrone and J. Llorente Bousquets, 173–177. Mexico, D.F.: Las Prensas de Ciencias, UNAM.

Nelson, G. and N. I. Platnick. 1978. The perils of plesiomorphy: Widespread taxa, dispersal, and phenetic biogeography. *Systematic Zoology* 27:474–477.

Nelson, G. and N. I. Platnick. 1980. A vicariance approach to historical biogeography. *Bioscience* 30:339–343.

Nelson, G. and N. I. Platnick. 1981. *Systematics and biogeography: Cladistics and vicariance*. New York: Columbia University Press.

Nelson, G. and N. I. Platnick. 1991. Three taxon statements: A more precise use of parsimony? *Cladistics* 7:351–366.

Nersting, L. G. and P. Arctander. 2001. Phylogeography and conservation of impala and greater kudu. *Molecular Ecology* 10:711–719.

Nihei, S. S. 2006. Misconceptions about parsimony analysis of endemicity. *Journal of Biogeography* 33:2099–2106.

Nihei, S. S. and C. J. B. de Carvalho. 2004. Taxonomy, cladistics and biogeography of *Coenosopia* Malloch (Diptera, Anthomyiidae) and its significance to the evolution of anthomyiids in the Neotropics. *Systematic Entomology* 29:260–275.

Nihei, S. S. and C. J. B. de Carvalho. 2005. Distributional patterns of the Neotropical fly genus *Polietina* Schnabl & Dziedzicki (Diptera: Muscidae): A phylogeny-supported analysis using panbiogeographic tools. *Papeis Avulsos de Zoologia* 45:313–326.

Nihei, S. S. and C. J. B. de Carvalho. 2007. Systematics and biogeography of *Polietina* Schnabl & Dziedzicki (Diptera, Muscidae): Neotropical area relationships and Amazonia as a composite area. *Systematic Entomology* 32:477–501.

Nixon, K. C. 1999. *About Winclada*. Ithaca, N.Y.: Author. Retrieved May 25, 2008, from http://www.cladistics.com/about_winc.htm.

Noonan, G. R. 1979. The science of biogeography with relation to carabids. In *Carabid beetles: Their evolution, natural history, and classification*, ed. T. L. Erwin, G. E. Ball, and D. R. Whitehead. The Hague: Junk.

Noonan, G. R. 1988. Biogeography of North American and Mexican insects, and a critique of vicariance biogeography. *Systematic Zoology* 37:366–384.

Nur, A. and Z. Ben-Avraham. 1980. Lost Pacifica continent: A mobilistic speculation. In *Vicariance biogeography: A critique*, ed. D. E. Rosen and G. Nelson, 341–358. New York: Columbia University Press.

Ochoa, L., B. Cruz, G. García, and A. Luis Martínez. 2003. Contribución al atlas panbiogeográfico de México: Los géneros *Adelpha* y *Hamadryas* (Nymphalidae), y *Dismorphia, Enantia, Leinix* y *Pseudopieris* (Pieridae) (Papilionoidea; Lepidoptera). *Folia Entomológica Mexicana* 42:65–77.

O'Donnell, K., E. Cigelnik, and H. I. Nirenberg. 1998. Molecular systematics and phylogeography of the *Gibberella fujikuroi* species complex. *Mycologia* 90:465–493.

O'Hara, R. J. 1988. Homage to Clio, or, toward an historical philosophy for evolutionary biology. *Systematic Zoology* 37(2):142–155.

Olmstead, R. G. and J. D. Palmer. 1997. Implications for the phylogeny, classification, and biogeography of *Solanum* from cpDNA restriction site variation. *Systematic Botany* 36:1–17.

Orange, D. I., B. R. Riddle, and D. C. Nickle. 1999. Phylogeography of a wide-ranging desert lizard, *Gambelia wislizenii* (Crotaphytidae). *Copeia* 1999:267–273.

Ortega, J. and H. T. Arita. 1998. Neotropical–Nearctic limits in Middle America as determined by distributions of bats. *Journal of Mammalogy* 79:772–781.

Ortmann, A. E. 1896. *Grundzuge der marinen Tiergeographie.* Jena: Gustav Fischer.

Osentoski, M. F. and T. Lamb. 1995. Intraspecific phylogeography of the gopher tortoise, *Gopherus polyphemus:* RFLP analysis of amplified mtDNA segments. *Molecular Ecology* 4:709–718.

Oyama, S. 2000. *Evolution's eye: A systems view of the biology–culture divide.* Durham, N.C.: Duke University Press.

Page, R. D. M. 1987. Graphs and generalized tracks: Quantifying Croizat's panbiogeography. *Systematic Zoology* 36:1–17.

Page, R. D. M. 1988. Quantitative cladistic biogeography: Constructing and comparing area cladograms. *Systematic Zoology* 37:254–270.

Page, R. D. M. 1989a. Comments on component-compatibility in historical biogeography. *New Zealand Journal of Zoology* 16:471–483.

Page, R. D. M. 1989b. *Component user's manual.* Release 1.5. Auckland: Author.

Page, R. D. M. 1990a. Component analysis: A valiant failure? *Cladistics* 6:119–136.

Page, R. D. M. 1990b. Temporal congruence and cladistic analysis of biogeography and cospeciation. *Systematic Zoology* 39:205–226.

Page, R. D. M. 1990c. Tracks and trees in the antipodes. *Systematic Zoology* 39:288–299.

Page, R. D. M. 1993a. *Component user's manual.* Release 2.0. London: The Natural History Museum.

Page, R. D. M. 1993b. Genes, organisms, and areas: The problem of multiple lineages. *Systematic Biology* 42:77–84.

Page, R. D. M. 1994a. Maps between trees and cladistic analysis of historical associations among genes, organisms, and areas. *Systematic Biology* 43:58–77.

Page, R. D. M. 1994b. Parallel phylogenies: Reconstructing the history of host–parasite assemblages. *Cladistics* 10:155–173.

Page, R. D. M. 1994c. *TreeMap.* Release 3.1. Oxford: University of Oxford.

Page, R. D. M. and M. A. Charleston. 1998. Trees within trees: Phylogeny and historical associations. *Tree* 13:356–359.

Page, R. D. M. and E. C. Holmes. 1998. *Molecular evolution: A phylogenetic approach.* Oxford: Blackwell Science.

Palma, R. E., P. A. Marquet, and D. Boric-Bargetto. 2005. Inter- and intraspecific phylogeography of small mammals in the Atacama desert and adjacent areas of northern Chile. *Journal of Biogeography* 32:1931–1941.

Papavero, N. 1990. *Introdução histórica à biologia comparada, com especial referência à biogeografia. II. A Idade Média.* Rio de Janeiro: Universidade Santa Úrsula. (Spanish translation: N. Papavero, G. J. Scrocchi, and J. Llorente Bousquets, 1995, *Historia de la biología comparada desde el Génesis hasta el Siglo de las Luces. II: La Edad Media,* Mexico, D.F.: Facultad de Ciencias, UNAM.)

Papavero, N. 1991. *Introdução histórica à biologia comparada, com especial referência à biogeografia. III. De Nicolau de Cusa a Francis Bacon.* Rio de Janeiro: Universidade Santa Úrsula. (Spanish translation: N. Papavero, J. Llorente Bousquets, and D. Espinosa, 1995, *Historia de la biología comparada desde el Génesis hasta el Siglo de las Luces. III: De Nicolás de Cusa a Francis Bacon,* Mexico, D.F.: Facultad de Ciencias, UNAM.)

Papavero, N. and J. Balsa. 1985. Introdução histórica e epistemológica à biologia comparada, com especial referência à biogeografia. I: Do Gênesis à queda do Império

Romano do Ocidente. Belo Horizonte: Biótica and Sociedade Brasileira de Zoologia. (Spanish translation: N. Papavero, J. Llorente Bousquets, and D. Espinosa, 1995, *Historia de la biología comparada desde el Génesis hasta el Siglo de las Luces. I: Del Génesis a la caída del Imperio Romano de Occidente,* México, D.F.: Universidad Nacional Autónoma de México.)

Papavero, N., D. M. Teixeira, and J. Llorente Bousquets. 1997. *História da biogeografia no período pré-evolutivo.* São Paulo: Ed. Pleiade. (Spanish translation: N. Papavero, D. M. Teixeira, J. Llorente Bousquets, and A. Bueno, 2004, *Historia de la biogeografía: I. El periodo preevolutivo,* Mexico, D.F.: Fondo de Cultura Económica.)

Parenti, L. R. 1981. Discussion. In *Vicariance biogeography: A critique,* ed. G. Nelson and D. E. Rosen, 490–497. New York: Columbia University Press.

Parenti, L. R. 1991. Ocean basins and the biogeography of freshwater fishes. *Australian Systematic Botany* 4:137–149.

Parenti, L. R. 2007. Common cause and historical biogeography. In *Biogeography in a changing world,* ed. M. C. Ebach and R. S. Tangney, 61–82. Boca Raton, Fla.: CRC Press.

Parenti, L. R. and C. J. Humphries. 2004. Historical biogeography, the natural science. *Taxon* 53(4):899–903.

Patterson, C. 1981. Methods of paleobiogeography. In *Vicariance biogeography: A critique,* ed. G. Nelson and D. E. Rosen, 446–489. New York: Columbia University Press.

Patterson, C. 1983. Aims and methods in biogeography. In *Evolution, time and space: The emergence of the biosphere,* ed. R. W. Sims, J. H. Price, and P. E. S. Whalley, 1–28. New York: Academic Press.

Patton, J. L., M. N. F. da Silva, and J. R. Malcolm. 2000. Mammals of the Rio Juruá and the evolutionary and ecological diversification of Amazonia. *Bulletin of the American Museum of Natural History* 244:1–306.

Pauly, G. B., O. Piskurek, and H. B. Shaffer. 2007. Phylogeographic concordance in the southeastern United States: The flatwoods salamander, *Ambystoma cingulatum,* as a test case. *Molecular Ecology* 16:415–429.

Peck, S. B. and J. Kukalová-Peck. 1990. Origin and biogeography of the beetles (Coleoptera) of the Galapagos archipelago, Ecuador. *Canadian Journal of Zoology* 68:1617–1638.

Perdices, A. and M. M. Coelho. 2006. Comparative phylogeography of *Zacco platypus* and *Opsariichthys bidens* (Teleostei, Cyprinidae) in China based on cytochrome b sequences. *Journal of Zoological Systematics and Evolutionary Research* 44(4):330–338.

Perret, M., A. Chautems, and R. Spichiger. 2006. Dispersal–vicariance analyses in the tribe Sinningieae (Gesneriaceae): A clue to understanding biogeographical history of the Brazilian Atlantic forest. *Annals of the Missouri Botanical Garden* 93:340–358.

Philippe, H. 1993. MUST: A computer package of management utilities for sequences and trees. *Nucleic Acids Research* 21:5264–5272.

Picard, D., T. Sempere, and O. Plantard. 2007. A northward colonization of the Andes by the potato cyst nematode during geological times suggests multiple host-shifts from wild cultivated potatoes. *Molecular Phylogenetics and Evolution* 42:308–316.

Pielou, E. C. 1992. *Biogeography.* Malabar, India: Krieger.

Pierrot-Bults, A. C., S. van der Spoel, B. J. Zahuranec, and R. K. Johnson. 1986. Pelagic biogeography. *UNESCO Technical Papers in Marine Science* 49:1–295.

Pindell, J. L. 1993. Regional synopsis of the Gulf of Mexico and Caribbean evolution. *GCSSEPM Proceedings,* July 1, 1993:251–274.

Pinto-da-Rocha, R. and M. B. da Silva. 2005. Faunistic similarity and historic biogeography of the harvestmen of southern and southeastern Atlantic rain forest of Brazil. *The Journal of Arachnology* 33:290–299.

Pizarro Araya, J. and V. Jerez. 2004. Distribución geográfica del género *Gyriosomus* Guérin-Méneville, 1834 (Coleoptera: Tenebrionidae): Una aproximación biogeográfica. *Revista Chilena de Historia Natural* 77:491–500.

Platnick, N. I. 1976. Concepts of dispersal in historical biogeography. *Systematic Zoology* 25:294–295.

Platnick, N. I. 1981. Widespread taxa and biogeographic congruence. In *Advances in cladistics, 1, Proceedings of the first meeting of the Willi Hennig Society,* ed. V. A. Funk and D. R. Brooks, 223–227. Bronx: New York Botanical Garden.

Platnick, N. I. 1988. Systematics, evolution and biogeography: A Dutch treat. *Cladistics* 4:308–313.

Platnick, N. I. 1991. On areas of endemism. *Australian Systematic Botany* 4:xi–xii.

Platnick, N. I. and G. Nelson. 1978. A method of analysis for historical biogeography. *Systematic Zoology* 27:1–16.

Platnick, N. I. and G. Nelson. 1988. Spanning-tree biogeography: Shortcut, detour, or dead-end? *Systematic Zoology* 37:410–419.

Popper, K. R. 1959. *The logic of scientific discovery.* London: Hutchinson.

Popper, K. R. 1963. *Conjectures and refutations: The growth of scientific knowledge.* London: Routledge.

Porzecanski, A. L. and J. Cracraft. 2005. Cladistic analysis of distributions and endemism (CADE): Using raw distributions of birds to unravel the biogeography of the South American aridlands. *Journal of Biogeography* 32:261–275.

Posada, D. and K. A. Crandall. 1998. MODELTEST: Testing the model of DNA substitution. *Bioinformatics* 14:817–818.

Posada, D., K. A. Crandall, and A. R. Templeton. 2000. GeoDis: A program for the cladistic nested analysis of the geographical distribution of genetic haplotypes. *Molecular Ecology* 9:487–488.

Posadas, P. 1996. Distributional patterns of vascular plants in Tierra del Fuego: A study applying parsimony analysis of endemicity (PAE). *Biogeographica* 72:161–177.

Posadas, P. E., J. M. Estévez, and J. J. Morrone. 1997. Distributional patterns and endemism areas of vascular plants in the Andean subregion. *Fontqueria* 48:1–10.

Posadas, P. and D. R. Miranda-Esquivel. 1999. El PAE (parsimony analysis of endemicity) como una herramienta en la evaluación de la biodiversidad. *Revista Chilena de Historia Natural* 72:539–546.

Posadas, P. and J. J. Morrone. 2003. Biogeografía histórica de la familia Curculionidae (Coleoptera) en las subregiones Subantártica y Chilena Central. *Revista de la Sociedad Entomológica Argentina* 62:75–84.

Poux, C., P. Chevret, D. Huchon, W. W. de Jong, and E. J. P. Douzery. 2006. Arrival and diversification of caviomorph rodents and platyrrhine primates in South America. *Systematic Biology* 55:228–244.

Poux, C., D. Madsen, E. Marquard, D. R. Vieites, W. W. de Jong, and M. Vences. 2005. Asynchronous colonization of Madagascar by the four endemic clades of primates, tenrecs, carnivores, and rodents as inferred from nuclear genes. *Systematic Biology* 54:719–730.

Powell, J. R. and R. DeSalle. 1995. *Drosophila* molecular phylogenies and their uses. *Evolutionary Biology* 28:87–138.

Prado, D. E. and P. E. Gibbs. 1993. Patterns of species distributions in the dry seasonal forests of South America. *Annals of the Missouri Botanical Garden* 80:902–927.

Presa, P., B. G. Pardo, P. Martínez, and L. Bernatchez. 2002. Phylogeographic congruence between mtDNA and rDNA ITS markers in brown trout. *Molecular Biology and Evolution* 19:2161–2175.

Presch, W. 1993. The family–subfamily taxon in lizards: Effects of plate tectonics on biogeographic patterns. In *Advances in cladistics*, Vol. 2: *Proceedings of the second meeting
of the Willi Hennig Society*, ed. N. I. Platnick and V. A. Funk, 191–198. New York:
Columbia University Press.

Puig, H. 1989. Análisis fitogeográfico del bosque mesófilo de montaña de Gomez Farías.
Biotam 1:34–53.

Quijano-Abril, M. A., R. Callejas-Posada, and D. R. Miranda-Esquivel. 2006. Areas of endemism and distribution patterns for Neotropical *Piper* species (Piperaceae). *Journal of
Biogeography* 33:1266–1278.

Racheli, L. and T. Racheli. 2003. Historical relationships of Amazonian areas of endemism
based on raw distributions of parrots (Psittacidae). *Tropical Zoology* 16:33–46.

Racheli, L. and T. Racheli. 2004. Patterns of Amazonian area relationships based on raw
distributions of papilionid butterflies (Lepidoptera: Papilioninae). *Biological Journal
of the Linnean Society* 82:345–357.

Raherilalao, M. J. and S. M. Goodman. 2005. Modeles d'endémisme des oiseaux forestiers
des hautes terres de Madagascar. *Revue d'Écologie (Terre Vie)* 60:355–368.

Rambaut, A. 2001. *RHINO*. Version 1.1. Retrieved May 25, 2008, from http://evolve.zoo.
ox.ac.uk/software.html.

Rambaut, A. and L. Bromham. 1998. Estimating divergence dates from molecular sequences. *Molecular Biology and Evolution* 15:442–448.

Rambaut, A. and M. Charleston. 2002. *Phylogenetic tree editor and manipulator v1.0 alpha 10*.
Oxford: Department of Zoology, University of Oxford.

Rapoport, E. H. 1968. Algunos problemas biogeográficos del nuevo mundo con especial
referencia a la región Neotropical. In *Biologie de l'Amerique Australe*, Vol. 4, ed. R.
Delamare Debouteville and E. H. Rapoport, 55–110. Paris: CNRS.

Rapoport, E. H. 1975. *Areografía: Estrategias geográficas de las especies*. Mexico, D.F.: Fondo
de Cultura Económica. (English translation: 1982, *Areography: Geographical strategies
of species*, Oxford: Pergamon.)

Rauchenberger, M. 1988. Historical biogeography of poecilid fishes in the Caribbean. *Systematic Zoology* 37:356–365.

Real, R., J. M. Vargas, and J. C. Guerrero. 1992. Análisis biogeográfico de clasificación de
áreas y especies. *Monografías Herpetológicas* 2:73–84.

Recuero, E., I. Martínez-Solano, G. Parra-Olea, and M. García-París. 2006. Phylogeography
of *Pseudacris regilla* (Anura: Hylidae) in western North America, with a proposal for
a new taxonomic rearrangement. *Molecular Phylogenetics and Evolution* 39:293–304.

Reig, O. A. 1962. Las interacciones cenogenéticas en el desarrollo de la fauna de vertebra-
dos tetrápodos de América del Sur. *Ameghiniana* 1:131–140.

Reig, O. A. 1981. *Teoría del origen y desarrollo de la fauna de mamíferos de América del Sur*. Mar
del Plata, Argentina: Museo Municipal de Ciencias Naturales Lorenzo Scaglia.

Renner, S. S., G. Clausing, and K. Meyer. 2001. Historical biogeography of Melastomata-
ceae: The roles of Tertiary migration and long-distance dispersal. *American Journal
of Botany* 88:1290–1300.

Retana-Salazar, A. P. 2005. Tras las huellas del hombre americano: Un enfoque parasi-
tológico. *Revista de Antropología Experimental* 5:1–10.

Reyes-Castillo, P., G. Amat-García, and C. Vasconcelos de Fonseca. 2005. Análisis de par-
simonia de endemismos de Passalidae (Coleoptera: Scarabaeoidea) de la subregión
Amazónica. In *Regionalización biogeográfica en Iberoamérica y tópicos afines: Primeras
Jornadas Biogeográficas de la Red Iberoamericana de Biogeografía y Entomología Sistemáti-
ca (RIBES XII.I–CYTED)*, ed. J. Llorente Bousquets and J. J. Morrone, 461–467.
Mexico, D.F.: Las Prensas de Ciencias, UNAM.

Ribas, C. C., R. Gaban-Lima, C. Y. Miyaki, and J. Cracraft. 2005. Historical biogeography and diversification within the Neotropical parrot genus *Pionopsitta* (Aves: Psittaciformes). *Journal of Biogeography* 32:1409–1427.

Ribera, I., D. T. Bilton, and A. P. Vogler. 2003. Mitochondrial DNA phylogeography and population history of *Meladema* diving beetles on the Atlantic islands and in the Mediterranean basin (Coleoptera, Dytiscidae). *Molecular Ecology* 12:153–167.

Ribichich, A. M. 2002. El modelo clásico de la fitogeografía de Argentina: Un análisis crítico. *Interciencia* 27:669–675.

Ribichich, A. M. 2005. From null community to non-randomly structured actual plant assemblages: Parsimony analysis of species co-occurrences. *Ecography* 28:88–98.

Richards, M. B., V. A. Macaulay, H. J. Bandelt, and B. C. Sykes. 1998. Phylogeography of mitochondrial DNA in western Europe. *Annals of Human Genetics* 62:241–260.

Riddle, B. R. 2005. Is biogeography emerging from its identity crisis? *Journal of Biogeography* 32:185–186.

Riddle, B. R. and D. J. Hafner. 2004. The past and future roles of phylogeography in historical biogeography. In *Frontiers of biogeography: New directions in the geography of nature,* ed. M. V. Lomolino and L. R. Heaney, 93–110. Sunderland, Mass.: Sinauer.

Riddle, B. R. and D. J. Hafner. 2006. A step-wise approach to integrating phylogeographic and phylogenetic biogeographic perspectives on the history of a core North American warm deserts biota. *Journal of Arid Environments* 66:435–461.

Riddle, B. R., D. J. Hafner, and L. F. Alexander. 2000a. Comparative phylogeography of Bailey's pocket mouse (*Chaetodipus baileyi*) and the *Peromyscus eremicus* species group: Historical vicariance of the Baja California Peninsular Desert. *Molecular Phylogenetics and Evolution* 17:161–172.

Riddle, B. R., D. J. Hafner, and L. F. Alexander. 2000b. Phylogeography and systematics of the *Peromyscus eremicus* species group and the historical biogeography of North American warm regional deserts. *Molecular Phylogenetics and Evolution* 17:145–160.

Riddle, B. R., D. J. Hafner, L. F. Alexander, and J. R. Jaeger. 2000c. Cryptic vicariance in the historical assembly of a Baja California Peninsular Desert biota. *Proceedings of the National Academy of Sciences* 97:14438–14443.

Ridley, M. 1996. *Evolution,* 2nd ed. Cambridge, Mass.: Blackwell Science.

Rieppel, O. 1991. Things, taxa and relationships. *Cladistics* 7:93–100.

Rieppel, O. 2004. The language of systematics, and the philosophy of "total evidence." *Systematics and Biodiversity* 2:9–19.

Ringuelet, R. A. 1957. Biogeografía de los arácnidos argentinos del orden Opiliones. *Contribuciones Científicas de la Facultad de Ciencias Exactas y Naturales, Serie Zoología* 1:1–33.

Ringuelet, R. A. 1961. Rasgos fundamentales de la zoogeografía de la Argentina. *Physis* (Buenos Aires) 22:151–170.

Robalo, J. I., V. C. Almada, A. Levy, and I. Doadrio. 2007. Re-examination and phylogeny of the genus *Chondrostoma* based on mitochondrial and nuclear data and the definition of 5 new genera. *Molecular Phylogenetics and Evolution* 42:362–372.

Robalo, J. I., C. S. Santos, V. C. Almada, and I. Doadrio. 2006. Paleobiogeography of two Iberian endemic cyprinid fishes (*Chondrostoma arcasii–Chondrostoma macrolepidotus*) inferred from mitochondrial DNA sequence data. *Journal of Heredity* 97:143–149.

Rode, A. L. and B. S. Lieberman. 2005. Integrating evolution and biogeography: A case study involving Devonian crustaceans. *Journal of Paleontology* 79:267–276.

Roderick, G. K. and R. G. Gillespie. 1998. Speciation and phylogeography of Hawaiian terrestrial arthropods. *Molecular Ecology* 7:519–531.

Roig-Juñent, S. 1992. Insectos de América del Sur, su origen a través del enfoque de la biogeografía histórica. *Multequina* (Mendoza) 1:107–114.

Roig-Juñent, S. 1994. Historia biogeográfica de América del Sur austral. *Multequina* (Mendoza) 3:167–203.

Roig-Juñent, S. 2002. Nuevas especies de *Cnemalobus* (Coleoptera: Carabidae) y consideraciones filogenéticas y biogeográficas sobre el género. *Revista de la Sociedad Entomológica Argentina* 61:51–72.

Roig-Juñent, S. 2004. Los Migadopini (Coleoptera: Carabidae) de América del Sur: Descripción de las estructuras genitales masculinas y femeninas y consideraciones filogenéticas y biogeográficas. *Acta Entomológica Chilena* 28:7–29.

Roig-Juñent, S. 2005. Las ideas biogeográficas de René Jeannel y su impacto en el conocimiento de la biogeografía de América del Sur. In *Regionalización biogeográfica en Iberoamérica y tópicos afines: Primeras Jornadas Biogeográficas de la Red Iberoamericana de Biogeografía y Entomología Sistemática (RIBES XII.I–CYTED)*, ed. J. Llorente Bousquets and J. J. Morrone, 55–66. Mexico, D.F.: Las Prensas de Ciencias, UNAM.

Roig-Juñent, S., J. V. Crisci, P. Posadas, and S. Lagos. 2002. Áreas de distribución y endemismo en zonas continentals. In *Proyecto de Red Iberoamericana de Biogeografía y Entomología Sistemática PrIBES 2002, Monografías Tercer Milenio*, Vol. 2, ed. C. Costa, S. A. Vanin, J. M. Lobo, and A. Meliá, 247–266. Saragossa, Spain: Sociedad Entomológica Aragonesa.

Roig-Juñent, S., M. C. Domínguez, G. E. Flores, and C. Mattoni. 2006. Biogeographic history of South American arid lands: A view from its arthropods using TASS analysis. *Journal of Arid Environments* 66:404–420.

Roig-Juñent, S. and G. E. Flores. 2001. Historia biogeográfica de las áreas áridas de América del Sur austral. In *Introducción a la biogeografía en Latinoamérica: Teorías, conceptos, métodos y aplicaciones*, ed. J. Llorente Bousquets and J. J. Morrone, 257–266. Mexico, D.F.: Las Prensas de Ciencias, UNAM.

Roig-Juñent, S., G. Flores, S. Claver, G. Debandi, and A. Marvaldi. 2001. Monte desert (Argentina): Insect biodiversity and natural areas. *Journal of Arid Environments* 47:77–94.

Roig-Juñent, S., G. Flores, and C. Mattoni. 2003. Consideraciones biogeográficas de la Precordillera (Argentina), con base en artrópodos epígeos. In *Una perspectiva latinoamericana de la biogeografía*, ed. J. J. Morrone and J. Llorente Bousquets, 275–288. Mexico, D.F.: Las Prensas de Ciencias, UNAM.

Rojas Parra, C. A. 2007. Una herramienta automatizada para realizar análisis panbiogeográficos. *Biogeografía* 1:31–33.

Rojas Soto, O. R., O. Alcántara Ayala, and A. G. Navarro. 2003. Regionalization of the avifauna of the Baja California peninsula, Mexico: A parsimony analysis of endemicity and distributional modelling approach. *Journal of Biogeography* 30:449–461.

Ron, S. R. 2000. Biogeographic area relationships of lowland Neotropical rainforest based on raw distributions of vertebrate groups. *Biological Journal of the Linnean Society* 71:379–402.

Ronquist, F. 1994. Ancestral areas and parsimony. *Systematic Biology* 43:267–274.

Ronquist, F. 1995. Ancestral areas revisited. *Systematic Biology* 44:572–575.

Ronquist, F. 1996. *DIVA*, version 1.0: Computer program for MacOS and Win32. Retrieved May 25, 2008, from http://www.ebc.uu.se/systzoo/research/diva/manual/dmanual.html.

Ronquist, F. 1997a. Dispersal–vicariance analysis: A new approach to the quantification of historical biogeography. *Systematic Biology* 46:195–203.

Ronquist, F. 1997b. Phylogenetic approaches in coevolution and biogeography. *Zoologica Scripta* 26:313–322.

Ronquist, F. 1998. Dispersal–vicariance analysis: A new approach to the quantification of historical biogeography. *Cladistics* 14:167–172.

Ronquist, F. 2002. *TreeFitter*. Version 1.3. Retrieved May 25, 2008, from http://www.ebc.uu.se/systzoo/research/treefitter/treefitter.html.

Ronquist, F. and S. Nylin. 1990. Process and pattern in the evolution of species associations. *Systematic Zoology* 39:323–344.

Rosa, D. 1918. *Ologenesi: Nuova teoria dell'evoluzione e della distribuzione geografica dei viventi.* Florence: Bemporad & Figlio Editori.

Rosas-Valdez, R. and G. Pérez-Ponce de León. 2005. Biogeografía histórica de helmintos parásitos de ictalúridos en América del Norte: Una hipótesis preliminar utilizando el método panbiogeográfico. In *Regionalización biogeográfica en Iberoamérica y tópicos afines: Primeras Jornadas Biogeográficas de la Red Iberoamericana de Biogeografía y Entomología Sistemática (RIBES XII.I–CYTED)*, ed. J. Llorente Bousquets and J. J. Morrone, 217–226. Mexico, D.F.: Las Prensas de Ciencias, UNAM.

Rosen, B. R. 1985. Long-term geographical controls on regional diversity. *Journal of the Open University Geological Society* 6:25–30.

Rosen, B. R. 1988a. Biogeographic patterns: A perceptual overview. In *Analytical biogeography: An integrated approach to the study of animal and plant distributions*, ed. A. A. Myers and P. S. Giller, 23–55. London: Chapman and Hall.

Rosen, B. R. 1988b. From fossils to Earth history: Applied historical biogeography. In *Analytical biogeography: An integrated approach to the study of animal and plant distributions*, ed. A. A. Myers and P. S. Giller, 437–481. London: Chapman and Hall.

Rosen, B. R. 1988c. Progress, problems and patterns in the biogeography of reef corals and other tropical marine organisms. *Helgolländer Meeresunters* 42:269–301.

Rosen, B. R. and A. B. Smith. 1988. Tectonics from fossils?: Analysis of reef-coral and sea-urchin distributions from late Cretaceous to Recent, using a new method. In *Gondwana and Tethys*, ed. M. G. Audley-Charles and A. Hallam, 275–306. London: Geological Society Special Publication no. 37.

Rosen, D. E. 1974. Space, time, form: The biological synthesis. *Systematic Zoology* 23:288–290.

Rosen, D. E. 1976. A vicariance model of Caribbean biogeography. *Systematic Zoology* 24:431–464.

Rosen, D. E. 1978. Vicariant patterns and historical explanation in biogeography. *Systematic Zoology* 27:159–188.

Rosen, D. E. 1979. Fishes from the uplands and intermontane basins of Guatemala: Revisionary studies and comparative geography. *Bulletin of the American Museum of Natural History* 162:267–376.

Rosen, D. E. 1981. Introduction. In *Vicariance biogeography: A critique*, ed. G. Nelson and D. E. Rosen, 1–5. New York: Columbia University Press.

Rosen, D. E. 1985. Geological hierarchies and biogeographic congruence in the Caribbean. *Annals of the Missouri Botanical Garden* 72:636–659.

Rosen, D. E. and G. Nelson, eds. 1980. *Vicariance biogeography: A critique.* New York: Columbia University Press.

Rotondo, G. M., V. G. Springer, G. A. J. Scott, and S. O. Schlanger. 1981. Plate movement and island integration: A possible mechanism in the formation of endemic biotas, with special reference to the Hawaiian islands. *Systematic Zoology* 30:12–21.

Rousseau, D. D. 1992. Is causal ecological biogeography a progressive research program? *Quarterly Science Review* 11:593–601.

Rovito, S. M., M. T. K. Arroyo, and P. Pliscoff. 2004. Distributional modelling and parsimony analysis of endemicity of *Senecio* in the Mediterranean-type climate area of central Chile. *Journal of Biogeography* 31:1623–1636.

Ruggiero, A. and C. Ezcurra. 2003. Regiones y transiciones biogeográficas: Complementariedad de los análisis en biogeografía histórica y ecológica. In *Una perspectiva latinoamericana de la biogeografía*, ed. J. J. Morrone and J. Llorente, 141–154. Mexico, D.F.: Las Prensas de Ciencias, UNAM.

Ruggiero, A., J. H. Lawton, and T. M. Blackburn. 1998. The geographic ranges of mammalian species in South America: Spatial patterns in environmental resistance and anisotropy. *Journal of Biogeography* 25:1093–1103.

Rundle, S. D., D. T. Bilton, and D. K. Shiozawa. 2000. Global and regional patterns in lotic meiofauna. *Freshwater Biology* 44:123–134.

Russell, A. L., R. A. Medellín, and G. F. McCraken. 2005. Genetic variation in the Mexican free-tailed bat (*Tadarida brasiliensis mexicana*). *Molecular Ecology* 14:2207–2222.

Russell, D. A. 1993. The role of central Asia in dinosaurian biogeography. *Canadian Journal of Earth Sciences* 30:2001–2012.

Rutschmann, F. 2006. Molecular dating of phylogenetic trees: A brief review of current methods that estimate divergence times. *Diversity and Distributions* 12:35–48.

Rzedowski, J. 1978. *Vegetación de México*. México D.F.: Limusa.

Salinas, M. and P. Ladrón de Guevara. 1993. Riqueza y diversidad de los mamíferos marinos. *Ciencias* 7:85–93.

Salisbury, B. A. 1999. *SECANT: Strongest evidence compatibility analysis tool*. Version 2.2. New Haven, Conn.: Department of Ecology and Evolutionary Biology, Yale University.

Sampson, S. D., L. M. Witmer, C. A. Foster, D. M. Krause, P. M. O'Connor, P. Dodson, and F. Ravoavy. 1998. Predatory dinosaur remains from Madagascar: Implications for the Cretaceous biogeography of Gondwana. *Science* 280:1048–1051.

Sanderson, M. J. 1998. Estimating rate and time in molecular phylogenies: Beyond the molecular clock? In *Molecular systematics of plants II: DNA sequencing*, ed. D. E. Soltis, P. S. Soltis, and J. J. Doyle, 242–264. Boston: Kluwer.

Sanderson, M. J. 2002. Estimating absolute rates of molecular evolution and divergence times: A penalized likelihood approach. *Molecular Biology and Evolution* 19:101–109.

Sanderson, M. J. 2003. R8s: Inferring absolute rates of molecular evolution and divergence times in the absence of a molecular clock. *Bioinformatics* 19:301–302.

Sanmartín, I. 2003. Dispersal vs. vicariance in the Mediterranean: Historical biogeography of the Palearctic Pachydeminae (Coleoptera, Scarabaeoidea). *Journal of Biogeography* 30:1883–1897.

Sanmartín, I., H. Enghoff, and F. Ronquist. 2001. Patterns of animal dispersal, vicariance and diversification in the Holarctic. *Biological Journal of the Linnean Society* 73: 345–390.

Sanmartín, I. and F. Ronquist. 2002. New solutions to old problems: Widespread taxa, redundant distributions and missing areas in event-based biogeography. *Animal Biodiversity and Conservation* 25:75–93.

Sanmartín, I. and F. Ronquist. 2004. Southern Hemisphere biogeography inferred by event-based models: Plant versus animal patterns. *Systematic Biology* 53:216–243.

Sanmartín, I., L. Wanntorp, and R. C. Winkworth. 2007. West wind drift revisited: Testing for directional dispersal in the Southern Hemisphere using event-based tree fitting. *Journal of Biogeography* 34:398–416.

Santos, C. M. D. 2005. Parsimony analysis of endemicity: Time for an epitaph? *Journal of Biogeography* 32:1284–1286.

Santos, C. M. D. 2007. On basal clades and ancestral areas. *Journal of Biogeography* 34: 1470–1471.

Savage, J. M. 1982. The enigma of the Central American herpetofauna: Dispersals or vicariance? *Annals of the Missouri Botanical Garden* 69:464–547.

Schaal, B. A., J. F. Gaskin, and A. L. Caicedo. 2003. Phylogeography, haplotype trees, and invasive plant species. *Journal of Heredity* 94:197–204.

Schäuble, C. S. and C. Moritz. 2001. Comparative phylogeography of two open forest frogs from eastern Australia. *Biological Journal of the Linnean Society* 74:157–170.

Schmidt, K. P. 1955. Animal geography. In *A century of progress in the natural sciences 1853–1953, published in celebration of the centennial of the California Academy of Sciences*, ed. E. L. Kessel, 767–794. San Francisco: California Academy of Sciences.

Schmidt, K. P. and R. F. Inger. 1951. Amphibians and reptiles of Hopkins–Branner expedition to Brazil. *Fieldiana, Zoology* 31:439–465.

Schuh, R. T. and G. M. Stonedahl. 1986. Historical biogeography in the Indo-Pacific: A cladistic approach. *Cladistics* 2:337–355.

Sclater, P. L. 1858. On the general geographical distribution of the members of the class Aves. *Journal of the Linnean Society, Zoology* 2:130–145.

Sclater, P. L. 1864. The mammals of Madagascar. *The Quarterly Journal of Science* 1:213–219.

Scotland, R. W. 2000. Taxic homology and the three-taxon statement analysis. *Systematic Biology* 49:480–500.

Seberg, O. 1986. A critique of the theory and methods of panbiogeography. *Systematic Zoology* 35:369–380.

Seberg, O. 1991. Biogeographic congruence in the South Pacific. *Australian Systematic Botany* 4:127–136.

Seeling, J. J. P. and G. Fauth. 2004. Global Campanian (Upper Cretaceous) ostracod palaeobiogeography. *Palaeogeography, Palaeoclimatology, Palaeoecology* 213:379–398.

Sequeira, A. S., A. A. Lanteri, M. A. Scataglini, V. A. Confalonieri, and B. D. Farrell. 2000. Are flightless *Galapaganus* weevils older than the Galápagos Islands they inhabit? *Heredity* 85:20–29.

Sereno, P. C. 1997. The origin and evolution of dinosaurs. *Annual Review of Earth and Planetary Sciences* 25:435–489.

Sereno, P. C. 1999. The evolution of dinosaurs. *Science* 284:2137–2147.

Sereno, P. C., A. L. Beck, D. B. Dutheil, B. Gado, H. C. E. Larsson, G. H. Lyon, J. D. Marcot, O. W. M. Rauhut, R. W. Sadleir, C. A. Sidor, D. B. Varricchio, G. P. Wilson, and J. A. Wilson. 1998. A long-snouted predatory dinosaur from Africa and the evolution of spinosaurids. *Science* 282:1298–1302.

Sfenthourakis, S. and S. Giokas. 1998. A biogeographical analysis of Greek Oniscidean endemism. *Israel Journal of Zoology* 44:273–282.

Shields, O. 1979. Evidence for the initial opening of the Pacific Ocean in the Jurassic. *Palaeogeography, Palaeoclimatology, Palaeoecology* 26:181–220.

Shields, O. 1991. Pacific biogeography and rapid earth expansion. *Journal of Biogeography* 18:583–585.

Shields, O. 1996. Plate tectonics or an expanding Earth? *Journal of the Geological Society of India* 47:399–408.

Siddall, M. E. 2005. Bracing for another decade of deception: The promise of secondary Brooks parsimony analysis. *Cladistics* 21:90–99.

Siddall, M. E. and S. L. Perkins. 2003. Brooks parsimony analysis: A valiant failure. *Cladistics* 19:554–564.

Siebert, D. J. and D. M. Williams. 1998. Recycled. *Cladistics* 14:339–347.

Silva, H. M. A. and V. Gallo. 2007. Parsimony analysis of endemicity of enchodontoid fishes from the Cenomanian. *Carnets de Géologie/Notebooks on Geology Letter* 2007/01:1–8.

Simpson, G. G. 1940. Mammals and land bridges. *Journal of the Washington Academy of Sciences* 30:137–163.

Simpson, G. G. 1950. History of the fauna of Latin America. *American Scientist* 38:361–389.

Simpson, G. G. 1953. *Evolution and geography: An essay on historical biogeography with special reference to mammals.* Eugene: Condon Lecture Series, Oregon State System of Higher Education. (Spanish translation: 1964, *Evolución y geografía: Historia de la fauna de América Latina*, Buenos Aires: Eudeba.)

Simpson, G. G. 1965. *The geography of evolution.* Philadelphia: Chilton.

Simpson, G. G. 1980. *Splendid isolation: The curious history of South American mammals.* New Haven, Conn.: Yale University Press.

Smith, A. B. 1988. Late Paleozoic biogeography of East Asia and paleontological constraints on plate tectonic reconstructions. *Philosophical Transactions of the Royal Society of London* A236:189–227.

Smith, A. B. 1992. Echinoid distribution in the Cenomanian: An analytical study in biogeography. *Palaeogeography, Palaeoclimatology, Palaeoecology* 92:263–276.

Smith, A. B. and J. Xu. 1988. Palaeontology of the 1985 Tibet Geotraverse, Lhasa to Golmud. *Philosophical Transactions of the Royal Society (Series A)* 327:53–105.

Smith, C. H. 1989. Historical biogeography: Geography as evolution, evolution as geography. *New Zealand Journal of Zoology* 16:773–785.

Soares, E. D. G. and C. J. B. de Carvalho. 2005. Biogeography of *Palpibrachus* (Diptera: Muscidae): An integrative study using panbiogeography, parsimony analysis of endemicity, and component analysis. In *Regionalización biogeográfica en Iberoamérica y tópicos afines: Primeras Jornadas Biogeográficas de la Red Iberoamericana de Biogeografía y Entomología Sistemàtica (RIBES XII.I–CYTED)*, ed. J. Llorente Bousquets and J. J. Morrone, 485–494. Mexico, D.F.: Las Prensas de Ciencias, UNAM.

Sober, E. 1988. The conceptual relationship of cladistic phylogenetics and vicariance biogeography. *Systematic Zoology* 37:245–253.

Soderstrom, T. R., E. J. Judziewicz, and L. G. Clark. 1988. Distribution patterns of Neotropical bamboos. In *Proceedings of a workshop on neotropical distribution patterns*, ed. P. E. Vanzolini and W. Ronald Heyer, 121–157. Rio de Janeiro: Academia Brasileira de Ciencias.

Solervicens, J. 1987. Filogenia y biogeografía del género *Eurymetopum* Blanchard, 1844 (Coleoptera: Cleridae: Phyllobaeninae). *Acta Entomológica Chilena* 14:127–154.

Solomon, J. C. 1982. The systematics and evolution of *Epilobium* (Onagraceae) in South America. *Annals of the Missouri Botanical Garden* 69:239–335.

Soltis, D. E., P. S. Soltis, and B. G. Milligan. 1992. Intraspecific chloroplast DNA variation: Systematic and phylogenetic implications. In *Molecular systematics of plants*, ed. P. S. Soltis, D. E. Soltis, and J. J. Doyle, 117–150. New York: Chapman and Hall.

Spellerberg, I. F. and J. W. D. Sawyer. 1999. *An introduction to applied biogeography.* Cambridge: Cambridge University Press.

Springer, M. S., W. J. Murphy, E. Eizirik, and S. J. O'Brien. 2003. Placental mammal diversification and the Cretaceous–Tertiary boundary. *Proceedings of the National Academy of Sciences* 100:1056–1061.

SPSS Inc. 2001. *Statistica for Windows 5.1.* Tulsa, Okla.: Author.

Stace, C. A. 1989. Dispersal versus vicariance: No contest. *Journal of Biogeography* 16: 201–202.

Steele, C. A. and A. Storfer. 2007. Phylogeographic incongruence of codistributed amphibian species based on small differences in geographical distribution. *Molecular Phylogenetics and Evolution* 43:468–479.

Strimmer, K. and A. von Haeseler. 1996. Quartet puzzling: A quartet maximum-likelihood method for reconstructing tree topologies. *Molecular Biology and Evolution* 13: 964–969.

Sukachev, V. N. 1958. On the principles of genetic classification in biocenology. *Ecology* 39:364–367.

Sullivan, J., J. A. Markert, and C. W. Kilpatrick. 1997. Phylogeography and molecular systematics of the *Peromyscus aztecus* species group (Rodentia: Muridae) inferred using parsimony and likelihood. *Systematic Biology* 46:426–440.

Swenson, U., A. Backlund, S. McLoughlin, and R. S. Hill. 2001. *Nothofagus* biogeography revisited with special emphasis on the enigmatic distribution of subgenus *Brassospora* in New Caledonia. *Cladistics* 17(1):28–47.

Swenson, U. and K. Bremer. 1996. Pacific biogeography of the Asteraceae genus *Abrotanella* (Senecioneae, Blemnospermatinae). *Systematic Botany* 22:493–508.

Swofford, D. L. 1999. *PAUP*: Phylogenetic analysis using parsimony (*and other methods)*. Version 4.0 beta. Sunderland, Mass.: Sinauer.

Swofford, D. L. 2003. *PAUP*: Phylogenetic analysis using parsimony (*and other methods)*. Version 4. Sunderland, Mass.: Sinauer. Retrieved May 25, 2008, from http://paup.csit. fsu.edu/.

Szumik, C. A., D. Casagranda, and S. Roig-Juñent. 2006. Manual de NDM/VNDM: Programas para la identificación de areas de endemismo. *Instituto Argentino de Estudios Filogenéticos* 5:1–26.

Szumik, C. A., F. Cuezzo, P. A. Goloboff, and A. E. Chalup. 2002. An optimality criterion to determine areas of endemism. *Systematic Biology* 51:806–816.

Szumik, C. A. and P. Goloboff. 2004. Areas of endemism: An improved optimality criterion. *Systematic Biology* 53:968–977.

Szumik, C. and S. Roig-Juñent. 2005. Criterio de optimación para áreas de endemismo: El caso de América del Sur austral. In *Regionalización biogeográfica en Iberoamérica y tópicos afines: Primeras Jornadas Biogeográficas de la Red Iberoamericana de Biogeografía y Entomología Sistemática (RIBES XII.I–CYTED)*, ed. J. Llorente Bousquets and J. J. Morrone, 495–508. Mexico, D.F.: Las Prensas de Ciencias, UNAM.

Taberlet, P., L. Fumagalli, A. G. Wust-Saucey, and J. F. Cosson. 1998. Comparative phylogeography and postglacial colonization routes in Europe. *Molecular Ecology* 7:453–464.

Takhtajan, A. 1969. *Flowering plants: Origin and dispersal*. Edinburgh: Oliver and Boyd.

Tassy, P. and P. Deleporte. 1999. Hennig XVII, a time for integration, 21–25 Septembre 1998, Sao Paulo (Brésil). *Bulletin de la Societé Française de Systematique* 21:13–14.

Tavares, E. E., A. J. Baker, S. L. Pereira, and C. Y. Miyaki. 2006. Phylogenetic relationships and historical biogeography of Neotropical parrots (Psittaciformes: Psittacidae: Arini) inferred from mitochondrial and nuclear DNA sequences. *Systematic Biology* 55:454–470.

Taylor, D. W. 1960. Distribution of the freshwater clam *Pissidium ultramontanum:* A zoogeographical enquiry. *American Journal of Science (A)* 258:1–35.

Taylor, F. B. 1910. Bearing of the Tertiary Mountain Belt on the origin of the earth's plan. *Bulletin of the Geological Society of America* 21:179–226.

Taylor, M. S. and M. E. Hellberg. 2003. Genetic evidence for local retention of pelagic larvae in a Caribbean reef fish. *Science* 299:107–109.

Tchaicka, L., E. Eizirik, T. G. De Oliveira, J. F. Candido, and T. R. O. Freitas. 2007. Phylogeography and population history of the crab-eating fox (*Cerdocyon thous*). *Molecular Ecology* 16:819–838.

Templeton, A. R. 1998. Nested clade analysis of phylogenetic data: Testing hypotheses about gene flow and population biology. *Molecular Ecology* 7:381–397.

Templeton, A. R. 2004. Statistical phylogeography: Methods of evaluating and minimizing inference errors. *Molecular Ecology* 13:789–809.

Templeton, A. R., E. Boerwinkle, and C. F. Sing. 1987. A cladistic analysis of phenotypic associations with haplotypes inferred from restriction endonuclease mapping. I. Basic theory and an analysis of alcohol dehydrogenase activity in drosophila. *Genetics* 117:343–351.

Templeton, A. R., K. A. Crandall, and C. F. Sing. 1992. A cladistic analysis of phenotypic associations with haplotypes inferred from restriction endenuclease mapping and DNA sequence data. III. Cladogram estimation. *Genetics* 132:619–633.

Templeton, A. R. and N. J. Georgiadis. 1996. A landscape approach to conservation genetics: Conserving evolutionary processes in the African Bovidae. In *Conservation genetics: Case histories from nature*, ed. J. C. Avise and J. L. Hamrick, 398–430. New York: Chapman and Hall.

Templeton, A. R., E. Routman, and C. A. Phillips. 1995. Separating population structure from population history: A cladistics analysis of the geographical distribution of mitochondrial DNA haplotypes in the tiger salamander, *Ambystoma tigrinum. Genetics* 140:767–782.

Thompson, J. D., T. J. Gibson, F. Plewniak, F. Jeanmougin, and D. G. Higgins. 1997. The ClustalX windows interface: Flexible strategies for multiple sequence alignment aided by quality analysis tools. *Nucleic Acids Research* 25:4876–4882.

Thorne, J. L. and H. Kishino. 2002. Divergence time and evolutionary rate estimation with multilocus data. *Systematic Biology* 51:689–702.

Thorne, J. L., H. Kishino, and I. S. Painter. 1998. Estimating the rate of molecular evolution. *Molecular Biology and Evolution* 15:1647–1657.

Torres-Miranda, A. and I. Luna-Vega. 2006. Análisis de trazos para establecer áreas de conservación en la Faja Volcánica Transmexicana. *Interciencia* 31:849–855.

Trejo-Torres, J. C. 2003. Biogeografía ecológica de las Antillas: Ejemplos de las orquídeas y las selvas cársticas. In *Una perspectiva latinoamericana de la biogeografía*, ed. J. J. Morrone and J. Llorente Bousquets, 199–208. Mexico, D.F.: Las Prensas de Ciencias, UNAM.

Trejo-Torres, J. C. and J. D. Ackerman. 2001. Biogeography of the Antilles based on a parsimony analysis of orchid distributions. *Journal of Biogeography* 28:775–794.

Trejo-Torres, J. C. and J. D. Ackerman. 2002. Composition patterns of Caribbean limestone forests: Are parsimony, classification, and ordination analyses congruent? *Biotropica* 34:502–515.

Tribsch, A. 2004. Areas of endemism of vascular plants in the eastern Alps in relation to Pleistocene glaciation. *Journal of Biogeography* 31:747–770.

Udvardy, M. D. F. 1969. *Dynamic biogeography.* New York: Van Nostrand.

Unmack, P. J. 2001. Biogeography of Australian freshwater fishes. *Journal of Biogeography* 28:1053–1089.

Upchurch, P. and C. A. Hunn. 2002. "Time": The neglected dimension in cladistic biogeography? In *International conference "Paleobiogeography and Paleoecology 2001," Piacenza and Castell'Arquato 2001*, ed. P. Monegatti, F. Cecca, and S. Raffi. *Geobios* 35 (*mémoire spéciale* 24):277–286.

Upchurch, P., C. A. Hunn, and D. B. Norman. 2002. An analysis of dinosaurian biogeography: Evidence for the existence of vicariance and dispersal patterns caused by geological events. *Proceedings of the Royal Society of London,* Series B 269:613–621.

van Soest, R. W. M. 1993. Affinities of the marine demosponge fauna of the Cape Verde islands and tropical West Africa. *Courier Forschung-Institut Senckenberg* 159:205–219.

van Soest, R. W. M. 1996. Recoding widespread distributions for general area cladogram construction. *Vie et Milieu* 46:155–161.

van Soest, R. W. M. and E. Hajdu. 1997. Marine area relationships from twenty sponge phylogenies: A comparison of methods and coding strategies. *Cladistics* 13(1–2): 1–20.

van Steenis, C. G. G. J. 1934–1935. On the origin of the Malaysian mountain flora. *Bulletin du Jardin Botanique Buitenzorg* 13:135–262, 289–417.

van Veller, M. G. P. 2004. Methods for historical biogeographical analyses: Anything goes? *Journal of Biogeography* 31:1552–1553.

van Veller, M. G. P. and D. R. Brooks. 2001. When simplicity is not parsimonious: *A priori* and *a posteriori* methods in historical biogeography. *Journal of Biogeography* 28:1–11.

van Veller, M. G. P., D. R. Brooks, and M. Zandee. 2003. Cladistic and phylogenetic biogeography: The art and the science of discovery. *Journal of Biogeography* 30:319–329.

van Veller, M. G. P., D. J. Kornet, and M. Zandee. 2000. Methods in vicariance biogeography: Assessment of the implementations of assumptions 0, 1, and 2. *Cladistics* 16:319–345.

van Veller, M. G. P., M. Zandee, and D. J. Kornet. 1999. Two requirements for obtaining valid common patterns under different assumptions in vicariance biogeography. *Cladistics* 15:393–406.

van Veller, M. G. P., M. Zandee, and D. J. Kornet. 2001. Measures for obtaining inclusive sets of area cladograms under assumptions zero, 1, and 2 with different methods for vicariance biogeography. *Cladistics* 17:248–259.

van Welzen, P. C. 1992. Interpretation of historical biogeographic results. *Acta Botanica Neerlandica* 41:75–87.

van Welzen, P. C., H. Turner, and P. Hovenkamp. 2003. Historical biogeography of Southeast Asia and the West Pacific, or the generality of unrooted area networks as historical biogeographic hypotheses. *Journal of Biogeography* 30:181–192.

van Welzen, P. C., H. Turner, and M. C. Roos. 2001. New Guinea: A correlation between accreting areas and dispersing Sapindaceae. *Cladistics* 17:242–247.

Vargas, J. M. 1992a. Un ensayo en torno al concepto de biogeografía. *Monografías en Herpetología* 2:7–20.

Vargas, J. M. 1992b. Escuelas y tendencias en biogeografía histórica. *Monografías en Herpetología* 2:107–136.

Vargas, J. M. 1993. Siete pecados capitales en biogeografía. *Zoologica Baetica* 4:39–56.

Vargas, J. M. 2002. *Proyecto docente de zoogeografía: Presentación para concurso de plaza de Catedrático.* Málaga, Spain: Universidad de Málaga.

Vargas, J. M., J. Olivero, A. L. Márquez, J. C. Guerrero, and R. Real. 2003. Relaciones biogeográficas de los sistemas montañosos de la península Ibérica: El caso de los micromamíferos. *Graellsia* 59:319–329.

Vargas, J. M., R. Real, and J. C. Guerrero. 1998. Biogeographical regions of the Iberian peninsula based on freshwater fish and amphibians distributions. *Ecography* 21:371–382.

Vergara, O. E., V. Jerez, and L. E. Parra. 2006. Diversidad y patrones de distribución de coleópteros en la Región del Bíobío, Chile: Una aproximación preliminar para la conservación de la biodiversidad. *Revista Chilena de Historia Natural* 79:369–388.

Vetter, J. 2006. Wallace's other line: Human biogeography and field practice in the eastern colonial tropics. *Journal of the History of Biology* 39:89–123.

Vila, C., I. R. Amorim, J. A. Leonard, D. Posada, J. Castroviejo, F. Petrucci-Fonseca, K. A. Crandall, H. Ellegren, and R. K. Wayne. 1999. Mitochondrial DNA phylogeography and population history of the grey wolf *Canis lupus*. *Molecular Ecology* 8:2089–2103.

Viloria, Á. 2005. Las mariposas (Lepidoptera: Papilionoidea) y la regionalización biogeográfica de Venezuela. In *Regionalización biogeográfica en Iberoamérica y tópicos afines: Primeras Jornadas Biogeográficas de la Red Iberoamericana de Biogeografía y Entomología Sistemática (RIBES XII.I–CYTED)*, ed. J. Llorente Bousquets and J. J. Morrone, 441–459. Mexico, D.F.: Las Prensas de Ciencias, UNAM.

Voelker, G. 1999. Dispersal, vicariance, and clocks: Historical biogeography and speciation in a cosmopolitan passerine genus (*Anthus:* Motacillidae). *Evolution* 53:1536–1552.

von Ihering, H. 1927. *Die Geschichte des Atlantisches Ozeans*. Jena, Germany: Gustav Fischer.

Vuilleumier, F. 1999. Biogeography on the eve of the twenty-first century: Towards an epistemology of biogeography. *Ostrich* 70:89–103.

Vuilleumier, F. and D. Simberloff. 1980. Ecology versus history as determinants of patchy and insular distribution in high Andean birds. *Evolutionary Biology* 12:235–379.

Waggoner, B. M. 1999. Biogeographic analyses of the Ediacara biota: A conflict with paleotectonic reconstructions. *Paleobiology* 25:440–458.

Waggoner, B. M. 2003. The Ediacaran biotas in space and time. *Integrative and Comparative Biology* 43:104–113.

Wallace, A. R. 1855. On the law which has regulated the introduction of new species. *Annals and Magazine of Natural History, Series 2*, 16:184–196.

Wallace, A. R. 1863a. On the physical geography of the Malay archipelago. *Journal of the Royal Geographical Society* 33:217–234.

Wallace, A. R. 1863b. On the varieties of man in the Malay archipelago. In *Report of the 33rd meeting of the British Association for the Advancement of Science*, 147–148. London: British Association for the Advancement of Science.

Wallace, A. R. 1864. On some anomalies in zoological and botanical geography. *Edinburgh New Philosophical Journal* 19:1–15.

Wallace, A. R. 1869. *The Malay archipelago: The land of the orang-utan*. New York: Dover.

Wallace, A. R. 1876. *The geographical distribution of animals, with a study of the relations of living and extinct faunas as elucidating the past changes of the earth's surface*. London: Macmillan.

Wallace, A. R. 1890. *The Malay archipelago: The land of the orang-utan and the bird of paradise; a narrative of travel, with studies of man and nature*, 10th ed. New York: Dover.

Wallace, C. C., J. M. Pandolfi, A. Young, and J. Wolstenholme. 1991. Indo-Pacific coral biogeography: A case study from the *Acropora selago* group. *Australian Systematic Botany* 4:199–210.

Walter, H. S. 2004. Understanding places and organisms in a changing world. *Taxon* 53:905–910.

Warren, B. R. and B. I. Crother. 2001. Métodos en biogeografía cladística: El ejemplo del Caribe. In *Introducción a la biogeografía en Latinoamérica: Teorías, conceptos, métodos y aplicaciones*, ed. J. Llorente Bousquets and J. J. Morrone, 233–243. Mexico, D.F.: Las Prensas de Ciencias, UNAM.

Watanabe, K. 1998. Parsimony analysis of the distribution patterns of Japanese primary freshwater fishes, and its application to the distribution of the bagrid catfishes. *Ichthyological Research* 45:259–270.

Waters, J. M., J. A. López, and G. P. Wallis. 2000. Molecular phylogenetics and biogeography of Galaxiid fishes (Osteichthyes: Galaxiidae): Dispersal, vicariance, and the position of *Lepidogalaxias salamandroides*. *Systematic Biology* 49:777–794.

Waters, J. M. and M. S. Roy. 2004. Out of Africa: The slow train to Australasia. *Systematic Biology* 53:18–24.

Webb, C. O., D. D. Ackerly, M. A. McPeek, and M. J. Donoghue. 2002. Phylogenies and community ecology. *Annual Review of Ecology and Systematics* 3:475–505.

Wegener, A. 1912. Die Entstehung der Kontinente und Ozeane. *Geologische Rundschau* 3: 276–292.

Wegener, A. 1929. *The origin of continents and oceans*. Dover: Dover Publications.

Weisrock, D. W. and F. J. Janzen. 2000. Comparative molecular phylogeography of North American softshell turtles (*Apalone*): Implications for regional and wide-scale historical evolutionary forces. *Molecular Phylogenetics and Evolution* 14:152–164.

Westermann, G. E. C. 2000. Biochore classification and nomenclature in paleobiogeography: An attempt at order. *Palaeogeography, Palaeoclimatology, Palaeoecology* 163:49–68.

Weston, P. H. and M. D. Crisp. 1994. Cladistic biogeography of waraths (Proteaceae: Embothriae) and their allies across the Pacific. *Australian Systematic Botany* 7:225–249.

Whittaker, R. J. 1998. *Island biogeography: Ecology, evolution, and conservation*. Oxford: Oxford University Press.

Wiens, J. J. and M. J. Donoghue. 2004. Historical biogeography, ecology and species richness. *Trends in Ecology and Evolution* 19:639–644.

Wiles, J. S. and V. M. Sarich. 1983. Are the Galapagos iguanas older than the Galapagos? In *Patterns of evolution in Galapagos organisms*, ed. R. I. Bowman, M. Berson, and A. E. Levington, 177–186. San Francisco: California Academy of Sciences.

Wiley, E. O. 1980. Phylogenetic systematics and vicariance biogeography. *Systematic Botany* 5:194–220.

Wiley, E. O. 1981. *Phylogenetics: The theory and practice of phylogenetic systematics*. New York: Wiley-Interscience.

Wiley, E. O. 1987. Methods in vicariance biogeography. In *Systematics and evolution: A matter of diversity*, ed. P. Hovenkamp, E. Gittenberger, E. Hennipman, R. de Jong, M. C. Roos, R. Sluys, and M. Zandee, 283–306. Utrecht: Institute of Systematic Botany, Utrecht University.

Wiley, E. O. 1988a. Parsimony analysis and vicariance biogeography. *Systematic Zoology* 37:271–290.

Wiley, E. O. 1988b. Vicariance biogeography. *Annual Review of Ecology and Systematics* 19:513–542.

Wiley, E. O. and R. L. Mayden. 1985. Species and speciation in phylogenetic systematics, with examples from the North American fish fauna. *Annals of the Missouri Botanical Garden* 72:596–635.

Willdenow, C. L. 1792. *Grundriß der Kräuterkunde zu Vorlesungen*. Berlin: Haude und Spener.

Williams, D. M. 2004. Homologues and homology, phenetics and cladistics: 150 years of progress. In *Milestones in systematics*, ed. D. M. Williams and P. L. Forey, 191–224. Boca Raton, Fla.: CRC Press.

Williams, D. M. 2007a. Ernst Haeckel and Louis Aggasiz: Trees that bite and their geographical dimension. In *Biogeography in a changing world*, ed. M. C. Ebach and R. S. Tangney, 1–59. Boca Raton, Fla.: CRC Press.

Williams, D. M. 2007b. Otto Kleinschmidt (1870–1954), biogeography and the "origin" of species: From *Formenkreis* to progression rule. *Biogeografía* 1:3–9.

Williams, D. M. and M. C. Ebach. 2004. The reform of palaeontology and the rise of bio-geography: 25 years after "Ontogeny, phylogeny, paleontology and the biogenetic law" (Nelson, 1978). *Journal of Biogeography* 31:685–712.

Williams, D. M. and C. J. Humphries. 2003. Component coding, three-item coding, and consensus methods. *Systematic Biology* 52:255–259.

Willis, J. C. 1915. The endemic flora of Ceylon, with reference to geographical distribution and evolution in general. *Philosophical Transactions of the Royal Society of London, Series B* 206:307–342.

Willis, J. C. 1922. *Age and area: A study in geographic distribution and origin of species.* Cambridge: Cambridge University Press.

Wilson, E. O. 1961. The nature of the taxon cycle in the Melanesian ant fauna. *The American Naturalist* 95:169–193.

Wilson, R. J. and L. W. Beinke, eds. 1979. *Applications of graph theory.* London: Academic Press.

Winfield, I., E. Escobar-Briones, and J. J. Morrone. 2006. Updated checklist and identification of areas of endemism of benthic amphipods (Caprellidea and Gammaridea) from offshore habitats in the SW Gulf of Mexico. *Scientia Marina* 70:99–108.

Wojcicki, M. and D. R. Brooks. 2004. Escaping the matrix: A new algorithm for phylogenetic comparative studies of coevolution. *Cladistics* 20:341–361.

Wojcicki, M. and D. R. Brooks. 2005. PACT: An efficient and powerful algorithm for generating area cladograms. *Journal of Biogeography* 32:755–774.

Won, H. and S. S. Renner. 2006. Dating dispersal and radiation in the gymnosperm *Gnetum* (Gnetales): Clock calibration when outgroup relationships are uncertain. *Systematic Biology* 55:610–622.

Xiang, Q. Y., S. J. Brunsfeld, D. E. Soltis, and P. S. Soltis. 1996. Phylogenetic relationships in *Cornus* based on chloroplast DNA restriction sites: Implications for biogeography and character evolution. *Systematic Botany* 21:515–534.

Xiang, Q. Y., D. J. Crawford, A. D. Wolfe, Y. C. Tang, and C. W. DePamphilis. 1998. Origin and biogeography of *Aesculus* L. (Hippocastanaceae): A molecular phylogenetic perspective. *Evolution* 52:988–997.

Xiang, Q. Y. and D. E. Soltis. 2001. Dispersal–vicariance analyses of intercontinental disjuncts: Historical biogeographical implications for angiosperms in the Northern Hemisphere. *International Journal of Plant Sciences* 162:S29–S39.

Xu, S. Q. 2005. Distribution and area of endemism of Catantopidae grasshopper species endemic to China. *Acta Zoologica Sinica* 51:624–629.

Yang, Z. 1997. PAML: A program package for phylogenetic analysis by maximum likelihood. *Computer Applications in the Biosciences* 13:555–556.

Yeates, D. K., P. Bouchard, and G. B. Monteith. 2001. Patterns and levels of endemism in the Australian wet tropics: Evidence from flightless insects. *Invertebrate Systematics* 16:605–619.

Young, G. C. 1987. Devonian palaeontological data and the Armorica problem. *Palaeogeography, Palaeoclimatology, Palaeoecology* 60:283–304.

Young, G. C. 1995. Application of cladistics to terrane history: Parsimony analysis of qualitative geological data. *Journal of Southeast Asian Earth Sciences* 11:167–176.

Yuan, Y. M., S. Wohlhauser, M. Möller, J. K. Lackenberg, M. W. Callmander, and P. Küpfer. 2005. Phylogeny and biogeography of *Exacum* (Gentianaceae): A disjunctive distribution in the Indian Ocean basin resulted from long distance dispersal and extensive radiation. *Systematic Biology* 54:21–34.

Zakharov, E. V., M. S. Caterino, and F. A. H. Sperling. 2004. Molecular phylogeny, historical biogeography, and divergence time estimates for swallowtail butterflies of the
genus *Papilio* (Lepidoptera: Papilionidae). *Systematic Biology* 53:193–215.

Zandee, M. and M. C. Roos. 1987. Component-compatibility in historical biogeography.
Cladistics 3:305–332.

Zhang, M. L. 2002. An improved method for PAE and its application in endemicity analysis of genus *Caragana*, sect. *Caragana* (Fabaceae). *Acta Botanica Yunnanica* 24:147–154.

Zhang, M. L., P. Y. Ladiges, and G. Nelson. 2002. Subtrees, TASS and an analysis of the
genus *Caragana*. *Acta Botanica Sinica* 44:1213–1218.

Zimmermann, E. A. W. 1778–1783. *Geographische Geschichte des Menschen und der vierfüssin-
gen Thiere*, 3 vols. Leipzig: Weygand.

Zink, R. M. 1996. Comparative phylogeography in North American birds. *Evolution*
50:308–317.

Zink, R. M. 2002. Methods in comparative phylogeography, and their application to studying evolution in the North American aridlands. *Integrative and Comparative Biology*
42:953–959.

Zink, R. M., R. C. Blackwell-Rago, and F. Ronquist. 2000. The shifting roles of dispersal
and vicariance in biogeography. *Proceedings of the Royal Society of London B*, 267:
497–503.

Zink, R. M., S. V. Drovetski, and S. Rohwer. 2002. Phylogeographic patterns in the great
spotted woodpecker *Dendrocops major* across Eurasia. *Journal of Avian Biology*
33:175–178.

Zuckerland, E. and L. Pauling. 1962. Molecular disease, evolution and genetic heterogeneity. In *Horizons in Biochemistry*, ed. M. Kasha and B. Pullman, 189–225. New York:
Academic Press.

Zunino, M. 1992. Per rileggere Croizat. *Biogeographia* 16:11–23.

Zunino, M. 2003. Nuevos conceptos en la biogeografía histórica: Implicancias teóricas y
metodológicas. In *Una perspectiva latinoamericana de la biogeografía*, ed. J. J. Morrone
and J. Llorente, 159–162. Mexico, D.F.: Las Prensas de Ciencias, UNAM.

Zunino, M. 2005. Corotipos y biogeografía sistemática en el Euromediterráneo. In *Regionalización biogeográfica en Iberoamérica y tópicos afines: Primeras Jornadas Biogeográficas
de la Red Iberoamericana de Biogeografía y Entomología Sistemática (RIBES XII.I–CYT-
ED)*, ed. J. Llorente Bousquets and J. J. Morrone, 181–187. Mexico, D.F.: Las Prensas
de Ciencias, UNAM.

Zunino, M., E. Barbero, C. Palestrini, E. Buffa, and A. Roggero. 1996. Ipotesi biogeogra-
fiche versus ipotesi filogenetiche: Il genere *Typhaeus* Leach (Coleoptera: Geotrupi-
dae) e il popolamento dell' area Sarda. *Biogeographia* 18:455–476.

Zunino, M. and A. Zullini. 1995. *Biogeografia: La dimensione spaziale dell'evoluzione*. Milan:
Casa Editrice Ambrosiana. (Spanish translation: 2003, *Biogeografía: La dimensión
espacial de la evolución*, Mexico, D.F.: Fondo de Cultura Económica.)

Author Index

Subject Index